Withdrawn
University of Waterloo

The Extractive Metallurgy of Gold

Withdrawn
University of Waterloo

The Extractive Metallurgy of Gold

J. C. Yannopoulos

VNR
VAN NOSTRAND REINHOLD
New York

To my wife and sons

Copyright © 1991 by Van Nostrand Reinhold

Library of Congress Catalog Card Number 90-12439

ISBN 0-442-31797-2

All rights reserved. No part of this work covered by the copyright hereon may be reproduced or used in any form by any means—graphic, electronic, or mechanical, including photocopying, recording, taping, or information storage and retrieval systems—without written permission of the publisher.

Printed in the United States of America

Van Nostrand Reinhold
115 Fifth Avenue
New York, New York 10003

Chapman and Hall
2-6 Boundary Row
London SE1 8HN, England

Thomas Nelson Australia
102 Dodds Street
South Melbourne 3205
Victoria, Australia

Nelson Canada
1120 Birchmount Road
Scarborough, Ontario MIK 5G4, Canada

16 15 14 13 12 11 10 9 8 7 6 5 4 3 2 1

Library of Congress Cataloging-in-Publication Data

Yannopoulos, J. C. (John C.)
The extractive metallurgy of gold / J. C. Yannopoulos.
p. cm.
Includes bibliographical references and index.
ISBN 0-442-31797-2
1. Gold—Metallurgy. 2. Gold ores. I. Title.
TN760.Y36 1990
669'.22—dc20

90-12439
CIP

Contents

	Preface	**ix**
Chapter 1	**Gold Ores**	**1**
	Geochemistry of Gold and Auriferous Deposits	**1**
	Auriferous Deposits	**4**
	Effects of Biological Systems on Metallic Gold	**7**
	Gold in Sea Water	**8**
	Exploration for Gold	**8**
Chapter 2	**Physical and Chemical Properties of Gold**	**11**
	Physical Properties of Gold	**11**
	Chemical Properties of Gold	**12**
	Gold Assaying	**21**
Chapter 3	**Treatment of Placer Deposits**	**25**
	Placer Environment and Formation	**25**
	Dredging or Hydraulic Mining	**31**
	Gold Recovery from Placer Deposits	**38**
	Gravity Concentration of Alluvial Gold	**41**
Chapter 4	**Milling of Amenable Gold Ores**	**55**
	Comminution of the Ore	**55**
	Partial Direct Recovery of Gold	**63**
	Leaching of Pulverized Gold Ores	**67**
	Gold Milling Flow Sheets	**68**
	Recovery of Gold from Solution	**70**
Chapter 5	**Treatment of Refractory Gold Ores**	**79**
	Mineralogy of Refractory Gold Ores	**79**
	Flotation of Some Refractory Ores	**86**

	High-Temperature Oxidation: Roasting	**87**
	High-Pressure Oxidation	**98**
	Biological Oxidation	**105**
	Chemical Oxidation	**107**
	Silica-Locked Gold	**110**
Chapter 6	**Leaching Low-Grade Gold Ores**	**115**
	Ore Testing	**116**
	Heap Leaching and Dump Leaching	**120**
	Vat Leaching	**133**
Chapter 7	**Recovery of Secondary Gold**	**137**
	Base Metals Present in Gold Scrap	**137**
	Special Cases of Gold Scrap	**138**
Chapter 8	**Cyanidation of Gold Ores**	**141**
	Theories on Gold Cyanidation	**143**
	The Mechanism of Cyanidation	**145**
	The Kinetics of Gold Cyanidation	**158**
	Cyanide Regeneration	**165**
	Destruction of Cyanide	**168**
Chapter 9	**Alternative Leaching Reagents for Gold**	**171**
	Thiourea Leaching of Gold	**171**
	Thiosulfate Leaching of Gold	**180**
	Leaching with Halogens and Halides	**182**
Chapter 10	**Recovery of Gold from Solutions**	**185**
	Zinc Cementation	**186**
	Activated Carbon Adsorption	**193**
	Electrowinning of Gold from Cyanide Solutions	**206**
	Electrowinning of Gold from Pregnant Gold-Mill Solutions	**209**
	Staged Heap Leaching and Direct Electrowinning	**214**
	Industrial Uses of Activated Carbon	**216**
	Ion-Exchange Resins	**230**
	Solvent Extraction	**235**
	Metal Chelating Agents	**235**

Chapter 11 **Melting and Refining of Gold** 241
The Wohlwill Electrorefining Process 243
Gold Refining by Dissolution/Precipitation 243

Chapter 12 **Gold Mill Tailings** 245
Disposal of Tailings 245
Water/Cyanide Recovery from Tailings Slurries 246
Destruction of Cyanide in Gold-Mill Effluents 248
Arsenic Removal from Gold-Mine Wastes 253
Recovery of Gold from Accumulated Old Tailings 253

Appendix A: Conversion Factors 257
Appendix B: Gold Production Statistics 259
Appendix C: Comparative Long-Term Total Production Costs in Selected Countries 261
Appendix D: Primary Gold Deposits and Mines in the U.S. 262
Appendix E: Gold Mill Sampling and Metallurgical Balance 264
Appendix F: Data Commonly Collected During Environmental Baseline Studies for an Environmental Assessment 269
Appendix G: Flowchart of Gold Recovery from Copper Refinery Slimes 271
Appendix H: Solubility of Minerals and Metals in Cyanide Solutions 272

Index 273

Preface

The history of gold begins in antiquity. Bits of gold were found in Spanish caves that were used by Paleolithic people around 40,000 B.C. Gold is the "child of Zeus," wrote the Greek poet Pindar. The Romans called the yellow metal *aurum* ("shining dawn"). Gold is the first element and first metal mentioned in the Bible, where it appears in more than 400 references.

This book provides the most thorough and up-to-date information available on the extraction of gold from its ores, starting with the mineralogy of gold ores and ending with details of refining. Each chapter concludes with a list of references including full publication information for all works cited. Sources preceded by an asterisk (*) are especially recommended for more in-depth study.

Nine appendices, helpful to both students and operators, complement the text. I have made every attempt to keep abreast of recent technical literature on the extraction of gold. Original publications through the spring of 1989 have been reviewed and cited where appropriate.

This book is intended as a reference for operators, managers, and designers of gold mills and for professional prospectors. It is also designed as a textbook for extractive metallurgy courses.

I am indebted to the Library of Engineering Societies in New York, which was the main source of the references in the book. The assistance of my son, Panos, in typing the manuscript is gratefully acknowledged.

The Extractive Metallurgy of Gold

CHAPTER 1

Gold Ores

Geochemistry of Gold and Auriferous Deposits

Gold, along with silver and copper, is a member of the IB group of the periodic table, the coinage metals (Table 1-1). Its principal oxidation states are +1 (aurous) and +3 (auric). Gold is soluble in cyanide solutions and in *aqua regia*, forming complexes of the types $[Au(CN)_2]^-$, $[AuCl_2]^-$, and $[AuCl_4]^-$. Gold solubility is highly influenced by redox reactions.

$$Au^0 + H^+ + 0.25O_2 \rightleftharpoons Au^+ + 0.5H_2O$$

In geochemical terms, gold will be leached and transported by oxidizing hydrothermal fluids, and it will be precipitated when the fluid enters a reducing environment.

$$AuCl_2^- + Fe^{2+} \rightleftharpoons \underline{Au} + Fe^{3+} + 2Cl^-$$

The abundance of gold in the upper lithosphere is estimated to be about 0.005 ppm, ranging from 0.003 ppm in limestone and granite-rhyolite to 0.03 ppm in sedimentary rocks (Boyle, 1987). Silver, then copper, are the elements most frequently associated mineralogically with gold. As, Sb, Bi, Fe, Pb, and Zn are commonly found in gold minerals. Petrovskaya (1973) depicted the tendency of the chemical elements to associate with gold on the periodic table, as shown in Figures 1-1 and 1-2.

Two types of auriferous deposits are generally recognized, lode (vein) deposits and placers. Quartz-pebble conglomerate deposits—which supply approximately 50% of the world's gold production—have been generally classified as modified paleo-placers. Henley (1975) classified gold-containing geological environments into seven broad groups.

TABLE 1-1. Numerical properties of the coinage metals.

Property	***Copper***	***Silver***	***Gold***
Atomic number	29	47	79
Outer electronic configuration	$3d^{10}4s^1$	$4d^{10}5s^1$	$5d^{10}6s^1$
Mass numbers, natural isotopes	63, 65	107, 109	197
Atomic weight	63.54	107.880	197.2
Density of solid at 20°C, grams/cc	8.92	10.5	19.3
Atomic volume of solid, cc	7.12	10.27	10.22
Melting point, °C	1083	960.5	1063
Boiling point, °C	2310	1950	2600
Ionization potential, ev	7.723	7.574	9.223
E^0_{298} for			
$M \rightleftharpoons M^+ + e^-$	−0.522	−0.799	−1.68
$M \rightleftharpoons M^{+2} + 2e^-$	−0.3448	−1.389	
Radii, A			
M	1.173	1.339	1.336
M^+	0.96	1.26	1.37

Reprinted by permission. T. Moeller, *Inorganic Chemistry*. John Wiley & Sons (New York), 1958.

Period	Subgroup																	
	Ia	IIa	IIIa	IVa	Va	VIa	VIIa	VIIIa			Ib	IIb	IIIb	IVb	Vb	VIb	VIIb	0
1	H																(H)	He
2	Li	Be	B	C											N	O	F	Ne
3	Na	Mg	Al	Si											P	S	Cl	Ar
4	K	Ca	Sc	Ti	V	Cr	Mn	Fe	Co	Ni	Cu	Zn	Ga	Ge	As	Se	Br	Kr
5	Rb	Sr	Y	Zr	Nb	Mo	Tc	Ru	Rh	Pd	Ag	Cd	In	Sn	Sb	Te	I	Xe
6	Cs	Ba	La	Hf	Ta	W	Re	Os	Ir	Pt	Au	Hg	Tl	Pb	Bi		At	Rh
7	Fr	Ra	Ac	Ku														

FIGURE 1-1. Geochemical table of elements associated with gold. 1. Elements universally associated with gold. 2. Elements typical of minerals commonly associated with gold. 3. Elements concentrated in gold-bearing mineral associations of individual orebodies. 4. Elements characteristic of gold ores only. 5. Trace elements commonly found in gold and its compounds (including artificial impurities).

Reprinted by permission. N. V. Petrovskaya, An outline of gold chemistry. In R. W. Boyle, *Gold: History and Genesis of Deposits*. Van Nostrand Reinhold (New York), 1987, p. 137. Article translated by Translation Bureau, Multilingual Translation Directorate, Secretary of State of Canada.

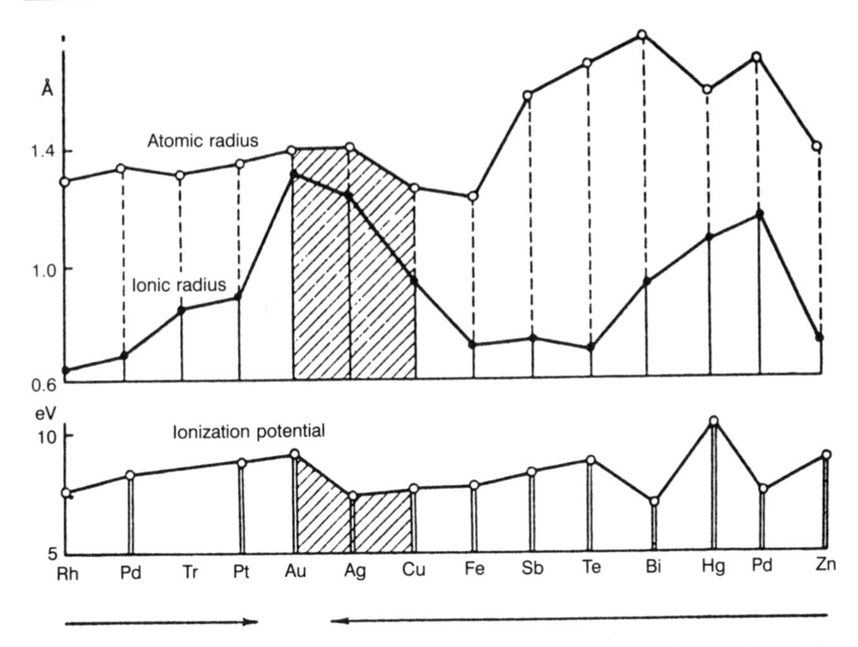

FIGURE 1-2. Comparison of the properties of elements associated with gold. The shaded areas denote the properties of the constant associates of gold. The arrows indicate the directions of intensification of elemental bond formation with gold.

Reprinted by permission. N. V. Petrovskaya, An outline of gold chemistry. In R. W. Boyle, *Gold: History and Genesis of Deposits*. Van Nostrand Reinhold (New York), 1987. Article translated by Translation Bureau, Multilingual Translation Directorate, Secretary of State of Canada.

1. Gold-quartz lodes
2. Epithermal deposits
3. Young placers
4. Fossil placers
5. Deposits with disseminated gold
6. Gold in nonferrous metal ores
7. Gold in sea water

McQuiston and Shoemaker (1975) proposed the following extraction-oriented classification of gold ores.

- *Native gold ores*, in which the precious metal can be removed by gravity separation, amalgamation, and/or cyanidation.
- *Gold associated with sulfides*, occurring either as free particles or disseminated in sulfides. Auriferous pyrite with gold finely disseminated in its matrix is rather common. (Pyrite is relatively stable in

cyanide, but pyrrhotite, if present, dissolves and increases the cyanide consumption.)

- *Gold tellurides*, which usually occur along with native gold and sulfides. Calaverite and krennerite are minerals containing about 40% gold, and sylvanetite contains about 25% gold (in addition to 13% silver).
- *Gold in other minerals*, as with arsenic and/or antimony (e.g., aurostibnite, $AuSb_2$), with copper porphyries (as selenide and telluride), with lead and zinc minerals, and with carbonaceous materials.

Auriferous Deposits

Boyle (1979) distinguishes ten types of auriferous deposits. A geochemical diagram appears in Figure 1-3.

1. *Auriferous porphyry dykes, sills, and stocks; auriferous pegmatites; coarse-grained granitic bodies; aplites and albitites*
The gold content of these granitic rocks is, as a rule, low, in the order of 3 ppb. Certain albitites, quartz-feldspar porphyry dykes, and rocks with indigenous pyrite and/or pyrrhotite may contain up to 0.10 ppm gold.

2. *Carbonatites and carbonatite-related bodies*
Most of the rocks comprising carbonatites are low in gold and silver (0.005 ppm Au, 0.1 ppm Ag). Very few carbonatites are enriched enough in gold and silver to be considered economic orebodies. Only sulfide phases enriched in gold and/or silver—if contained in carbonatites—may be explored as possible gold deposits.

3. *Auriferous skarn-type*[1] *deposits*
These contain early developed Ca-Fe-Mg silicate and oxide minerals, along with later silicate, carbonate, sulfide, and arsenide minerals. Gold is contained as native or in the form of gold tellurides. The elements most frequently enriched with gold in skarn deposits are Fe, S, Cu, Ag, Zn, Pb, Mo, As, Bi, and Te. Tungsten is a common trace element in gold-bearing skarn deposits (Boyle, 1987).

4. *Gold-silver and silver-gold veins, lodes, mineralized pipes, and irregular silicified bodies in fractions, faults, and zones*
These deposits occur in rocks of all ages, but mostly in those of the Precambrian and Tertiary eras. A few deposits occur in or near the throats of extinct (or active) hot springs and/or in siliceous hot-spring aprons. The mineralization of these particular deposits is distinctively composed of

---

[1]Any lime-bearing siliceous rock produced by metamorphism (especially of limestone or dolomite) and by the introduction of new elements.

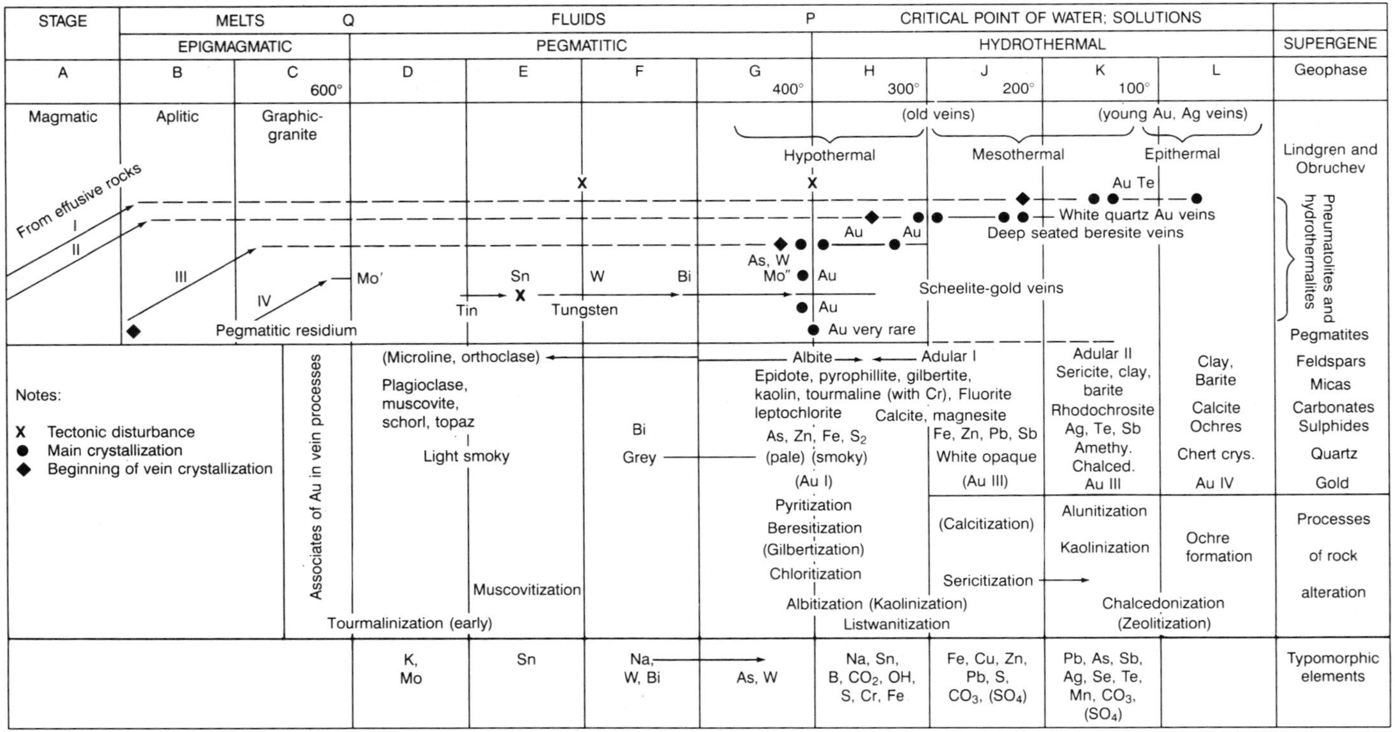

STAGE	MELTS		Q	FLUIDS			P	CRITICAL POINT OF WATER; SOLUTIONS			
	EPIGMAGMATIC		PEGMATITIC				HYDROTHERMAL				SUPERGENE
A	B	C 600°	D	E	F	G 400°	H 300°	J 200°	K 100°	L	Geophase
Magmatic	Aplitic	Graphic-granite					(old veins) Hypothermal	Mesothermal	(young Au, Ag veins) Epithermal		Lindgren and Obruchev
From effusive rocks I II	III	IV Pegmatitic residium	Mo′ Tin	Sn X Tungsten	W X	Bi	As, W Mo″ Au Au Au very rare	Au Au X	Deep seated beresite veins Scheelite-gold veins	White quartz Au veins Au Te	Pneumatolites and hydrothermalites Pegmatites
Notes: X Tectonic disturbance ● Main crystallization ◆ Beginning of vein crystallization		Associates of Au in vein processes	(Microline, orthoclase) Plagioclase, muscovite, schorl, topaz Light smoky		Bi Grey	Albite Epidote, pyrophillite, gilbertite, kaolin, tourmaline (with Cr), Fluorite leptochlorite As, Zn, Fe, S_2 (pale) (smoky) (Au I)		Adular I Calcite, magnesite Fe, Zn, Pb, Sb White opaque (Au III)	Adular II Sericite, clay, barite Rhodochrosite Ag, Te, Sb Amethy. Chalced. Au III	Clay, Barite Calcite Ochres Chert crys. Au IV	Feldspars Micas Carbonates Sulphides Quartz Gold
			Tourmalinization (early)	Muscovitization		Pyritization Beresitization (Gilbertization) Chloritization Albitization (Kaolinization)	Listwanitization	(Calcitization) Sericitization	Alunitization Kaolinization Chalcedonization	Ochre formation (Zeolitization)	Processes of rock alteration
			K, Mo	Sn	Na, W, Bi	As, W	Na, Sn, B, CO_2, OH, S, Cr, Fe	Fe, Cu, Zn, Pb, S, CO_3, (SO_4)	Pb, As, Sb, Ag, Se, Te, Mn, CO_3, (SO_4)		Typomorphic elements

FIGURE 1-3. Geochemical diagram.

Reprinted by permission. A. E. Fersman, Gold, *Geochemistry* (4): 262–71, 1939. In R. W. Boyle, *Gold: History and Genesis of Deposits*. Van Nostrand Reinhold (New York), 1987. Article translated by Translation Bureau, Multilingual Directorate, Secretary of State of Canada.

quartz, carbonate minerals, pyrite, arsenopyrite, base-metal sulfide, and sulfosalt minerals. Native gold and gold tellurides are the principal gold minerals; aurostibnite occurs in some deposits. Commonly concentrated elements in this class of deposits include Cu, Ag, Zn, Cd, Hg, B, Tl, Bp, As, Sb, Bi, V, Se, Te, S, Mo, W, Mn, and Fe, as carbonates and/or silicates.

5. *Auriferous veins, lodes, sheeted zones, and saddle reefs in faults and fractures*
These deposits are developed mainly in sequences of shale and sandstone of marine origin. A few economic deposits occur in the granitic batholiths that invade the graywacke-slate sequences. The principal gangue mineral in these deposits is quartz. Among the metallic minerals, pyrite and arsenopyrite are most common, but galena, sphalerite, chalcopyrite, and pyrrhotite may also occur. The valuable minerals in these ores are native gold (generally low in silver), auriferous pyrite, and auriferous arsenopyrite. Elements that occur frequently in these deposits are Cu, Ag, Mg, Ca, Zn, Cd, B, Si, Pb, As, Sb, S, W, Mn, and Fe.

6. *Gold-silver and silver-gold veins, lodes, stockworks, and silicified zones in a complex geological environment with sedimentary, volcanic, and various igneous intrusive and granitized rocks*
Quartz is a predominant gangue, with some deposits containing moderate developments of carbonates. These orebodies are mainly quartz veins, lodes, and silicified and carbonated zones. The contained gold is mainly free; it may also be present as tellurides and disseminated in pyrite and arsenopyrite. The Au/Ag ratio varies depending on the particular district.

7. *Disseminated and stockwork gold-silver deposits in igneous, volcanic, and sedimentary rocks*
Three general subgroups can be recognized.

- Disseminated and stockwork gold-silver deposits in igneous strata
- Disseminated gold-silver and silver-gold occurrences in volcanic flows and associated volcaniclastic rocks
- Disseminated gold-silver deposits in volcaniclastic and sedimentary beds

The grade of these deposits is highly variable, but most are relatively low grade (less than 15g/ton) with large tonnages. The elements commonly concentrated in these deposits are Cu, Ag, Au, Zn, Cd, Hg, B, Pb, As, Sb, Bi, V, S, Se, Te, Mo, W, Fe, Co, and Ni. In most deposits, the ratio of Au/Ag is greater than one.

Gold deposits that result from significant infiltration or replacement of favorable beds are mainly developed in calcareous and dolomitic pelites and psammites, and in thin-bedded carbonate rocks invaded by granitic stocks and porphyry dykes (Boyle, 1987). Most deposits are characterized by introductions of Au, Ag, Hg, Tl, B, Sb, As, Se, Te, and the base metals. The gold is disseminated through the altered rocks in very finely divided form ($< 5\ \mu$). Deposits of this type, referred to as "Carlin type," are widely distributed throughout the world. Ashton (1989) has reported As, Hg, and, to a lesser extent, Sb to be effective pathfinder elements, with consistent threshold levels, for disseminated gold deposits.

8. *Gold deposits in quartz-pebble conglomerates and quartzites*
These constitute the largest and most productive gold mines, yielding about 50% of the annual gold production in the world (in South Africa, Ghana, and Brazil). These deposits are marked by the presence of abundant pyrite (and/or hematite) along with minor to trace amounts of other sulfides, arsenides, and uranium minerals. Very fine native gold ($< 80\ \mu$) is present in the conglomerates or quartzites. In most orebodies, enrichments of Fe, S, As, Au, and Ag are to be found; some deposits are marked with U, Th, rare earths, Cu, Zn, Pb, Ni, Co, and platinum-group metals. The average ratio of Au/Ag is about 10.

9. *Eluvial and alluvial placers*
These produce gold nuggets and dust with a low silver content. Heavy minerals like monazite, scheelite, and cinnabar, as well as platinum-group metals, may accompany eluvial and alluvial gold. Pactolus, a small river in Lydia (Asia Minor), was famous for its placers in ancient times, and the name means "very rich."

10. *Miscellaneous sources of gold*
These include chalcopyrite, copper-nickel sulfides, pyrite, arsenopyrite, other base-metal sulfides, selenides, arsenides, and sulfosalts. Gold follows the base metal during smelting, and it is recovered from the slimes produced during electrorefining of the base metal.

Effects of Biological Systems on Metallic Gold

The transport of gold by underground water, which contains organic material from the decomposition of vegetable matter, was reported a long time ago (Freise, 1931). Freise noticed that completely exhausted alluvial gold deposits in Brazil could be reworked after a period of years. However, the new gold differed from the original placer in color and purity. It was, in effect, similar to "black gold," which consists of fine gold particles covered by a thin coating. The coating could usually be removed by washing with a 5% solution of potassium carbonate at 35°–45° C. In cases where the coating was thicker, washing under pressure at 300°–330° C had to be followed by dilute sulfuric-acid washing to remove the appreciable iron content in the organic coating.

Freise was able to prove experimentally that water containing humic materials could dissolve gold (and other metals such as iron, copper, and manganese) under anaerobic conditions. Gold and other metals precipitated when the humic solution was exposed to air.

Higher-than-normal concentrations of gold in a number of plant species collected over areas of gold mineralization have been reported by Girling and Peterson (1980). Evidence that gold dissolution occurs *in situ*, caused by plant substances and/or by microorganisms in the soil, has accumulated in recent years (Rapson, 1982).

Gold in Sea Water

The presence of very low concentrations of gold in sea water was established late in the 19th century. The accurate quantitative determination of gold traces in sea water is complicated by the necessity of using ultrapure analytical reagents and of avoiding concentration by evaporation of the sea-water sample. Hence, reported values of sea-water gold content vary widely, from 1 up to 4,000 ppt.[2]

Exploration for Gold

Geophysical magnetic and electromagnetic surveys may provide direct targets of testing for gold. Most gold deposits have no particular magnetic expression, but magnetometer surveys can determine some rock types. Such surveys may detect structural complexities such as faults and fold noses. Aerial photography can often delineate fractures, and patterns of fractures, that indicate a promising district.

Geological techniques of exploration include recording on maps the bedding, facing, and volcanic structures of the potential deposit. Attention is paid to chemical sediments like banded iron formation or chert. All mineral-alteration assemblages have to be recorded on maps along with any apparent silification, veining, sericitization, and appearance(s) of distinct minerals.

Geochemical exploration techniques include systematic chip sampling of outcrops and reverse-circulation drill cuttings, which can detect broad halos around orebodies. Fletcher (1981) summarizes the sample preparation techniques and suggested analytical methods in Figure 1-4. Using biogeochemical surveys has been valuable in some exploration campaigns.

---

[2]Parts per trillion, equivalent to kg/km^3. The dream of economic recovery of gold from sea water (where an average of five million gallons of water must be pumped to recover one gram of gold) is comparable to the alchemists' dream of transmutation of common matter into gold.

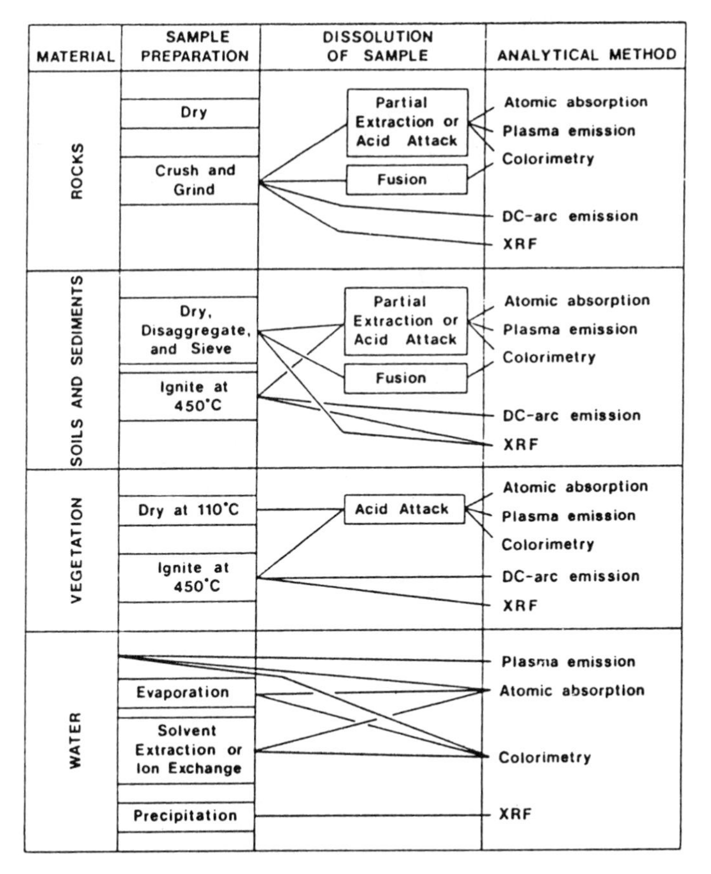

FIGURE 1-4. Suggestions for preparation, dissolution, and analysis of exploration samples.

Reprinted by permission. W. K. Fletcher, *Handbook of Exploration Geochemistry, Vol. 1.* Elsevier Science Publishers (New York), 1981.

References

Note: Sources with an asterisk (*) are recommended for further reading.

Ashton, L. W. 1989. Geochemical exploration guidelines to disseminated gold deposits. *Min. Eng.*, March: 169–74.

Boyle, R. W. 1979. The geochemistry of gold and its deposits. *Canada Geol. Survey Bull.* 280.

*Boyle, R. W. 1987. *Gold: History and Genesis of Deposits.* New York: Van Nostrand Reinhold.

Fersman, A. E. 1939. Gold, *Geochemistry* (4): 262–71. (In Boyle, 1987, pp. 105–16.)

Fletcher, W. K. 1981. Analytical methods in geochemical prospecting. *Handbook of Exploration Geochemistry*, Vol. 1. New York: Elsevier Scientific Publishers.

Freise, F. W. 1931. *Econ. Geol.* 26(4): 424–31.

Girling, C. A., and P. J. Peterson. 1980. *Gold Bull.* 13(4): 151–57.

Henley, K. J. 1975. Gold ore mineralogy and its relation to metallurgical treatment. *Mineral Sc. and Eng.* 17(4): 289–312.

McQuiston Jr., F. W., and R. S. Shoemaker. 1975. *Gold and Silver Cyanidation Plant Practice*. Vol. I, S.M.E.-A.I.M.E., New York.

Moeller, T. 1958. *Inorganic Chemistry*, pp. 818–43, New York: John Wiley & Sons.

*Petrovskaya, N. V. 1973. An outline of the geochemistry of gold. In *Native Gold*, pp. 8–20, Moscow. (In Boyle, 1987, pp. 135–50.)

Rapson, W. S. 1982. Effects of biological systems on metallic gold. *Gold Bull.* (15): 19–20.

CHAPTER 2

Physical and Chemical Properties of Gold

Physical Properties of Gold

Gold is a soft yellow metal, with the highest ductility and malleability of any metal. Gold crystallizes in the cubic system, although crystals of gold are very rare (it is usually found as irregular plates or grains). Gold has high thermal and electrical conductivities. The only natural isotope of gold is ^{197}Au; however, 19 isotopes—ranging from ^{185}Au to ^{203}Au—have been produced artificially. Those isotopes are radioactive, with half-lives ranging from a few seconds to 199 days (see Table 2-1).

Pure gold and many gold alloys are nonmagnetic. An alloy of gold and manganese is somewhat magnetic, and alloys of gold with iron, nickel, or cobalt are ferromagnetic.

The equilibria of numerous binary gold alloys are described by Hansen and Anderko (1958). Except for white golds (Au-Ag), the carat golds, used mainly in jewelry, are alloys of gold, silver, and copper. The carat is used to express the proportion of gold contained: 24 carats are pure gold, 18 carats are 75% gold, and so on.

Gold forms alloys with a number of metals (see Table 2-2). Mercury wets gold particles, forming amalgams, and it is used in gold extraction operations to selectively remove gold from ground ores. Gold has a very low solubility (0.13%) in mercury. Mercury forms a solid solution with gold up to about 16% Hg. Larger contents of mercury form intermetallic compounds like Au_3Hg and Au_2Hg. Molten lead is a good solvent for gold, and is used as such in fire assay and in some secondary smelting operations.

TABLE 2-1. The isotopes of gold.

Mass No.[a]	*Half life*	*Mode of decay*[b]
177	1.35 sec	α
178	2.65 sec	α
179	7.25 sec	α
181	11.55 sec	α, EC
183	45.5 sec	α
185	4.3 min	α, EC
186	12 min	EC, γ
187	8 min	α, EC
188	8 min	EC, γ
189	29.7 min	EC, γ
189m	4.7 min	EC, γ
190	39 min	EC, γ
191	3.2 hr	EC, γ
192	5.0 hr	EC, β+, γ
193m	3.9 sec	γ
193	17.5 hr	EC, γ
194	39.5 hr	EC, β+, γ
195m	31.0 sec	β+, γ
195	183 day	EC,γ
196m	9.7 hr	γ
196	6.2 day	EC, β, γ
197m	7.5 sec	γ
197	stable	
198	2.70 day	β, γ
199	3.15 day	β, γ
200	48.4 min	β, γ
201	26 min	β, γ
202	30 sec	β, γ
203	55 sec	β, γ
204	4.05 sec	β, γ

[a]m = metastable.
[b]α = alpha emission, EC = electron capture, β+ = positron emission,
β = beta particle emission, γ = gamma radiation.
Reprinted by permission. R. J. Puddephatt, *The Chemistry of Gold*. Elsevier Science Publishers, Physical Science and Engineering Division (New York), 1978, p. 8. (Originally published in L. Myerscough, *Gold Bull.*, (6): 62, 1973.)

Chemical Properties of Gold

Gold is the most inert, or the noblest, of the metallic elements. It exhibits great stability and resistance to corrosion. Simple mineral acids, with the exception of selenic acid, do not dissolve gold. Hydrochloric acid in the presence of oxidants (such as nitric acid, oxygen, cupric or ferric ions, and manganese dioxide) dissolves gold. The combination of hydrochloric and nitric acids, *aqua regia*, vigorously attacks gold.

$$Au + 4HCl + HNO_3 \rightarrow H[AuCl_4] + 2H_2O + NO$$

TABLE 2-2. Eutectic and peritectic temperatures of gold alloys.

Alloy	*Eutectic °C*	*% Second metal*	*Peritectic °C*	*% Second metal*
Au-Bi	241	82		
Au-Cd	625	25		
Au-Co	990	10		
Au-Cr			1,160	21
Au-Fe			1,168	40
Au-Ge	356	12		
Au-Ni	950	17.5		
Au-Pb	215	85		
Au-Sb	360	24		
Au-Si	370	6		
Au-Zn	642	15		

Reprinted by permission. R. J. Puddephatt, *The Chemistry of Gold*. Elsevier Scientific Publishers, Physical Science and Engineering Division (New York), 1978, p. 31.

Gold will dissolve, as above, in aqueous solutions containing an oxidizing agent and a ligand for gold.

$$2Au + 2HCl + 6FeCl_3 \rightarrow 2H[AuCl_4] + 6FeCl_2$$

The main reaction in the extraction of gold from its ores is its dissolution in cyanide, with oxygen as the oxidant (the Elsner reaction)—another example of the oxidant-ligand effect.

$$4Au + 8KCN + O_2 + 2H_2O \rightarrow 4K[Au(CN)_2] + 4KOH$$

Gold is the only metal not attacked by either oxygen or sulfur at any temperature. It does, however, react with tellurium at high temperatures.

$$Au + 2Te \rightarrow AuTe_2$$

Gold reacts with all halogens; it reacts exothermally with bromine at room temperature.

$$2Au + 3Br_2 \rightarrow Au_2Br_6$$

Oxidation States in the Aqueous Chemistry of Gold

The electronic arrangements, $(n - 1)d^{10}ns^1$, characteristic of periodic group IB permit the removal of more than a single electron, since the energy differences between the ns and $(n - 1)d$ electrons are not major. Oxidation states +1, +2, and +3 are generally encountered with the IB group of elements. The extranuclear structure (ns^1) of the atoms of the IB metals indicates +1 as their most characteristic state. This is true only for silver, however. The +2 state for copper and the +3 state for gold are more common and more resistant to reduction than their respective +1 states (refer to Table 1-1).

Oxidation State I. Among the group IB elements, silver alone forms a simple aqueous ion in oxidation state I. Gold(I) and copper(I) compounds disproportionate in aqueous solutions unless they are of very low solubility—e.g., Au(I)—or strongly complexed (e.g., $[Au(CN)_2]^-$, $[AuCl_2]^-$).

$$2Cu(I) \rightarrow Cu + Cu(II)$$
$$3Au(I) \rightarrow 2Au + Au(III)$$

The aurocyanide complex with a high stability constant (2×10^{38}) is by far more stable than the cuprous or silver cyanide complexes that are common in gold-extraction cyanide solutions. Other aurous complex ions of interest in the extractive metallurgy of gold are the chloride $[AuCl_2]^-$, the thiosulfate $[Au(S_2O_3)_2]^{3-}$, and the thiourea $[Au(NH_2CSNH_2)_2]^+$ ions.

Oxidation State II. Oxidation state II is very important in the chemistry of copper, but it is less common for silver and very rare for gold. There is evidence, however, of rare occurrences of Au(II):

- As a transient intermediate in redox reactions between Au(I) and Au(III)
- In compounds where there is a gold-to-gold bond
- In complexes with certain unsaturated sulfur ligands

Oxidation State III. This is the most important oxidation state for gold, and it is very rare for copper and silver. The auricyanide complex, $[Au(CN)_4]^-$, like the corresponding aurocyanide complex, is extremely stable (1×10^{56}). Examples of other stable gold(III) complexes are the chloride $[AuCl_4]^-$, the thiocyanate $[Au(SCN)_4]^-$, and the ammonia $[Au(NH_3)_4]^{+++}$ complexes.

Oxidation State V. The first gold(V) complex, $[AuF_6]^-$, was reported in the 1970s. The compound AuF_5 can also be prepared. These gold(V) compounds are powerful oxidizing agents. There are no similar compounds of copper and silver.

Summary of Oxidation States. The prevalent oxidation states of gold are III (auric) and I (aurous). Neither auric nor aurous ions are known to exist. Gold compounds are bound covalently and most often as complexes. Auric complexes are strong oxidizing agents.

There seem to be more differences than similarities among the properties of analogous compounds of the group IB metals. In some instances, similarities of gold complexes with complexes of its horizontal neighbors in the periodic table exist (e.g., $[AuCl_2]^{-1}$ *vs.* $[HgCl_2]^{-1}$ and $[AuCl_4]^{-1}$ *vs.* $[PtCl_4]^{-1}$).

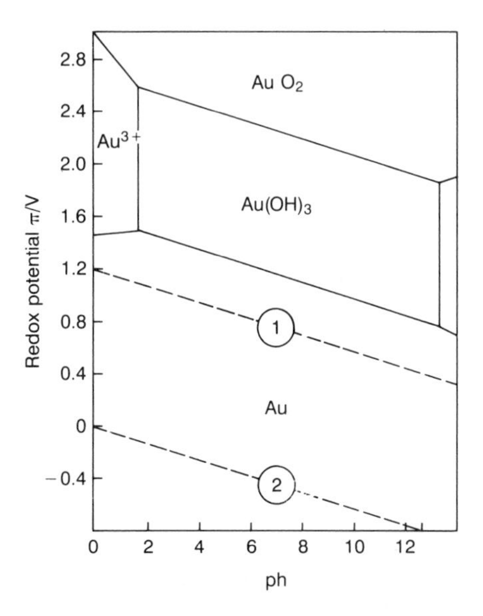

FIGURE 2-1. Pourbaix diagram for the system Au-H_2O at 250°C (all gold species at 10^{-4} M concentrations).

Reprinted by permission. R. J. Puddephatt, *The Chemistry of Gold*. Elsevier Scientific Publishers, Physical Science and Engineering Division (New York), 1978, p. 38.

Oxidation-Reduction Potentials

$$Au \rightarrow Au^{+} + e^{-}$$

The tendency of this reaction to occur, as given by the Nernst equation, is

$$q = q_0 + 2.303\frac{RT}{F}\log\frac{[Au^+]}{[Au]} = 1.7 + 0.059\log[Au^+] \text{ (in volts)}$$

For other oxidation properties

$$Au \rightarrow Au^{+++} + 3e^{-} \quad q = 1.50 + 0.0197\log[Au^{+++}]$$
$$Au + 3H_2O \rightleftharpoons Au(OH)_3 + 3H^{+} + 3e^{-} \quad q = 1.46 - 0.059pH$$
$$Au(OH)_3 \rightleftharpoons AuO_2 + H_2O + H^{+} + e \quad q = 2.63 - 0.059pH$$

These reactions in aqueous solution are limited by the stability of water, and can only take place if sufficiently strong oxidants are present. The system redox potential *vs*. pH is shown graphically in the Pourbaix diagram in Figure 2-1.

Binary Compounds of Gold

The complexes of gold(I) and gold(III) are the mainstays of its chemistry. The simple binary compounds of gold are often unstable in air. Gold forms

TABLE 2-3. Halides of gold and thermodynamic data at 298K for crystalline compounds.

Compound	*Color*	ρ_{20}^{pyk}/ g cm^{-3}	ΔH_f^θ/ kJ mol^{-1}	ΔG_f^θ/ kJ mol^{-1}
Oxidation state (I)				
(AuF)			−75 est.	
AuCl	yellow-white	7.8	−35	−16
AuBr	light yellow	7.9	−19 [20]	−15 [20]
AuI	lemon yellow	8.25	1.7 [22]	−3.3 [22]
Oxidation state (III)				
AuF_3	golden yellow	6.7	−360 [15]	
$AuCl_3$	red	4.3	−121	−54
$AuBr_3$	dark brown	—	−67.3 [20]	−36 [20]
Oxidation state (V)				
AuF_5	dark red	—	—	—

Reprinted by permission. R. J. Puddephatt, *The Chemistry of Gold*. Elsevier Science Publishers, Physical Science and Engineering Division (New York), 1978, p. 31.

hydrides, nitrides, and, indirectly, oxides, hydroxides, nitrates, and fluorosulfates. The following discussion, nevertheless, will be restricted to the more common binary compounds: halides, sulfides, selenides, and tellurides of gold.

Halides of gold. The known halides of gold and their thermodynamic data are presented in Table 2-3.

Gold Fluorides. AuF is unstable and disproportionates to AuF_3, which decomposes when heated to 500° C. AuF_3 is a powerful fluorinating agent. AuF_3 and AuF_5 are diamagnetic compounds.

Gold Chlorides. AuCl is metastable and disproportionates to Au and Au_2Cl_6 very slowly. This disproportionation to Au(III) species, including $AuCl_3 \cdot H_2O$, is very rapid in water. It dissolves in concentrated NaCl solution:

$$NaCl + AuCl \rightleftharpoons Na[AuCl_2]$$

The complex $[AuCl_2]^-$ is decomposed by water:

$$2Na[AuCl_2] + H_2O \rightarrow Au + Na[AuCl_4] + NaOH + 0.5H_2$$

AuCl, in the gas phase, exists as the dimer Au_2Cl_2. Gold(III) chloride—auric chloride—has a dimeric structure, as shown in Figure 2-2.

Au_2Cl_6 also dissolves in concentrated chloride solutions to form aurichlorides, salts of chloroauric acid ($HAuCl_4$):

$$2NaCl + Au_2Cl_6 \rightleftharpoons 2Na[AuCl_4]$$

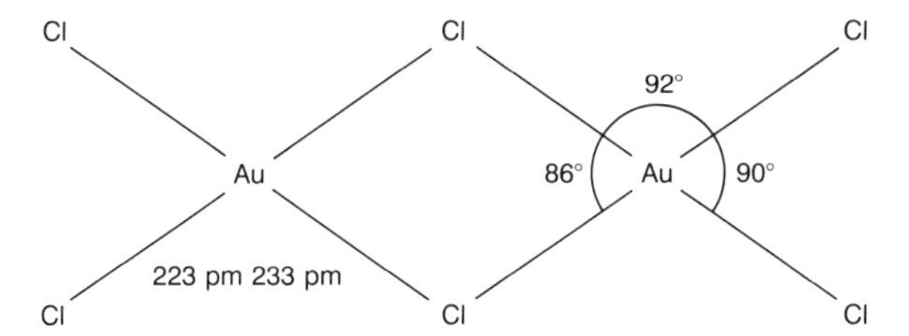

FIGURE 2-2. Dimeric structure of auric chloride.
Reprinted by permission. R. J. Puddephatt, *The Chemistry of Gold*. Elsevier Scientific Publishers, Physical Science and Engineering Division, 1978, p. 32.

The salts of chloroauric acid hydrolyze in water:

$$Na[AuCl_4] + H_2O \rightarrow [AuCl_3OH]^- + Na^+ + H^+ + Cl^-$$

The chlorine bridges of Au_2Cl_6 (see Figure 2-2) are easily broken by many neutral ligands (L) to give complexes, e.g., $[LAuCl_3]^-$. With easily oxidized ligands, Au_2Cl_6 may act as a chlorinating agent. Auric chloride is decomposed in air by heat. This decomposition is rapid at temperatures higher than 180° C.

Gold Bromides. Liquid bromine reacts exothermally with gold powder to form Au_2Br_6. Its structure is dimeric, like that of Au_2Cl_6, and its chemical properties resemble also those of Au_2Cl_6. For example, it forms complexes of the type $[LAuBr_3]^-$ with nitrogen-donor ligands. On heating Au_2Br_6 to about 200° C, it decomposes to bromine and AuBr. The monobromide of gold, like its corresponding monochloride, is sensitive to moisture decomposing to auri- and aurobromide complexes, such as $[LAuBr]^-$ and $[AuBr_2]^-$.

Gold Iodides. Many gold(III) complexes readily oxidize iodides to iodine. Compounds with the formula AuI_3 are actually compounds of gold with the complex I_3^- (I_2I^-) rather than gold(III) iodide. Gold(I) iodide may be obtained from the reaction of aqueous aurichloride with iodide.

$$H[AuCl_4] + KI \rightarrow AuI + KCl + HCl + Cl_2$$

Gold iodide may also be obtained from the rather slow, direct reaction of gold with iodine, at elevated temperature. AuI forms yellow crystals that are more stable in water than AuCl. It dissolves in aqueous iodide solution to form $[AuI_2]^-$, and in iodide-iodine solutions to give mixtures of auro- and auri-complexes (e.g., $[AuI_2]^-$ and $[AuI_4]^-$).

Pseudohalides of Gold. Yellow crystals, which have a polymeric structure with linear -AuCN-AuCN- chains, are formed by heating $K[Au(CN)_2]$ with hydrochloric acid. The polymeric chain is insoluble in water, but redissolves readily in the presence of cyanide solution, forming the complex ion $[Au(CN)_2]^-$.

Anhydrous gold(III) cyanide is not known, but its hydrated complex, $Au(CN)_3 \cdot 3H_2O$ can be prepared by dehydration of solutions of the auricyanide complex, $H[Au(CN)_4]$. The $Au(CN)_3 \cdot 3H_2O$ decomposes to gold(I) cyanide (AuCN) upon being heated.

Sulfides, Selenides, and Tellurides of Gold. Although gold does not react directly with sulfur, gold(I) and gold(III) sulfides can be prepared indirectly.

$$2\,K[Au(CN)_2] + H_2S \rightarrow Au_2S + 2KCN + 2HCN$$

Au_2S precipitates as black-brown powder that is slightly soluble in water but quite soluble in cyanide and polysulfide solutions.

Au_2S_3 can be prepared by reaction of H_2S with Au_2Cl_6 in anhydrous ether at a low temperature. It dissolves with decomposition to $[Au(CN)_2]^-$ in cyanide solution. Auric sulfide, Au_2S_3, is stable when dry up to 200° C, but it decomposes in water to form metallic gold and sulfuric acid.

Two selenides of gold are known, AuSe and Au_2Se_3. AuSe may be better represented as $Au(I)Au(III)Se_2$. Of the tellurides of gold, $AuTe_2$ occurs in the minerals calaverite and sylvanite, and can also be prepared by direct reaction of gold with tellurium. Au_2Te_3 occurs in the mineral montbrayite. Gold selenides and tellurides are often found in the slimes collected from cells of copper electrolytic refining.

Gold Complex Ions

The chemistry of gold complexes is quite extensive. A very brief overview is presented here. A thorough presentation of this subject is to be found in *The Chemistry of Gold* by R. J. Puddephatt (1978).

The stabilities of both Au(I) and Au(III) complexes tend to decrease with increasing electronegativity of the atom bound directly to the gold (Finkelstein, 1972), as shown in the diagram below:

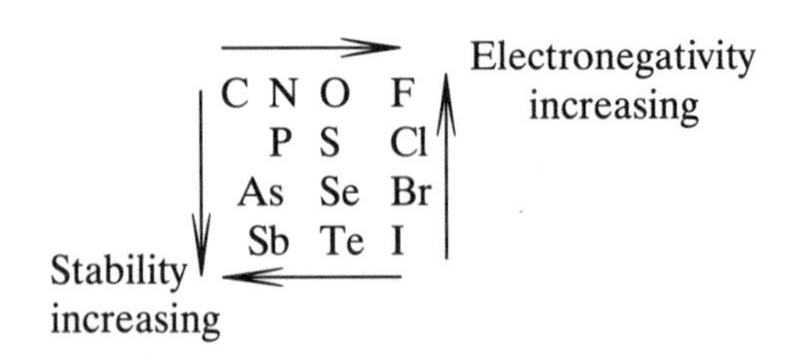

Some overall stability constants of Au(I) and Au(III) complexes, as given by Finkelstein (1972), can be found in Table 2-4.

Gold(I) Complexes. The complex ion $[Au(CN)_2]^-$—the backbone of the chemical extraction of gold from its ores—is perhaps the most stable complex formed by gold (Puddephatt, 1978).

$$Au^+ + 2CN^- \rightarrow [Au(CN)_2]^- \text{ Stab. const. k} \approx 10^{39}$$

On boiling an aqueous solution of colorless $K[Au(CN)_2]$ with 2M HCl, the complex decomposes and lemon-yellow AuCN crystallizes from solution. The acid $H[Au(CN)_2]$ can be prepared by ion exchange from $K[Au(CN)_2]$, but it decomposes to HCN and AuCN when heated to 100° C. Gold(I) also forms complexes with carbonyl, e.g., [AuCl(CO)]; isocyanide, e.g., $[Au(RNC)_2]^+$; nitrogen-donor complexes, e.g., $[Au(MeCN)_2]^+$; azides; isocyanates; phosphorous, arsenic, and antimony complexes; gold-oxygen bonds; and sulfur or selenium-donor ligands, including thiosulfate complexes, dialkylsulfide, and dithiocarbamate.

Halide complex ions $[AuX_2]^-$ are known. Their stability increases with the heavier halide ions.

$$AuI + I^- \rightleftharpoons [AuI_2]^-$$
$$[AuI_2^-] + I_2 \rightleftharpoons [AuI_4]^-$$

The existence of $[AuCl_2]^-$ and $[AuBr_2]^-$ in aqueous solutions has been debated.

Gold(II) Complexes. Many complexes of gold whose empirical formulas suggest the presence of gold(II) have been shown to be mixed gold(I) and gold(III) species.

Gold(III) Complexes. Adding a solution of $K[AuCl_4]$ to a concentrated solution of KCN results in the complex $K[Au(CN)_4]$. Ammine, nitrogen-

TABLE 2-4. Stability constants.

Au(I)			*Au(III)*
Complex	B_2	Complex	B_4
$Au(CN)_2^-$	2×10^{38}	$Au(CN)_4^-$	10^{56}
$Au(S_2O_3)_2^{-3}$	5×10^{28}	AuI_4^-	5×10^{47}
AuI_2^-	4×10^{19}		
$Au(SCN)_2^-$	1.3×10^{17}	$Au(SCN)_4^-$	10^{32}
$AuCl_2^-$	10^9	$AuCl_4^-$	10^{26}

Reprinted by permission. N. P. Finkelstein, Recovery of gold from solutions. In *Gold Metallurgy in South Africa*, edited by R. J. Adamson. Chamber of Mines of South Africa (Johannesburg), 1972.

donor, and related complexes can be prepared by displacing chloride from aurichloride, $H[AuCl_4]$, with a nitrogen-donor ligand, e.g., $[Au(NH_3)_4]NO_3$.

Gold(III) also forms complexes with phosphorus-, arsenic-, and antimony-donor ligands; nitrates, e.g., $[NO_2]^+[Au(NO_3)_4]^-$; and sulfur- and selenium-donor ligands. Complex halides of gold(III) include $K[AuF_4]$, $AuF_3{\cdot}BrF_3$, and $AuF_3{\cdot}SeF_3$; $H[AuBr_4]$; and $H[AuCl_4]$. Hydroxo species are formed in alkaline solutions of gold(III).

$$[AuCl_4]^- + H_2O \rightleftharpoons [AuCl_3(OH)]^- + H^+ + Cl^-$$
$$[AuCl_4]^- + 2H_2O \rightleftharpoons [AuCl_2(OH)_2]^- + 2H^+ + 2Cl^-$$

Stability of Gold Complexes. Stability constants of gold(I) complexes in aqueous solutions are difficult to obtain due to their tendency to disproportionate. The order of stability constants of gold(III) complexes is given below (Puddephatt, 1978).

- Anionic complexes $[AuX_2]^-$

$$NCO^- < NCS^- \sim Cl^- < Br^- < I^- \ll CN^-$$

- Cationic complexes $[AuL_2]^+$

$$Ph_3PO < Me_2S < py < AsPh_3 < NH_3 \ll PPh_3 \sim MeNC$$
$$MeNC < PMePh_2 < PMe_2Ph$$

- Neutral complexes $[AuXL]$

$$Me_2O \sim NPh_3 \ll Me_2S < Me_2Se < SbPh_3 < AsPh_3 < Me_2Te$$
$$Me_2Te < PPh_3 < PMePh_2 < PMe_2Ph$$

Gold Dissolution and Precipitation Reactions

As already stated, gold will dissolve in aqueous solutions containing an oxidizing agent and a ligand for gold. The dissolution of gold in *aqua regia* is used in analytical chemistry for either volumetric or gravimetric determinations of the soluble gold.

$$Au + 4HCl + HNO_3 \rightarrow H[AuCl_4] + 2H_2O + NO$$

The dissolution of gold in cyanide solution, in the presence of oxygen, is the main reaction in the extraction of gold from its ores.

$$4Au + 8NaCN + O_2 + 2H_2O \rightarrow 4Na[Au(CN)_2] + 4NaOH$$

Thiourea, $CS(NH_2)_2$, forms a soluble complex of gold, and is used as leaching reagent for gold under acidic conditions:

$$2Au + 4CS(NH_2)_2 + Fe_2(SO_4)_3 \rightarrow [Au(CS(NH_2)_2)_2]_2SO_4 + 2FeSO_4$$

Gold reacts with halogens, forming soluble halides:

$$2Au + 3Br_2 \rightarrow Au_2Br_6$$

Finely divided gold is soluble to a limited extent in solutions of sodium thiosulfate. Cementation of gold by zinc dust is widely used for the recovery of gold from cyanide solution.

$$Na[Au(CN)_2] + 2NaCN + H_2O + Zn \rightarrow$$
$$Au + Na_2[Zn(CN)_4] + + NaOH + O.5H_2$$

Aluminum can precipitate gold from cyanide solution in the presence of sodium hydroxide.

$$Al + 4OH^- + Na^+ + 3Au^+ \rightleftharpoons 3Au + Na^+ + Al0_2^- + 2H_2O$$

Gold Assaying

Sample Preparation

Samples of ores, concentrates, residues, slags, flue dusts, and any nonmetallic gold-containing materials are pulverized to -100 mesh, mixed very well, and split to smaller samples. Molten metals may be sampled by shotting or dipping, whereas solid metals have to be drilled.

Detection

Gold in solution may be detected by the "Purple of Cassius" color test. The traces of gold, if any, are precipitated by adding a small quantity of NaCN solution, zinc dust, and lead acetate solution to a large sample (1 l) of the solution to be tested. After shaking, a precipitate may form. After decanting the excess of solution, the precipitate is treated with *aqua regia* and cooled. Addition of drops of stannous chloride ($SnCl_2$) will produce a yellow to purple tint (for solutions originally containing 0.03 to 0.3 ppm of gold). Gold, in solids or in solutions, may also be detected by emission spectrography.

Gold Determination[1]

Gravimetric Methods. Fire assay is the traditional method employed for determination of gold in ores, concentrates, metals, refinery slimes, and other solid materials. One part of the finely ground solid is mixed with three parts of a flux containing litharge and carbon. The mixture is fused, at 900–1,100° C, for at least an hour. After cooling, a button of lead (the

[1]See Lenahan and Marray-Smith (1986) for a detailed description of methods.

product of the reduction of litharge), which has collected all the gold and other precious and base metals, is separated from the slag. The button is placed in a cupel (made of bone, ash, or magnesia); "cuppelation"—consisting of heating to 950–1,000° C—is carried out; and the lead oxides are absorbed by the cupel. The remaining bead is treated with nitric acid to dissolve silver, leaving the gold and any platinum metals to be weighed.

Sulfur dioxide will reduce gold in hydrochloric solution to the metal. Only selenium, tellurium, and traces of lead may be found with the gold precipitate, and they can be redissolved with nitric acid. Alternative, reliable gravimetric procedures for gold in solution involve using other reducing reagents such as sodium nitrite, oxalic acid, and hydroquinone.

Volumetric Determinations. Volumetric determinations of gold in solution are based on reduction of gold(III) to metallic gold with titrated solutions of reducing reagents. Reduction of gold(III) to Au(I) by excess of iodide and titration of the released iodine is another method of determination.

Spectrophotometric Methods. These methods are especially useful with a sample that is insufficient for gravimetric and volumetric determinations. The simplest method is to convert the gold to $(AuCl_4)^-$ or $(AuBr_4)^-$ and measure the absorbance due to these ions. Impurities may interfere, and it is recommended that the auricomplex be extracted into an organic solvent before measuring absorbance.

Atomic Absorption and Emission Spectrography. Atomic absorption is a valuable technique for the determination of small concentrations of gold. Using chromatographic concentration and emission spectroscopy, the very low content of gold in sea water (10^{-12} ppm) can be determined with $\pm 10\%$ accuracy.

X-ray Fluorescence. This has been used for the determination of gold. However, it has not become a routine for analytical determination.

Neutron Activation. This is the most sensitive analytical method for determining gold. It is the preferred technique for use with very low concentrations of gold, as in sea water.

References

Note: Sources with an asterisk (*) are recommended for further reading.

Finkelstein, N. P. 1972. The chemistry of the extraction of gold from its ores. Chapter 10 in *Gold Metallurgy in South Africa*, edited by R. J. Adamson. Capetown: Chamber of Mines of South Africa.

Hansen, M., and K. Anderko. 1958. *Constitution of Binary Alloys*, 2d ed. New York: McGraw-Hill.

Lenahan, W. C., and R. de L. Marray-Smith. 1986. *Assay and Analytical Practice in the South African Mining Industry.* Johannesburg, S. Afr.: I.M.M.

Myerscough, L. 1973. The isotopes of gold. *Gold Bull.* (6): 62.

*Puddephatt, R. J. 1978. *The Chemistry of Gold*. New York: Elsevier Scientific Publishers.

CHAPTER 3

Treatment of Placer Deposits

Alluvial, or placer, gold has been washed away by flowing water or carried by blowing wind. It has then been deposited along with earth, sand, gravel, and other transported matter, especially on river beds. Although there is no way of knowing exactly how long ago humans felt enamored of the shiny yellow metal, there is reason to presume that gold, specifically alluvial gold, has been known and appreciated for at least 12,000 years. The Egyptians knew how to melt gold and silver about 6,000 years ago, and records indicate that gravity concentration of gold was practiced around 4000 B.C.

Placer Environment and Formation

The Russian scientist M. V. Lomonosov (1711-1765) was among the first to recognize that placers result from "the fracturing of lodes and nowhere is it more hopeful to seek them as along rivers in upper reaches of [gold] ore mountains." Charles Lyell recognized the indomitable power of natural erosion, and stated in his *Principles of Geology* (1830) that "given enough time whole landscapes can be created or destroyed by the action of the slow, yet relentless, forces of Nature."

Contrary to the ancient Greek Neptunian theory, which ascribed the formation of all strata and rocks solely to water and to catastrophic floodings, modern scientists—starting with Lomonosov and then Lyell—recognized the effect of additional environmental factors. The task of rationally classifying the forces that have been shaping the skin of our planet has

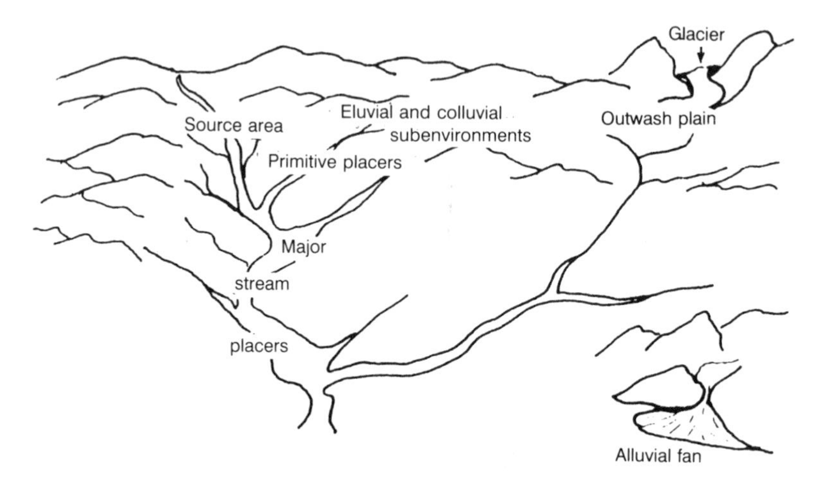

FIGURE 3-1. Sketch of a continental placer environment.
Reprinted by permission. E. H. MacDonald, *Alluvial Mining*. Chapman & Hall (London), 1983.

been going on for almost a century. A classification based on "techniques for exploration and mining" was proposed by Macdonald (1983), and includes three cardinal environments.

1. The continental placer environment (Figure 3-1), where common exploration techniques rely on drilling and/or pitting to obtain samples
2. The transitional placer environment (Figure 3-2), which starts offshore, where the waves first disturb sediments on the sea floor, and extends inland to the limits of the aeolian transportation
3. The marine placer environment (Figure 3-3), which is an underwater continuation of the adjacent land with the same bedrock geology

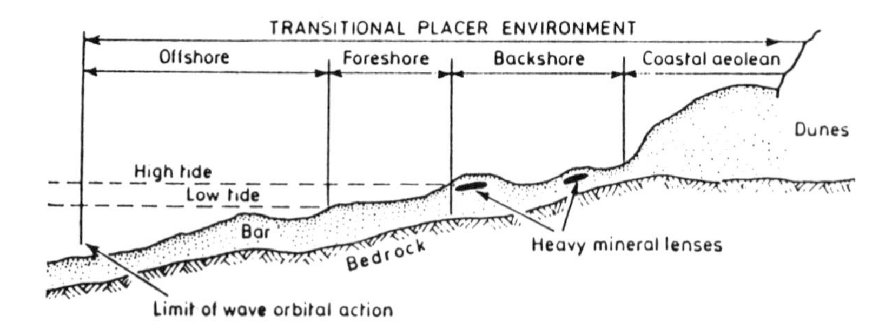

FIGURE 3-2. Typical section across transitional placer environment.
Reprinted by permission. E. H. MacDonald, *Alluvial Mining*. Chapman & Hall (London), 1983.

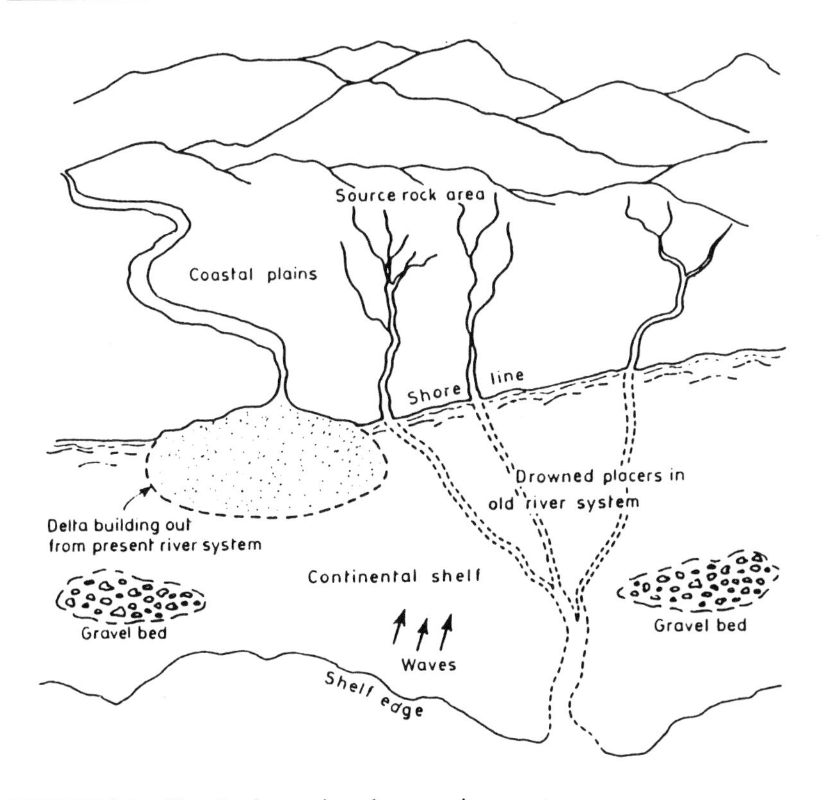

FIGURE 3-3. Sketch of a marine placer environment.
Reprinted by permission. E. H. MacDonald, *Alluvial Mining*. Chapman & Hall (London), 1983.

These main placer environments—all of interest for finding alluvial gold—are described more adequately in Table 3-1. The distinctive features of the continental and transitional subenvironments are listed respectively in Tables 3-2 and 3-3.

The following extract from E. H. Macdonald's (1983) book on alluvial mining gives an excellent description of placer gold formation.

> Placer gold is derived mainly from the weathering of quartz or calcite reefs and stringers within igneous and metamorphic rocks, although rich concentrations are not necessarily indicative of a central rich source. For example, the Raigarh Goldfield of Madhya Pradesh, India, covers an area of about 100 km; it has yielded large quantities of placer gold for many hundreds of years with no historical record of underground mining and recent investigations to discover economically viable reef deposits have been no more successful. Rich alluvial gold concentrations at Edie Creek and Bulolo in New Guinea were also probably derived from widespread stringers rather than

TABLE 3-1. The placer environments.

Environment	*Sub-environment*	*Main products*	*Environmental process elements*	*Exploration techniques*	*Mining methods*
Continental	Eluvial	Au, Pt, Sn, WO_3 Ta, Nb, gem stones (all varieties)	Percolating waters, chemical and biological reactions, heat, wind, and rain	Soil sampling, shallow pitting, churn drilling	Open cast, hydraulic sluicing, hand mining
	Colluvial	Au, Pt, Sn, WO_3 Ta, Nb, gem stones (all varieties)	Surface creep, wind, rainwash, elutriation; frost	Stream and soil sampling, shallow pitting and trenching	Hydraulic sluicing, bulldozer and loader—hand mining
	Fluvial	Au, Pt, Sn, rarely Ta, Nb, diamonds, and corundums	Flowing streams of water	Stream sampling, geophysics, pitting, churn auger and pitdigging drills, banka drills	Bucket dredging in active beds, bucket dredging, hydraulic sluicing, and dozer-loader operations in old stream beds
	Desert	Au, Pt, Sn, WO_3 Ta, Nb, gem stones (all varieties)	Wind with minor stream flow; heat and frost	Shallow pitting, churn and pit digging drills, geophysics	Various earth-moving combinations
	Glacial	Au (rarely)	Moving streams of ice and melt waters	Stream sampling and pitting	Hydraulic sluicing
Transitional	Strandline	Ti, Zr, Fe, ReO, Au, Pt, Sn	Waves, currents, wind, tides	Hand augering and sludging, sample splitting allowable	Suction-cutter dredging, bulldozer and loaders, bucket-wheel dredging

	Coastal Aeolean	Ti, Zr, Fe, ReO	Wind and rain splash	Power augers (hollow), sample splitting allowable	Suction-cutter dredging, bulldozers and buried loaders
	Deltaic	Ti, Zr, Fe, ReO	Waves, currents, wind, tides and channel flow	Hand augering and sludging, sample splitting allowable	Specially designed shallow depth dredges having great mobility
Marine	Drowned placers	Au, Pt, Sn, diamonds, minor Ti, Zr, Fe, ReO, industrial sand and gravel	Eustatic, isostatic, and tectonic movements—net rise in sea level	Geophysics (seismic refraction and reflection) bottom sampling, remote sensing, hammer, jet, vibro and banka drills, positioning	Bucket-line dredging, jetting, clamshell, rarely suction-cutter dredging

Reprinted by permission, E. H. MacDonald, *Alluvial Mining*. Chapman & Hall (London), 1983.

TABLE 3-2. Subenvironments of the continental placer environment.

Subenvironment	***Features of distinction***
Eluvial	Weathering *in situ*; upgrading largely through the removal of soluble minerals and colloids; some surface material removed by sheet flow, rivulets, and wind; all minerals may be represented in partially weathered sectors; end products at surface may be chemically stable.
Colluvial	Downslope movement of weathered rock controlled principally by gravity; all placer minerals may be represented but sorting poorly developed.
Fluvial	The most important subdivision of the environment characterized by wide range of depositional land forms; most deposits formed within few kilometers of source rocks; particle size reduces and sorting improves with distance from source; only chemically and physically stable particles persist; many deposits are relict from earlier times.
Desert	Somewhat similar characteristics to eluvial and colluvial subenvironments except for wind as principal transporting agency; flash flooding in some desert areas induces fluviation in gutters and channels; all types of placer minerals may be represented but principal varieties are tin, gold, and pegmatite minerals.
Glacial	Deposits unsorted and unstratified; rare economic concentrations; moraines and tills in outwash plains sometimes give rise to more important concentrations due to subsequent stream action; upgrading also occurs along shorelines where glaciers discharge into the sea.

Reprinted by permission. E. H. MacDonald, *Alluvial Mining*. Chapman & Hall (London), 1983.

from single primary bodies although none have been found. On the other hand, in other placer mining districts such as the famous old gold diggings at Ballarat, Victoria, Australia, many of the most important deposits of gold and other placer minerals have been formed from weathering over a large area of provenance and an abundance of mineralized veinlets, widely dispersed.

Evans (1981) has proposed an alternative provenance for some placer gold deposits suggesting that gold, in small amounts but evenly distributed throughout ultrabasic rocks, may be chemically dissolved and reprecipitated in possibly commercial placer concentrations by the normal process of lateritization. He cites as possible examples a French Guiana gold placer where the only possible source appeared to be the thick, red, lateritic soil cut by the stream channel; and some of the gold deposits in the higher Tertiary channels of California's Mother Lode country, and on the west slope of the Sierra Nevada. A similar environment to that of French Guiana, in Suriname, provides another possible example at Royal Hill

TABLE 3-3. Subenvironments of the transitional placer environment.

Subenvironment	***Features of distinction***
Strandline	Beach placers are formed by the action of waves, tides, currents, and wind; principal minerals are the most resistant of the low-density varieties such as rutile, ilmenite, zircon, and monazite; sometimes gold, platinum, tin, and diamonds; rarely others.
Aeolean	Placers formed in sand blown up from beaches; dunal systems developed from both stationary and transgressive dunes; mineral suites as for beaches but particle size generally finer.
Deltaic	Formed around mouths of rivers carrying large-sediment loads; seaward margins favor reworking of sediments and repetitive vertical sequences common; lower-density varieties common.

Reprinted by permission. E. H. MacDonald, *Alluvial Mining*. Chapman & Hall (London), 1983.

where the auriferous laterites, totalling some 5 million tonnes, have given rise to extensive placer workings. Evans makes reference to an article by Dr. Landsweert (1869) who wrote that:

> Everyone concurs in the belief that alluvial gold has been derived at some time or other from lodes; but seeing that the largest piece of gold ever found in the matrix is insignificant when compared with the nuggets that have sometimes been found in the alluvium, it has been a difficult matter to reconcile belief with experience. . . . The occurrence of larger nuggets in gravel deposits that have been found in quartz lodes with the fact that alluvial gold almost universally has a higher standard of fineness, would seem to imply a different origin for the two.

A significant fraction of alluvial gold is generally expected to be coarser than lode or reef gold. Both physical and chemical accretions have been advanced as the cause of the coarse size fraction. However, several experimental studies with detrital minerals in a simulated river bed have indicated that gold is degraded more than it is accreted. The fact is that the data on the particle size of alluvial gold are usually obtained from concentrates, from which the "fine gold tail" may have escaped (during the gravity concentration), and some accretions among malleable gold particles may have occurred.

Dredging or Hydraulic Mining

Dredging is a method of digging underwater placer deposits by rotating a cutterhead and suction line or by rotating a cutting-bucket line. Dredging equipment digs, scrapes, and raises gravel and/or mud from the bottom of rivers, lakes, or ponds and delivers it onto a floating platform (a pontoon or hull). The dredged material is washed and screened, and the contained gold is recovered by gravity concentration. The tailings are returned into the water or stacked on the bank, if there are no environmental objections.

Dredgers may be divided according to their means of digging and raising up the gravel, which may be raised to:

- Suction pumps or cutterhead and suction pumps (suction-cutter dredgers)
- Continuous bucket-line or bucket-wheel elevators
- Crane and bucket (clamshell) or mechanical shovel

Bucket-line and suction-cutter dredgers are the equipment mainly used in mining gold placer deposits. Clamshell dredgers have a limited scope for mining small deposits of underwater loose gravel.

Transitional and marine placer deposits can easily generate solid-water pulps that can be mined and transported more easily (and therefore at a lower cost) than rocky or solid earth deposits. The description of a device claiming the elevation of fluids through the use of a rotating core was first recorded in France in 1372. A centrifugal pump, in a configuration remotely close to modern designs, was invented by Papin in 1705. The first steam engine for dredge service was constructed in England in 1795.

The first recorded use of a centrifugal pump in a dredging barge is in the year 1855. Since the middle 1700s, primitive forms of dredging had been used along the Gold Coast of Africa. The idea of moving valuable ore as solid-liquid suspension has been very appealing because of its relatively low cost. The Suez Canal was built by the French engineer Bozin in 1867 mainly by hydraulic transportation of water-solid mixtures. The Chicago area was drained with suction-cutter dredgers.

A so-called "hydraulic dredge" built in Germany in 1855 was a barge with a steam engine and a pump assembly. A mechanical cutter head surrounding the suction nozzle was added later. The first suction-cutter dredger in the United States was developed in 1862. The conversion of a steamship for dredging in the United States was proposed in 1871, and cutterhead dredges were in extensive use by 1876. Electric power, generated by diesel engine, was first used for dredger propulsion and operation in 1922.

Bucket-Line (or Bucket-Elevator) Dredgers

These dredgers were introduced to alluvial mining in the second half of the 19th century. Bucket-line dredging is performed by a line of heavy steel buckets revolving on an endless chain that is supported by a heavy, strong dredging ladder. The dredging ladder is made of girders held together by cross members, and is pivoted from a central framework that also supports the driving pulleys. The supporting structures, the digging mechanism, and

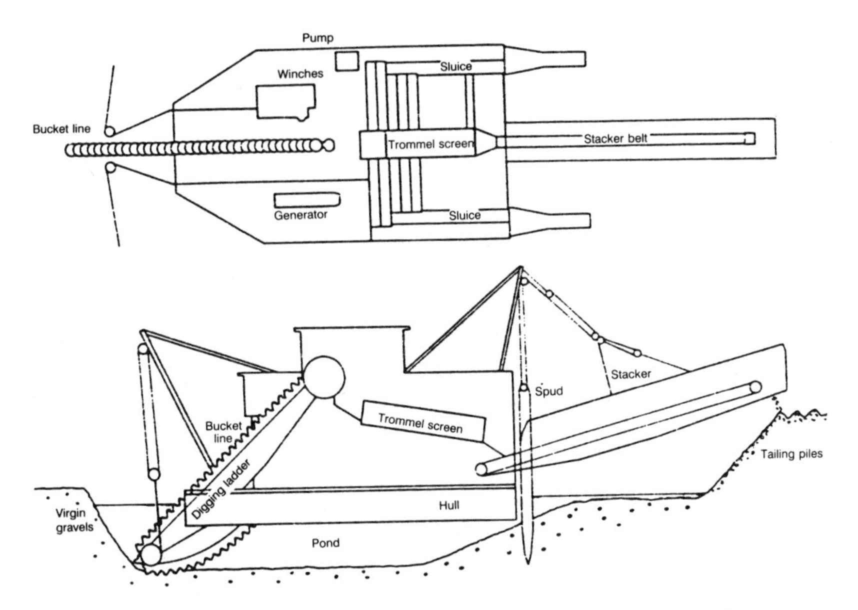

FIGURE 3-4. Plan view and section of a continuous bucket-line dredger.
Reprinted by permission. A. Woodsend, in *The Mining Journal Ltd.* (London), August 1984: 94.

(in most cases) the gravity concentration facilities are installed on a large pontoon (Figure 3-4). The boxlike pontoon has an elongated cut at its forward end to accommodate the dredging ladder and the moving bucket chain.

The forceful movement of the bucket-line dredger excavates the placer deposit and feeds it into the treatment plant. Great strength is needed for the continuous, multibucket excavation and ore transportation; hence, bucket-line dredgers are huge, very heavy structures. They weigh 7,000 to 8,000 tons, and are able to dredge 1,000 to 1,200 $m^3 h^{-1}$.

Bucket-line dredgers require high capital investment and therefore their selection for mining alluvial deposits has to be supported by

- large reserves (over 10,000,000 m^3 for small dredgers and over 120,000,000 m^3 of good-grade placer ore for large operations);
- an adequate supply of pond water, without inordinate numbers of tree roots and aquatic vines; and
- a favorable "floor" for dredging, not densely compacted, gently sloping, and without large boulders.

Most of the bucket-line dredgers currently in operation are used in mining alluvial tin. A good number are used in dredging gold, platinum, and gemstone deposits. Dredgers in the tin fields are the most advanced and yield the highest production rates (up to 5×10^6 m^3 per year). They dig as deep as 45 meters.

High-grade manganese steels have to be used for dredger buckets; their lips must be exceptionally hardened. Casting the lip as an integral part of the bucket is currently preferred over riveted or bolted replaceable lips. Advanced wear of lips (as compared to wear of the hood) should be corrected by welding inserts.

Although large buckets are expected to create high production rates, a number of specific local conditions (such as large boulders, rocky ground, and depth) may limit the bucket size. Buckets in actual practice range in volume between 400 and 680 liters.

Additional costs are involved in the construction of the dredger and in lifting the alluvial mineral from increasing depths. Deep deposits require higher productivity (larger buckets, at the optimum speed) to compensate for the added costs. Economic considerations will suggest the minimum bucket size allowable at the specific depth.

The design speed of a bucket line obviously depends on the conditions of the ground to be cropped. Most dredgers operate at a speed of 20 to 24 buckets per minute, regardless of local conditions. In good, sandy grounds, speeds up to 35 buckets per minute can be achieved. Heavy steel frames designed to carry the bucket line to the maximum anticipated depth are required for the dredging ladders.

Suction-Cutter Dredging

Suction-cutter dredging was carried out initially by bringing the suction pipe of a pump down to the bottom of the pond to be dredged, jetting water around the tip of the pipe to fluidize the sand, and sucking up the sand. This was a simple, portable, low-cost set-up, but it had the potential disadvantage of high losses in heavy minerals. Water jets cannot completely fluidize the heavy components of the sand, and cannot disintegrate compacted layers of materials.

Some powerful digging was required in most instances, and cutter heads were introduced to dig and disintegrate indurated and compacted materials. The mechanical action of the cutter assists in fluidizing the broken materials—including heavy minerals—around the end of a suction pipe. Revolving cutter heads with curved steel blades, drive-shafts, and driving mechanisms are mounted on a steel-beam ladder along with the dredge's suction pipe. The ladder is pivoted on the pontoon and lowered through a slot toward the alluvial deposit, to dig it and fluidize it. The closeness of the end of the suction pipe to the cutter head assures efficient recovery of the broken solids.

Dredges for offshore work have the required gravity-concentration equipment and the dredging ladder mounted on one pontoon (Figure 3-5). Dredgers for onshore work are preferably self-contained on a smaller hull

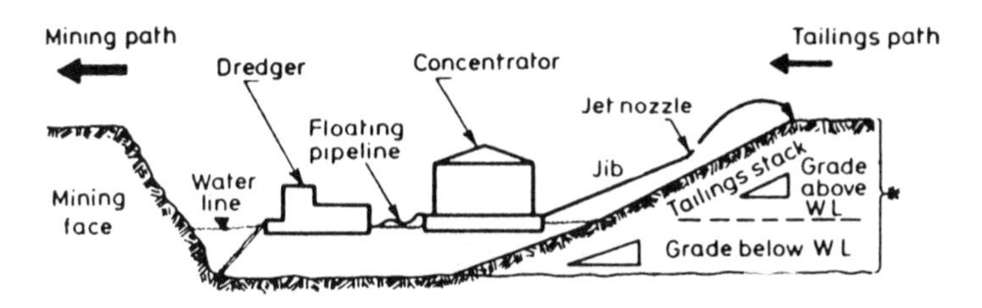

FIGURE 3-5. Suction dredger along with a concentration plant.
Reprinted by permission. E. H. MacDonald, *Alluvial Mining*. Chapman & Hall (London), 1983.

so they have more mobility in small, artificial ponds. Their dredged slurry is pumped into the onshore receiving bin of a treatment plant.

The suction lift of hull-mounted pumps is limited to about 5 meters, by practical design considerations, in most dredging operations. Large suction cutters have up to 7,000 HP connected to cutters and pumps. Suction-cutter dredgers are, of course, more power-intensive than bucket-line dredgers, mainly due to the amount of material cut and pumped. The stresses imposed on ladders, pivots, and hulls in suction-cutters are, however, lower than in bucket-line dredgers. Hence, suction-cutter dredgers have lower capital costs and generate higher operating costs than bucket-line dredgers of the same rating capacity. A relative comparison (1969 conditions) of capacities and costs for bucket-line and suction-cutter (hydraulic) dredgers is presented in Figure 3-6.

Suction-cutter dredgers work efficiently—to full capacity—only on fine, loose, or loosely consolidated alluvial deposits. Bucket-line dredgers can operate successfully even on tight formations with sizable gravel, cobbles, and small boulders.

Suction-cutter dredgers are ideal for large-scale stripping of sandy and/or clay overburden, since they can mine and transport large quantities of fine earth over considerable distances in a single operation. Huge capacities, in excess of 2,000 m^3h^{-1} of fine solids, can be achieved with suction dredgers operating with only one pump and delivering the spoil through a pipeline to distances greater than 400 meters.

Some suction-cutter dredgers working offshore (in Thailand) can operate at depths of up to 35 meters with specially designed pumps and drives installed near the bottom of the ladder. Hydraulic dredgers, using jet-lift pumps, dredge alluvial tin sands from depths of up to 100 meters in waters too deep for conventional suction dredging.

The cutting devices are cutting heads at the end of a shaft, rotating at 15 to 45 rpm depending on the nature of the deposit, the size of the cutter, and the length of the rotating shaft. The cutting devices are classified, according to the position of their shaft in relation to the suction line, into

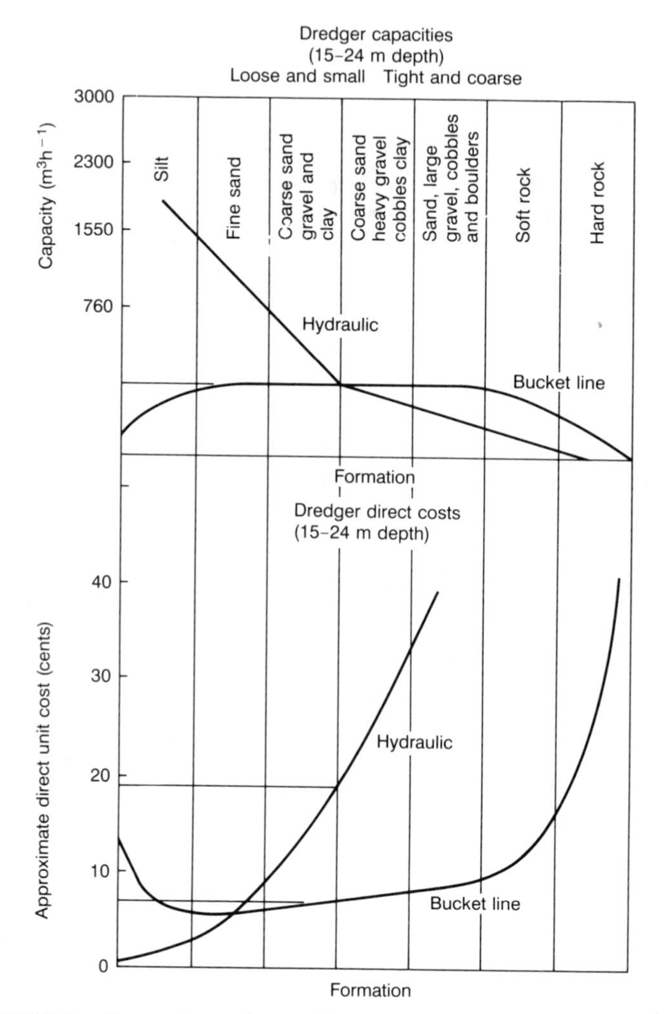

FIGURE 3-6. Comparison of capacities and relative (1969) costs for bucket-line and suction-cutter dredgers.

Reprinted by Permission, E. H. MacDonald, *Alluvial Mining*. Chapman & Hall (London), 1983.

those with a rotating shaft parallel to the suction line and those with a rotating shaft vertical to the suction line.

In addition to fluidizing fine solids close to the tip of the suction pipe, cutters must be designed to be able to dig into compacted or consolidated materials. Cutter-head assemblies must be designed to be able to chop tree roots and vines into pieces small enough to flow freely through pumps. The stainless steels used for cutting heads must be tough, hard, and resistant to breakage.

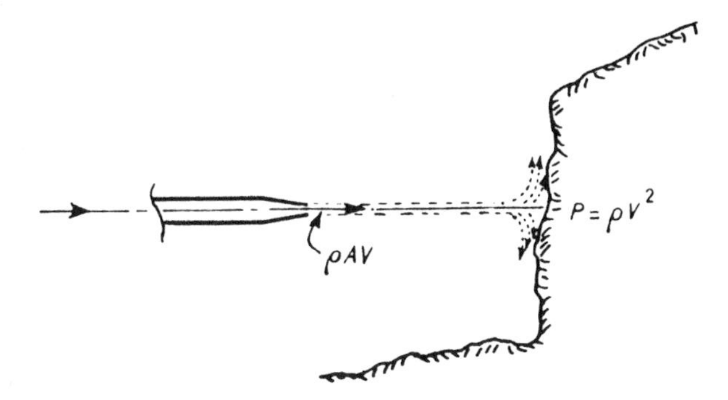

FIGURE 3-7. Hydraulic sluicing relies on the impact of high-velocity jets of water. Reprinted by permission. E. H. MacDonald, *Alluvial Mining*. Chapman & Hall (London), 1983.

Hydraulic Mining

Hydraulic mining (or hydraulic sluicing) is one of the oldest means of mining placers of not-too-coarse and rather uniform size distribution.[1] It was developed for use wherever water could be launched onto the face of a deposit at pressure heads of 30 meters or more.

Hydraulic sluicing operations rely heavily on the impact of high-velocity jets of water to break down the bank of ore deposit to be slurried (Figure 3-7). Since force equals momentum

$$F = \rho(AV)V = \rho AV^2$$

where A is the cross-sectional area of the jet (m^2) opening, F is the force, V is the velocity of water jet, ms^{-1}, and ρ is the density of fluid, Kgm^{-3}.

Hydraulic sluicing was first applied in regions with high hills or mountains, where a natural head could provide high-pressure water at the mining face without powerful pumps. With the advent of mechanization, powerful pumps supplied water to the monitors (Figure 3-8) at a sufficiently high pressure to be operated from a safe distance from the ore bank. Monitors (hydraulic giants) throw water under high pressure onto the alluvial bank to knock it down and direct the spoils, as slurry, to a pump sump, or often to a vibrating grizzly for the removal of barren oversize (Figure 3-9).

[1]"Hydraulic mining made it possible for the Romans to process a colossal volume of rock in northwestern Spain; estimates run as high as 500 million tons. Strabo writes more gold was won this way than from underground mines. It was practiced, according to Pliny, in central Gaul, in the Welsh tin mines, and other places as well." (Marx, 1978)

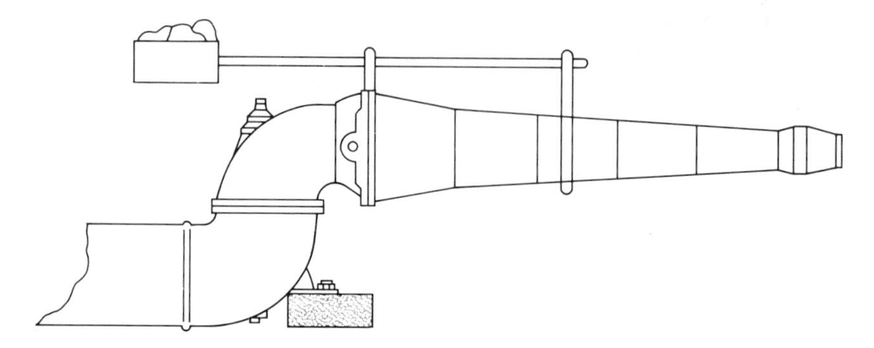

FIGURE 3-8. Giant monitor to eject water under pressure.

From T. K. Rose and W. A. C. Newman. *The Metallurgy of Gold.* J. B. Lippincott (Philadelphia), 1937. Reprinted by permission of Harper & Row, Publishers, Inc.

The successful application of hydraulic mining requires a plentiful and uninterrupted accumulation of auriferous gravel and/or sand that is no less than 30 feet thick and is not overlaid by any appreciable barren material. Actual data from a Russian operation indicate that the productivity of a hydraulic jet decreases and the water consumption (per meter of ground) increases with increasing distance traveled.

Gold Recovery from Placer Deposits

Alluvial gold can be found in continental-type placers. The gold particles may range from nuggets to micron-sized grains in the wash (partly weathered

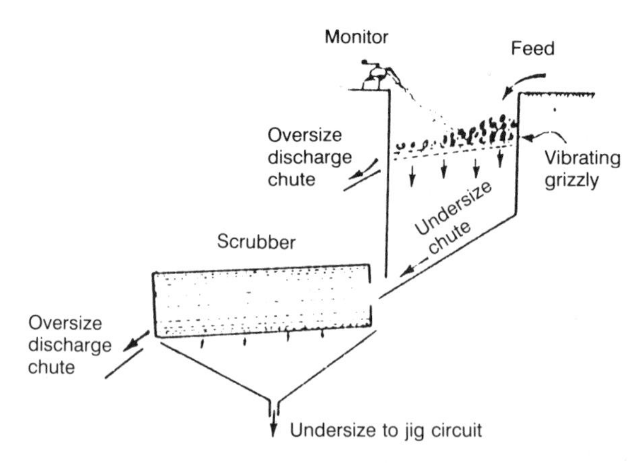

FIGURE 3-9. Feed preparation (scrubbing section) of a gravity-concentration plant.

Reprinted by permission. E. H. MacDonald, *Alluvial Mining.* Chapman & Hall (London), 1983.

mixtures of clays, sand, and gravel). Effective slurrying of such deposits is achieved by hydraulic mining. An autogenous scrubber (as in Figure 3-9) with high-pressure water jets for washing should complement the action of the hydraulic mining. Tests have to be conducted with any particular placer deposit to elucidate the characteristics of the ore, ascertain the degree of gold liberation, establish the size distributions of the ore and the free gold, and determine the optimum processing flow sheet.

Samples for Testing

The importance of selecting a representative ore sample for developing a mineral beneficiation procedure or a metal extraction process cannot be overemphasized.

- The alluvial-gold mining industry has enormous sampling problems. Large boulders, sometimes over 1 meter in their largest dimension, are often found along with miniscule grains of fine gold. Large trommels are commonly used to remove the washed barren oversize, but inefficient scrubbing may lead to gold losses in the oversize.
- If tests are run only with the feed material, that is most difficult to process, the resulting plant can be overdesigned and expensive. Testing with the easiest-to-process ore, however, may result in a plant design that is inadequate to meet the production targets.
- Testing with an "average-representative" sample and designing for average conditions will result in a plant that, in the absence of continuous ore blending, will have periods of inadequacies in both production rate and product quality.

Sampling for testing the beneficiation of placer gold can be inherently erratic due to some significant peculiarities of the alluvial gold deposits:

- Extremely low head assays, often less than 0.2 g/m^3
- Very large difference in specific gravity between the free gold (sp.gr. 15–19) and the gangue minerals (sp.gr. 2.5–5.0)
- Very high degree of gold liberation and very wide range of ore distribution

These conditions may call for very large samples and for sophisticated handling techniques in testing.

Sampling for a reliable process development should include representative samples of (a) the most-difficult-to-process feed material and (b) the easiest-to-process feed material (if such differentiation exists). Tests should be run with samples (a), samples (b), and samples (a + b), the last blended in proportions dictated by the actual alluvial deposit and its mining plan. The advisability of blending ores for testing should be considered only if such blending is feasible and affordable in industrial practice.

The additional cost of mining for blending, if any, must be taken into account. The cost of blending ores must be justified, of course, by savings in capital or operating costs and/or higher gold recoveries. Blending may have to be avoided for practical reasons, such as inaccessibility of all kinds of ore.

Laboratory Tests

Laboratory tests with a great number of samples involve assembling a complete description of the ore deposit under study, exploring the feasibility of gold concentration by lab-size equipment, and establishing, with designed experiments, the optimum gold-concentration process.

Screen sizing analyses, along with chemical assays and mineralogical examinations, will establish the gold-containing size-fractions, the degree of gold liberation, and the size distribution of the contained gold particles. Particular microscopic attention has to be paid to the gold particles, especially the finer ones, in regard to the shape and the appearance of their surfaces. Thus, information on the following ore characteristics becomes available:

- Size distribution of the gangue materials
- Maximum size and size distribution of the gold particles
- Shape and surface conditions of the finer gold particles
- Degree of gold liberation

The gold-containing fractions should also be submitted to X-ray diffraction and X-ray fluorescence radiometric tests. It pays to know the composition and properties of the alluvial deposit thoroughly, since such knowledge is bound to lead to the optimum gold-recovery process.

The investigator should have, by now, a feel for the potential approaches to gold concentration. The upper, coarse, no-gold-bearing fractions should be removed by screening. After desliming of the remaining gold-containing finer ore, gravity separation tests should be run before and after sizing with lab-size equipment (gigs, strakes, tables, spirals, cones, etc.). The products and tails of the meaningful tests have to be sized and assayed, in search for leads to enrich the product and minimize the gold loss in the tail. Further experimentation to optimize promising flowsheets should continue in the laboratory.

Pilot-Plant Testing

Results obtained with lab-sized gravity-separation equipment are very useful for guidance in further testing, but they should not be trusted as design parameters for the industrial plant. Time and money have to be spent for a thorough pilot-plant study to confirm the laboratory results. This is done to settle on the most promising technique and to refine the parameters and conditions required for the design of the industrial installation.

The pilot plant should be run with samples of (a) the most-difficult-to-process ore, (b) the easiest-to-process ore, and (c) blended feed material (a + b) in order to collect experimental data leading to the optimizations of the mining plan and of the commercial plant design. The pilot plant should be run, initially, as an open circuit to steady operation, so that all products (including middlings and tail fractions) can be automatically sampled. Automatic sampling of feed, middlings, products, and tails at predetermined time intervals is bound to provide information that adequately describes all the flows of the process.

The characteristics of all collected samples should be established with screening, assays, microscopic examinations, and lab-scale side tests. On the basis of this information, the recycling of certain middling and tail fractions has to be tested and optimized. The pilot plant must have a flexible design to accommodate and test changes in the flows of middling and tail fractions. The optimum pilot-plant operation has to simulate and conclusively test the prototype plant conditions.

Gravity Concentration of Alluvial Gold

Gravity concentration, has played a cardinal part in the recovery of alluvial gold since very ancient times. According to inscriptions on monuments of the Fourth Dynasty in Upper Egypt, gold was recovered by gravity at least as far back as 4000 B.C. Strabo, who lived in the first half of the ninth century, reported in his writings that "around 4000 B.C., in the country of Saones, the winter torrents brought down gold which the barbarians collected in troughs pierced with holes and lined with fleeces." The legend of the trip of the Argonauts (approximately 1200 B.C.) to find the "golden fleece" was probably no more than an expedition to raid a faraway river where fine alluvial gold was concentrated by gravity on sheepskins.

Gravity (partial) concentration of gold during milling for cyanidation of gold ores, is often feasible and highly advisable, and will be discussed in Chapter 4. Only gravity concentration of alluvial auriferous gravels and sands is discussed in the present section.

The pronounced difference between the high density of placer gold (sp.gr. 15–19) and the low density of the alluvial gangue (sp.gr. 2.5–5.0) makes gravity concentration the decidedly advantageous means of placer-gold recovery. Gravity separation of gold has distinct advantages over flotation and chemical dissolution: It requires significantly lower capital investment, less power, and lower handling costs per ton of feed material than either flotation or leaching. Gravity concentration results also in an effluent that is less harmful to the environment than effluent from the other two processes.

Although the high density of placer gold is a very strong advantage in gravity concentration, some of its other characteristics are rather detrimental. Flattened, flakelike shapes, cavities, pores, and irregularities, which may be filled with low-density soils, reduce the *effective density* of placer gold and may cause it to float rather than sink. The flat shapes of fine sizes may cause hydroplaning through the beneficiation system and impair gold recovery. Gold surfaces are hydrophobic; hydrophobicity causes flotability and antagonizes gravity concentration.

The shapes of placer gold particles are far from spherical, cylindrical, or elongated solids. Deviations from sphericity are often described with shape factors. The Corey shape factor is

$$S_f = \frac{T}{\sqrt{LB}}$$

where T, L, and B are thickness (measured parallel to the direction of motion), length, and breadth of the particle.

Most nonspherical particles will settle with their maximum cross section normal to the direction of the flow. Hence, T is their shortest dimension, whereas B controls the sieve size of the particle. The shape factor of a particle has a pronounced effect on its settling velocity in water, as can be seen in Figure 3-10.

The drag coefficient, C_D, is a description of the fluid resistance to the free-settling rates of particles in the fluid.

$$C_D = \frac{4}{3} \cdot \frac{gd(\rho_s - \rho)}{\rho w^2}$$

The Reynolds number is a dimensionless, hydrodynamic parameter describing the condition of the fluid flow.

$$Re = \frac{wd_\rho}{\mu}$$

where w is the particle-settling velocity (cm/sec), d is the particle diameter (cm), g is the gravity acceleration (cm/sec^2), μ is the fluid viscosity (g/cm · sec), and ρ_s and ρ are the densities of solid and fluid (g/cm^3). Figure 3-10 shows the pronounced effect of the shape factor. Free-settling rates of particles (w) in fluids of known properties can be calculated from the above equations and Figure 3-10.

The size of a particle also has a very significant effect on its settling velocity, since the effect of the gravity force is proportional to the mass of the particle. Therefore, the coarse barren gravel should be removed by screening from the alluvial feed before any attempt at gold separation. A grizzly followed by a trommel are often adequate for removing the barren

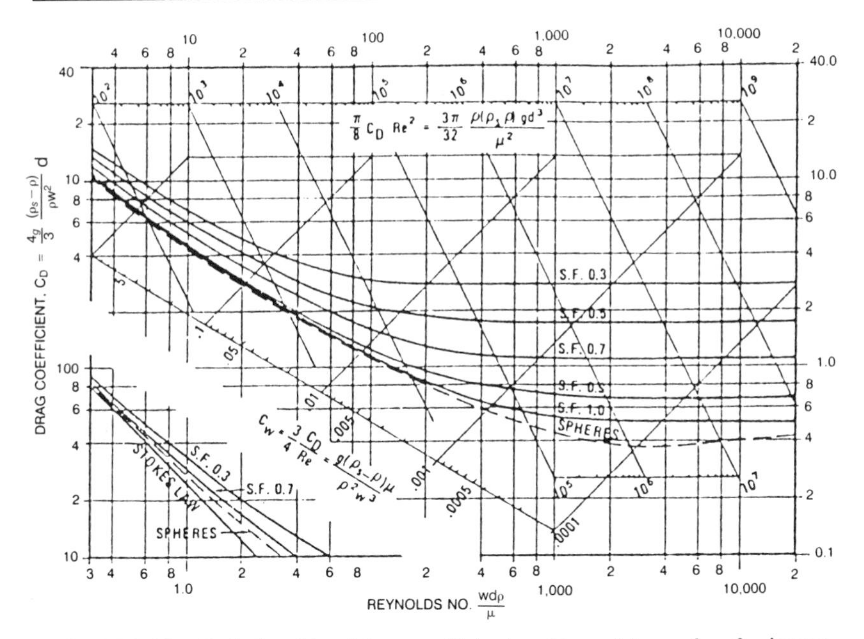

FIGURE 3-10. Relationship of drag coefficient *vs*. Reynolds number for irregularly shaped particles.

Reprinted by permission. Canadian Institute of Mining & Metallurgy (Montreal). W. Wengian and G. W. Poling, in *C.I.M. Bull.,* December 1983, 76 (860): 47–56.

oversize before scrubbing, desliming, and further sizing—if necessary—of the feed to the gravimetric concentration plant. Elaborate feed sizing may be required for auriferous sands with a wide size distribution, since the gravity-separation equipment has some rather distinct size limitations for maximum recovery results.

The moving river and glacier-bed loads are natural grinding mills. Gold will degrade to fine particles in a simulated river environment. Therefore, alluvial gold is expected to have a size distribution ending in a fine tail, although conventional sizing field methods may not detect fine gold (< 100 μm). This is especially true if the ore contains a substantial amount of clay.

Table 3-4 indicates that most gravity-separation equipment is not efficient in recovering very fine gold. In "Metallurgical Efficiency in the Recovery of Alluvial Gold" (1984 Fricker) concludes that "conventional gold saving (gravity separation) equipment does not recover gold finer than 200 μm very efficiently." Figure 3-11 indicates the inefficiency of riffled sluices, jigs, and even shaking tables for gold particles finer than 200 μm (0.2 mm).

TABLE 3-4. Particle size limitations of gravity-separation equipment.

Unit	*Particle size range*
Jigs	75 μm–25 mm
Spiral concentrators	75 μm–3.0 mm
Strakes and riffles	70 μm–25 mm
Reichert Cones	45 μm–20 mm
Hydrocyclones	40 μm–3.0 mm
Pinched sluices and cones	30 μm–3.0 mm
Shaking tables	15 μm–3.0 mm
Amalgamation	70 μm–1.5 mm
Cyanidation	-200 μ m

Reprinted by permission. Adapted from E. H. MacDonald, *Alluvial Mining*. Chapman & Hall (London), 1983.

Feed Preparation and Control Equipment (Sizing, Scrubbing, Washing, and Slurrying)

Effective liberation of the gold and size classification are very important for a successful concentration process. The effect of gravity force is proportional to the mass of the particle (volume · density), so the coarse, barren gravel has to be removed on a grizzly followed by a trommel, where

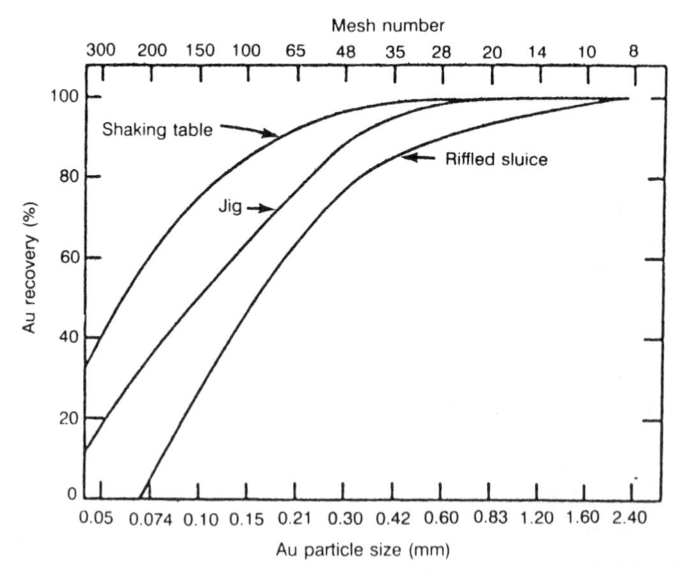

FIGURE 3-11. Expected recovery of gold particles by gravity as a function of particle size.

Reprinted by permission, Canadian Institute of Mining & Metallurgy (Montreal). W. Wengian and G. W. Poling, in *C.I.M. Bull.*, December 1983, 76 (860): 47–56.

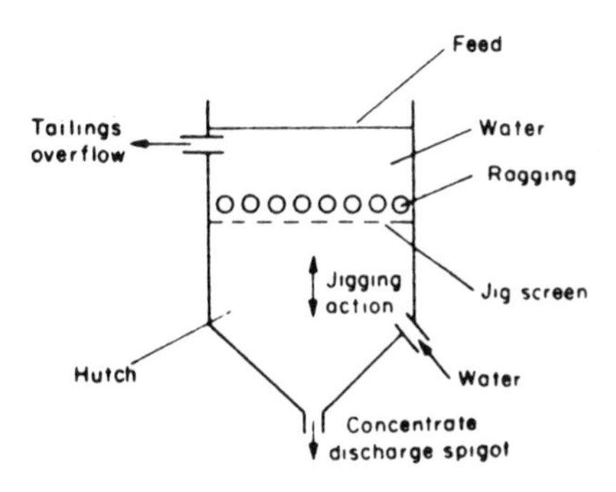

FIGURE 3-12. Sketch of jigging action.

Reprinted by permission. B. A. Wills, *Mineral Processing Technology*. Pergamon Press (Oxford), 1988.

scrubbing and desliming take place (Figure 3-9). Gravity separation of valuable minerals can be effective only with particles of the same size range, where density prevails as the criterion of separation. Attention must be paid so that the removed coarse, barren gravel does not include fines and slimes attached to or coagulated on the rock chunks. Persistent washing on a vibrating screen—or, preferably, scrubbing and washing in a trommel—will thoroughly clean the coarse fraction and thus minimize the gold-bearing fines discarded. Strong jets of water, in a trommel, break up clays and wash, screen, and distribute slime, sand, and fine gravel to the processing plant.

In addition to rock fragments and boulders, placer deposits may contain a lot of trash, such as weathered wood, plant roots, and fibers, that must be removed to protect the integrity and the smooth operation of the plant. A good portion of such trash is expected to be removed on the primary grizzly and trommel described earlier.

Gravity-Concentration Equipment

Once the gold is liberated, the sized ore should be charged to high-capacity gravity-concentration equipment (e.g., jigs and pinched sluices) in series or in parallel, while lower-capacity equipment (e.g., spirals and tables) should be used for upgrading primary concentrates or treating small-size, high-value materials. Table 3-4 lists the main gravity-concentration equipment in a sequence of diminishing recoverable particle size.

Jigs are hindered settling devices consisting of shallow, flat trays with perforated bottoms (Figure 3-12). The trays normally contain layers of thick, high-density material (e.g., pieces of haematite or steel punchings) through which water pulsates up and down. The direction of flow through the bed is reversed several times per second. The bed is dilated by the forward stroke of a plunger and compacted by the plunger's backward stroke (Figures 3-13 and 3-14).

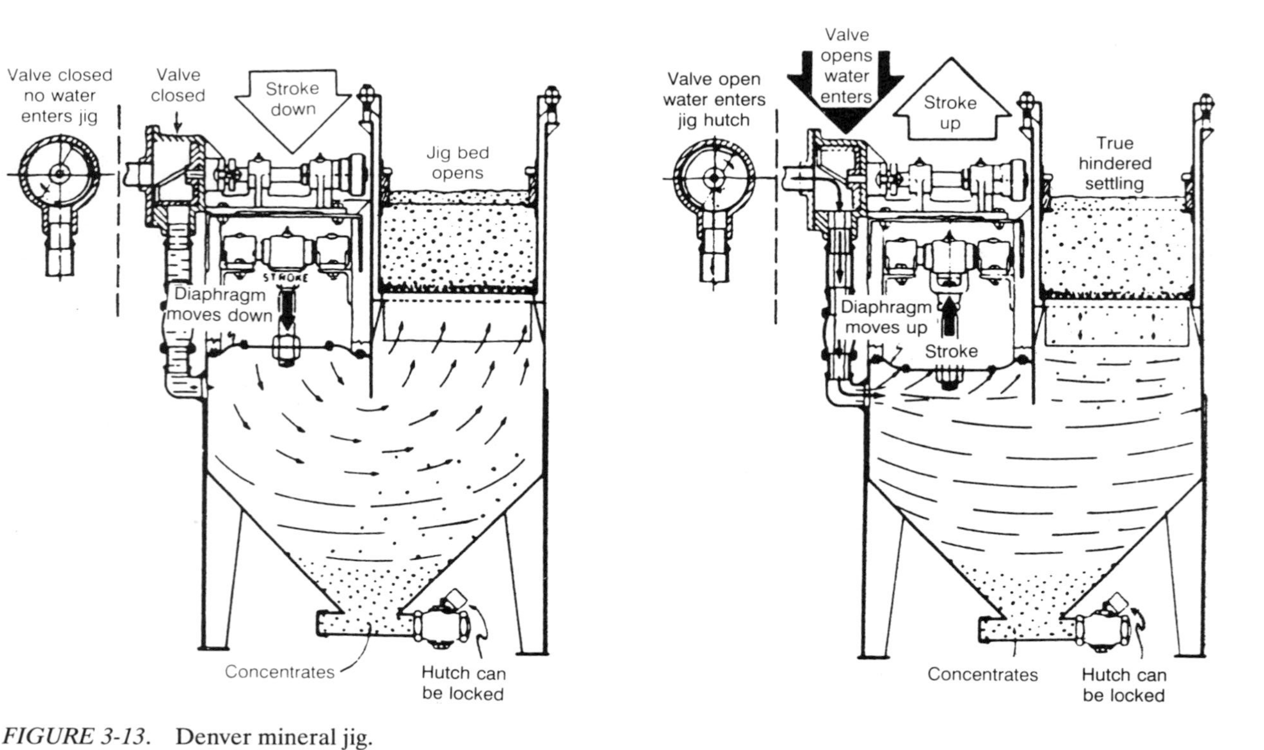

FIGURE 3-13. Denver mineral jig.

Reprinted by permission. B. A. Wills, *Mineral Processing Technology*. Pergamon Press (Oxford), 1988.

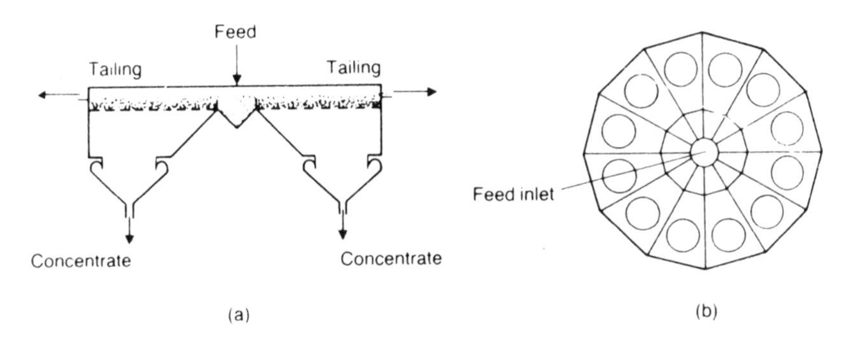

FIGURE 3-14. (a) Section of a circular jig. (b) Top view of a 12-module jig. Reprinted by permission. B. A. Wills, *Mineral Processing Technology*. Pergamon Press (Oxford), 1988.

The feed slurry flowing across the bed is submitted to conditions of hindered settling, which keep the lighter particles in suspension and make the heavier particles sink through the bed. The lighter particles continue to move over the end of the jig and are discharged as tailings. The heavier particles sink and can be drawn off from spigots at the bottom of each compartment of the jig.

High proportions of fine sand and slime interfere with jig performance. The finer fractions of jig tailings may have to be scavenged by sluices, spirals, or tables.

Spiral concentrators (Figure 3-15) were made originally of cast iron, but light-weight fiberglass spirals with polyurethane or rubber linings have been common since the late 1970s. The gravity separation is effected by centrifugal and gravitational forces (Figure 3-16). The water piles up and flows on the outer rim of the spiral channel, reaching the level where the centrifugal force is in equilibrium with the gravitational force. The bottom layer of the flowing stream is subjected to a smaller centrifugal force since it is retarded by friction, and it flows toward the inner part of the spiral channel.

Operating variables are the diameter and pitch of the spiral, the density of the pulp, the location of the take-off points, and the volume/pressure of the wash water added to the system. Spiral concentrators have some form of adjustable splitters that divert the stream of concentrate from the main flow and direct it, finally, to a collecting box.

Most spiral concentrators are constructed as two or three units around a common vertical pipe. This results in floor-space savings and very low cost supervision. One operator can service up to 200 multispirals if the feed and water are adequately controlled and screened. Spiral concentrators have been used to upgrade many minerals, including gold ores and coals. The feed rate depends on the type of material to be concentrated, and is usually one to two tons per hour, with up to 20 liters per minute of wash water.

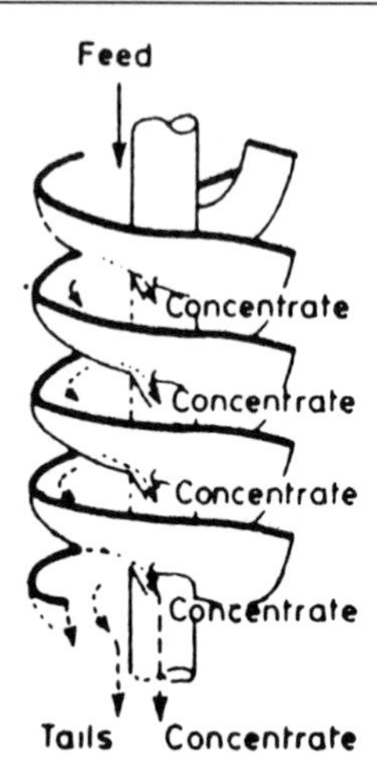

FIGURE 3-15. Sketch of a twin-spiral concentrator.
Reprinted by permission. E. H. MacDonald, *Alluvial Mining*. Chapman & Hall (London), 1983.

Spirals (Reichert Mark 7A) with no wash water have been available since the early 1980s, and have significantly affected the performances attainable, especially with feeds containing fines.

Riffled sluices are inclined launders, 20 to 60 cm wide with 20-cm-high side walls, in sections three to four meters long. Each sluice-box tapers to fit into the succeeding box. Transverse pieces of wood, metal, or rubber—

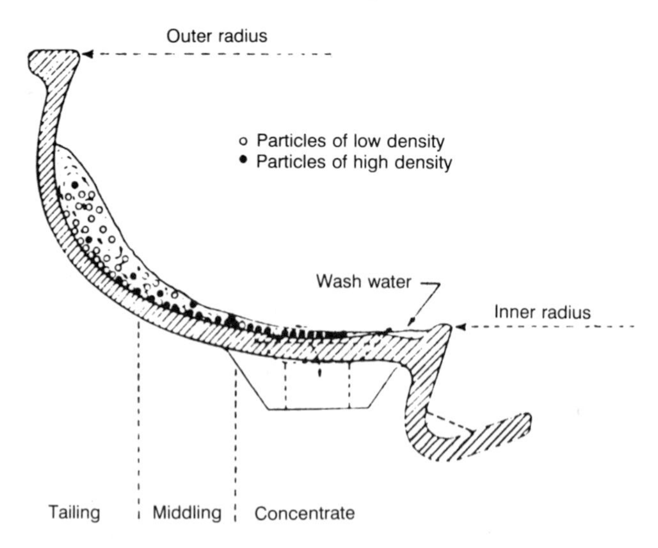

FIGURE 3-16. Cross-section of stream in spiral concentrator.
Reprinted by permission. B. A. Wills, *Mineral Processing Technology*. Pergamon Press (Oxford), 1988.

called riffles—with a cross section of 1.8 to 6 cm^2, are spaced from 2.5 to 6 cm apart on the bottom of the launder. Heavy (gold) particles are expected to concentrate ahead of the riffles. Intense turbulence in the spaces between riffles must be avoided so that gold particles can sink and not be uplifted by current eddies.

A *belt concentrator* is a riffled belt sloped downwards (at approximately 12°) but moving upward at a rate of 0.4 meters per minute. The mill pulp flows down over the riffles, which collect a concentrate of gold and any heavy minerals that are present. The concentrate moves counter-current
expected to concentrate ahead of the riffles. Intense turbulence in the spaces between riffles must be avoided so that gold particles can sink and not be uplifted by current eddies.

A *belt concentrator* is a riffled belt sloped downwards (at approximately 12°) but moving upward at a rate of 0.4 meters per minute. The mill pulp flows down over the riffles, which collect a concentrate of gold and any heavy minerals that are present. The concentrate moves counter-current
to the feed-pulp and is washed out of the riffles from under the head pulley.

Pinched sluices of multiple forms have been used for heavy mineral concentration since ancient times. *Fanning concentrators* (Figure 3-17) have the disadvantage of excessive circulating loads and need multistage operation for satisfactory upgrading. The most widely used pinched-sluice system is the *Reichert Cone Concentrator*, a high-capacity (60 to 100 tons per hour), low-cost gravity separator. A 60%- to 70%-solids pulp flows evenly around the top of the distribution cone (Figure 3-18) to a rubber "finger," which splits it between two concentrating cones that operate in parallel and are superimposed, one above the other. A Reichert Cone comprises, in effect, a number of pinched sluices arranged in circles. The pinched-sluice separation by stratification is depicted in detail in Figure 3-19.

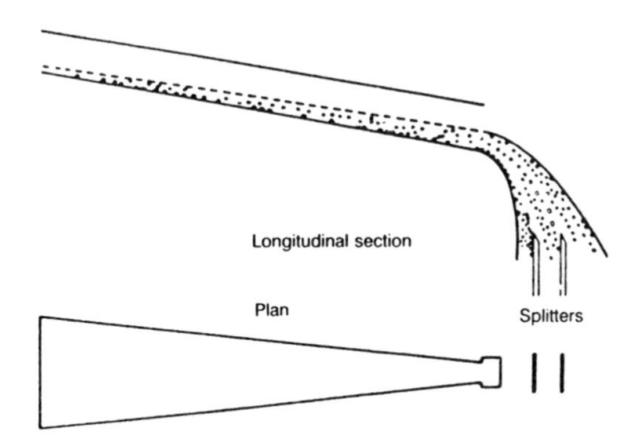

FIGURE 3-17. Pinched sluice.

Repinted by permission B. A. Wills, *Mineral Processing Technology*. Pergamon Press (Oxford), 1988.

FIGURE 3-18. Reichert Cones.

Reprinted by permission. B. A. Wills, *Mineral Processing Technology*. Pergamon Press (Oxford), 1988.

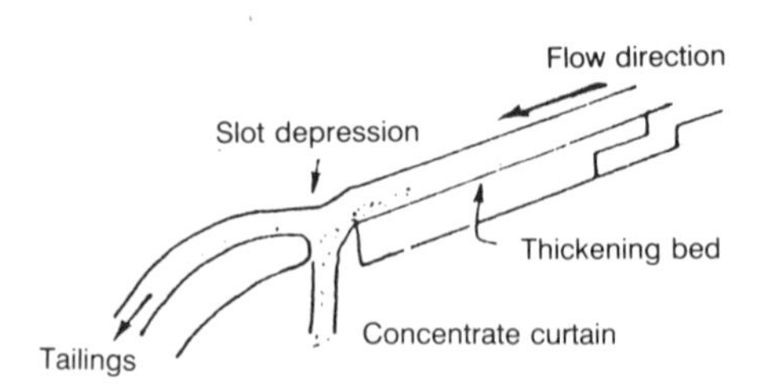

FIGURE 3-19. Stratification of a flow in Reichert Cone.

Reprinted by permission. E. H. MacDonald, *Alluvial Mining*. Chapman & Hall (London), 1983.

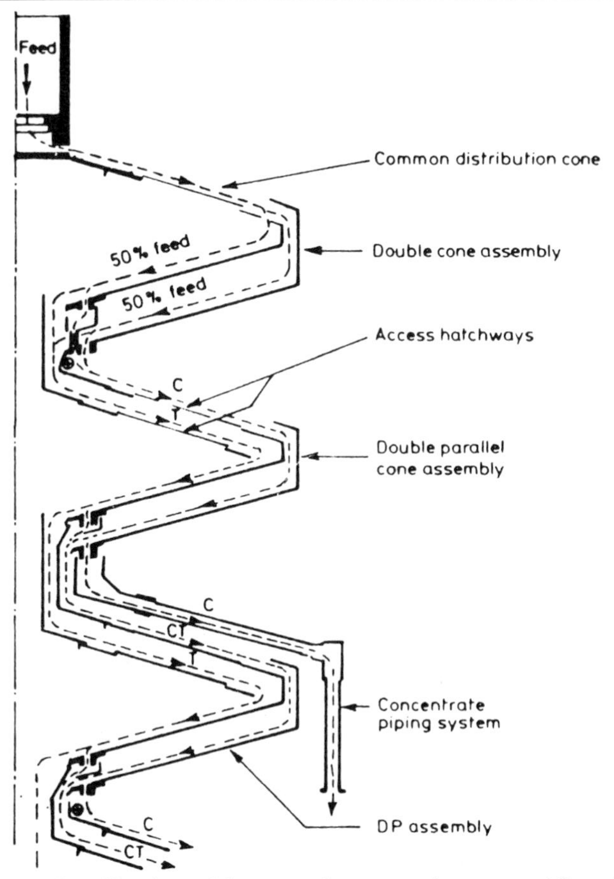

FIGURE 3-20. Stratification of flows on the upper three assemblies of a 3-m diameter Reichert Cone.

Reprinted by permission. E. H. MacDonald, *Alluvial Mining*. Chapman & Hall (London), 1983.

The Reichert Cone has been used extensively for upgrading mineral sands in Australia, with large units treating up to 300 tons per hour. It can be used for a complete gravity plant, with roughing, scavenging, and cleaning cones (Figure 3-20). In addition to the feed rate and the density, parameters that control the product grade and recovery are the slope of the cones and the cone-insert width.

Hydrocyclones (short-cone or flat-bottomed) have been reported to achieve satisfactory recoveries in gold concentration. *Shaking tables* (Figure 3-21) oscillate horizontally and concentrate the high-density particles due to inertia differences created by the brisk, reciprocating motion of the inclined deck of the table. Coarse feeds require long strokes at lower speeds; fine feeds require shorter strokes at faster speeds. Thin strips of metal on the deck, parallel to the motion, act as riffles.

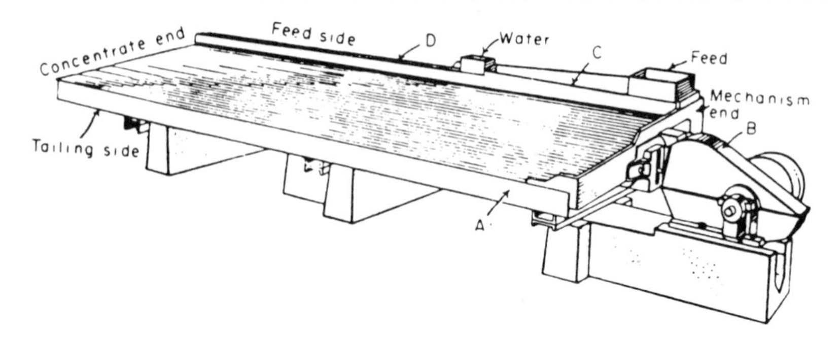

FIGURE 3-21. Shaking table. A: inclined deck; B: mechanism of vibration; C: feed distribution box; D: water distribution launder.

Reprinted by permission. B. A. Wills, *Mineral Processing Technology*. Pergamon Press (Oxford), 1988.

The feed slurry (about 25% solids) spreads on the deck of the table due to the reciprocating motion (280–325 strokes per minute) and the transverse flow of water (Figure 3-22). As the slurry flows on the deck, the lighter particles are washed over the riffles, whereas the heavier particles travel along the riffles activated by the reciprocating motion and discharge over the end of the inclined deck. The light particles of the tailings are entrained by the weak flow of wash water over the lower edge of the deck (Figure 3-23). A middling product flows off between the concentrate and the tailings.

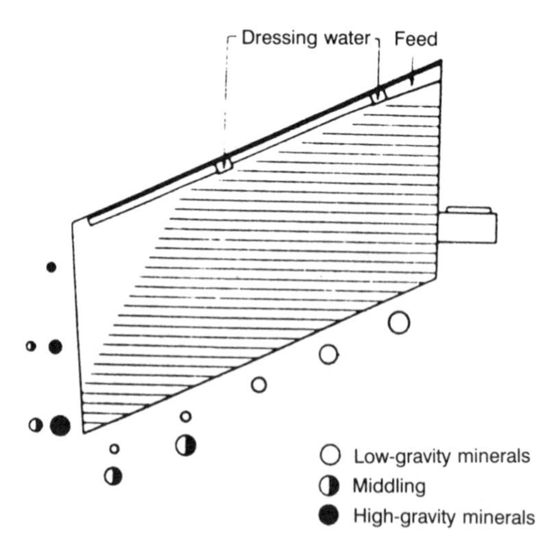

FIGURE 3-22. Distribution of shaking-table products.

Reprinted by permission. B. A. Wills, *Mineral Processing Technology*. Pergamon Press (Oxford), 1988.

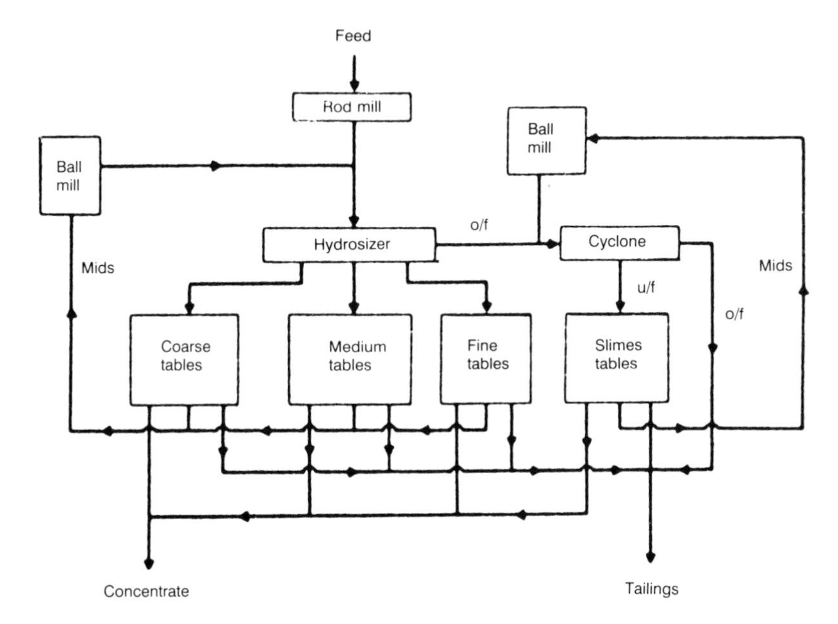

FIGURE 3-23. Typical flow sheet of concentration with shaking tables.
Reprinted by permission. B. A. Wills, *Mineral Processing Technology*. Pergamon Press (Oxford), 1988.

References

Note: Sources with an asterisk (*) are recommended for further reading.

Evans, D. L. 1981. Lateralization as possible contributor to gold placers. *Eng. & Min. J.*: 86–91.

Fricker, A. G. 1984. Metallurgical efficiency in the recovery of alluvial gold. *Proc. Australas. I.M.M.* 289, February: 59–67.

*Macdonald, E. H. 1983. *Alluvial Mining*. New York: Chapman & Hall, Ltd.

Marx, J. 1978. *The Magic of Gold*. New York: Doubleday & Co.

Rose, T. K., and W. A. C. Newman. 1937. *The Metallurgy of Gold*. Philadelphia: J. B. Lippincott, 105–32.Wengian, W., and G. W. Poling, 1983. Methods for recovering fine placer gold. *C.I.M. Bull.*, December: 47–55.

Wengian, W., and G. W. Poling, 1983. Methods for recovering fine placer gold. *C.I.M. Bull.*, December: 47–55.

Wills, B. A. 1988. *Mineral Processing Technology*, 4th ed. Elmsford, New York: Pergamon Press, 377–419.

Woodsend, A. 1984. Dredging for gold in the Yukon. *Min. Mag.*, August: 92–97.

CHAPTER 4

Milling of Amenable Gold Ores

The objective of milling gold ores is to extract the gold for the highest financial return. To recover the maximum amount of gold, the ore must be finely ground in order to liberate the gold particles for gravity separation and/or chemical extraction. Ores yielding acceptable gold recovery (more than 88%) when "normally ground" (60–75% −200 mesh), by direct cyanide leaching and defined as "amenable" (in contrast to the "refractory" ores, which require extremely fine grinding and/or pretreatment before cyanidation). A general flowchart of the processing steps required to extract gold from its ores is presented in Figure 4-1.

Comminution of the Ore

The lithology, geological characteristics, and mineralogical composition of the gold deposit must be known in detail. The hardness and friability of the deposit, its gold distribution, and cost considerations are the cardinal parameters that dictate the selection of the optimum comminution circuit.

Wide variations in comminution properties may occur within a large gold deposit. Hence, numerous ore samples must be tested in order to design the optimum comminution plant. The size and required number of ore samples have to be determined by the management of the operation in a compromise between the cost of the drilling-sampling program and the safety of the plant design. Pronounced geological differences within the ore deposit impose higher costs of sampling and testing for satisfactory safety and efficiency in plant design.

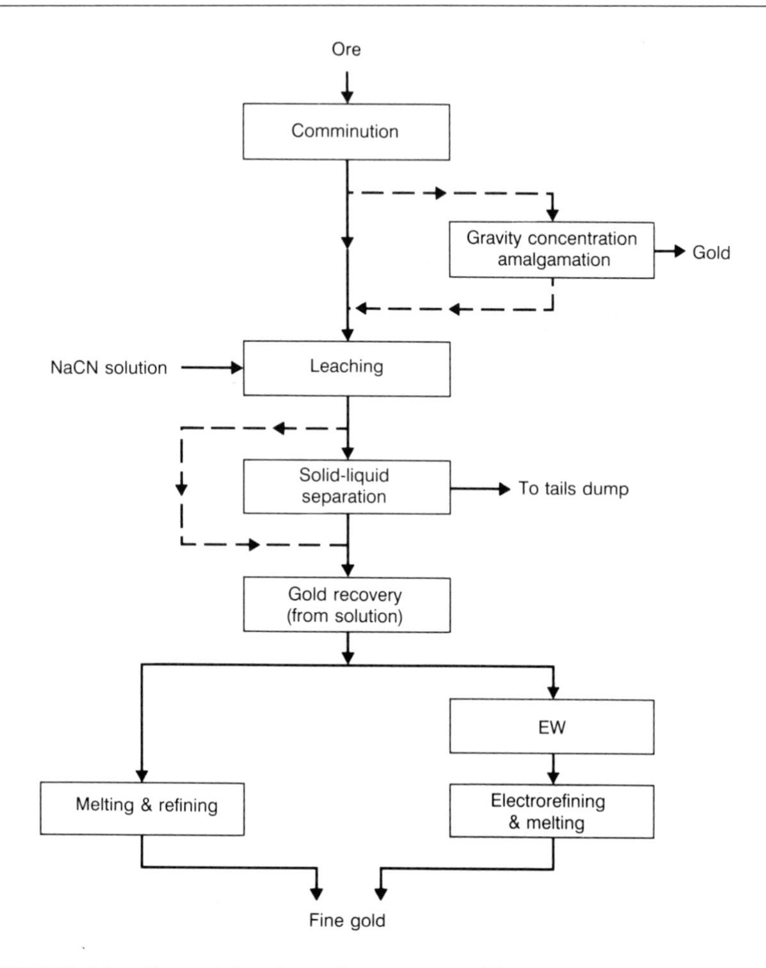

FIGURE 4-1. General flowchart: From ore to gold.

Each ore has its own gold-liberation characteristics. Tests and feasibility studies have to define the optimum comminution circuit that can deliver adequate mill feed throughout the year *under the specific climatic conditions of the mill site*. The three basic types of comminution circuits are

- Conventional three-stage closed-circuit crushing and rod-mill/ball-mill grinding
- Primary crushing and autogenous grinding
- Primary crushing with autogenous mill followed by fine grinding in a ball mill

The mineralogy of the gold ore dictates the required fineness of grinding for adequate gold liberation and the economically optimum extraction recovery.

Crushing

The ore often has to be first crushed at the mine, especially if it is mined underground. A large jaw crusher (48′ × 36′) or a large gyratory crusher can produce −5 inches material. This is transported and stored, preferably in a coarse-ore bin, ahead of the secondary crusher or the autogenous grinding mill.

If the deposit contains enough hard rock, and an autogenous or semi-autogenous grinding circuit is selected, a single large crusher can be adequate (Figure 4-2). The lithology of the ore, and/or the high cost of energy at the specific site, might exclude autogenous grinding from consideration. Then an elaborate, multistage crushing plant will be required ahead of a conventional grinding circuit.

Tests and cost studies are required to develop design criteria and recommendations that lead to the selection of primary, secondary, and tertiary crushers, with conventional (nonautogenous) grinding. Such feasibility studies have to take under serious consideration the climatic conditions of the specific site, especially if the ore is mined from open pits. A screen between primary and secondary crushers will save energy and alleviate potential flow problems caused by wet fines and clays in wintertime (Figure 4-3). A shorthead cone-crusher in closed circuit, by means of a screen

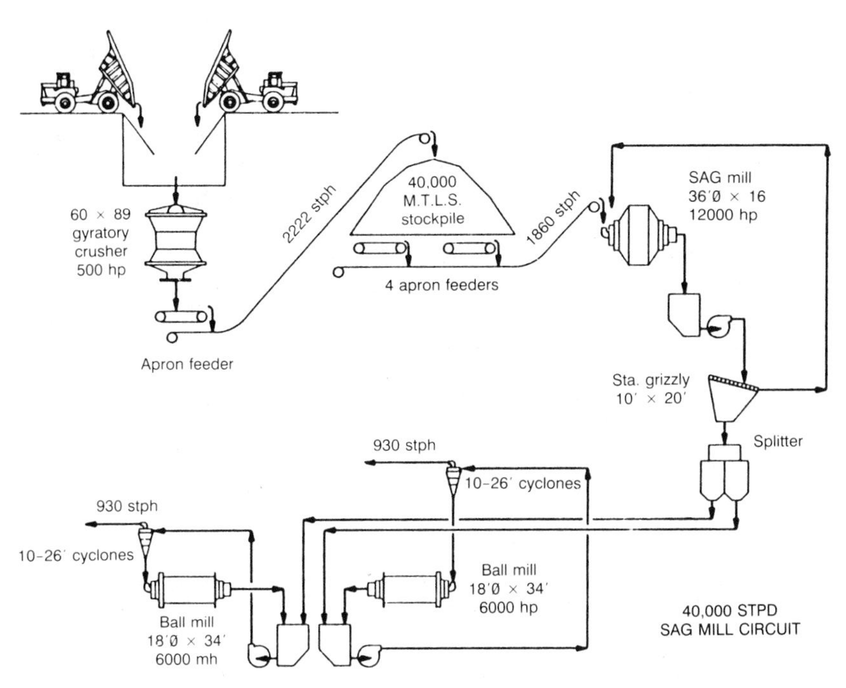

FIGURE 4-2. Grinding circuit with semi-autogenous (SAG) mill.

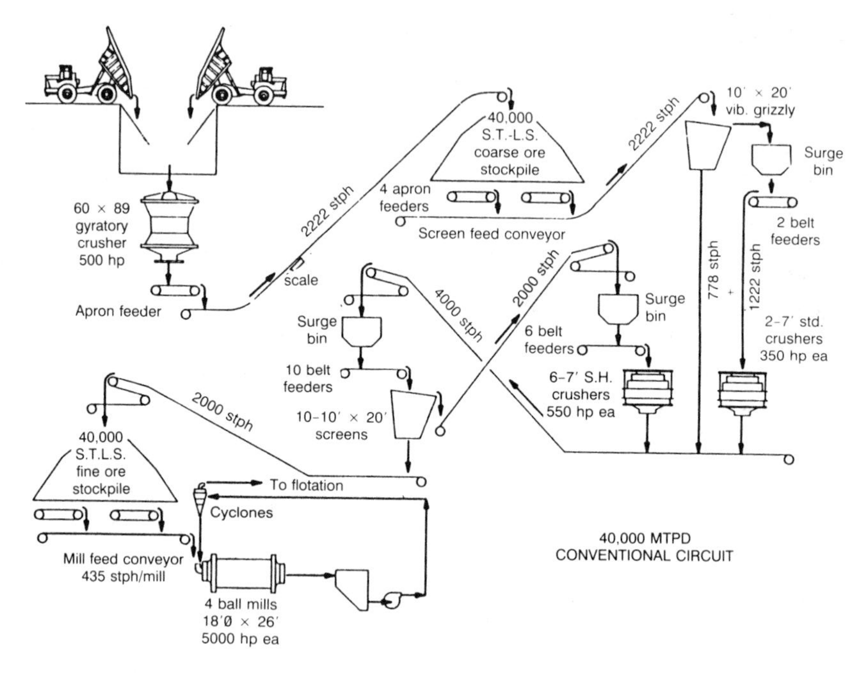

FIGURE 4-3. Conventional circuit with three-stage crushing and ball-mill grinding.

(preferably double-decked), should deliver the crushed material required for grinding by steel-media mills (about 80% −1/2″). A closed crushing circuit delivers the proper size of crushed ore to the grinding circuit, optimizes the energy consumption, and may alleviate wet-winter problems.

Nordberg Inc. (1989) has developed a wet crushing system called Waterflush (WF) Technology. The system incorporates a cone-crusher with special seals, components, and lubricants able to operate with large water flows. Wet crushing claims the following advantages.

- Up to 25% reduction in the energy consumed by the comminution circuits
- Up to 50% increase in crusher throughput capacity
- Maximum flakiness of the crushed ore, with significant improvement in ball grinding

Water is added to the WF cone-crusher in a proportion that produces a 30–50% solids slurry. The significant volume of water flushes fines through the crusher and prevents any buildup on liners.

A crushing plant should include the following design features, in order to maintain a high production rate at all times.

- A screen between primary and secondary crushers, to alleviate potential flow problems by wet fines and clays
- Magnetic protection for tramp-iron removal
- Sloping floors and pump sumps for easy cleanup
- Dust scrubbers to keep the working environment clean and comply with environmental emissions regulations

Dust collected from dry cyclones and wet scrubbers in the crushing plant is valuable, since the fines are generally enriched in gold content (often containing three times the gold in the head grade of the ore). The collected dust is sent to the cyanide leaching plant.

A multistage crushing system provides opportunities for manual, size, radiometric, or photometric sorting. Such sorting, if allowed by the ore geology and appearance, will result in upgrading the mill feed, thus reducing the milling cost per ounce of gold produced. As mentioned above, in a number of ores gold tends to concentrate in the fines of the crushing operation; hence, sorting may sometimes be carried out by screening off the coarse fraction of the crushed ore.

Grinding

In order to enhance gold extraction by cyanidation, the gold grains (or, in some ores, the gold micrograins) have to be wetted by the solvent so that the liquid-solid reaction of leaching takes place. Fine grinding of the ore is required to liberate the gold particles and maximize the reaction rate and efficiency of leaching.

The feasibility and advisability of *autogenous grinding*—or *semi-autogenous grinding (SAG)*—must be considered and, if attractive, seriously studied.[1] A comminution circuit with autogenous grinding involves, as a rule, significantly lower capital costs than multistage crushing and conventional grinding. However, autogenous grinding is energy intensive and might not be acceptable in areas of high energy costs.

The simplest comminution process is direct run-of-mine ore grinding, which is currently in use in some large gold plants. Huge (4.9 m dia. by 12.2 m) autogenous (or semi-autogenous, after the introduction of some

[1]An autogenous tumbling mill is, by definition, the mill that employs coarse lump ore as the grinding medium, while it is itself being ground. In practice, autogenous grinding often has to be changed to semi-autogenous grinding (SAG), when the lack of adequately hard ore dictates the addition of large steel balls to maintain the required grinding rate at the prescribed fineness. SAG involves, as a rule, lower (70–75%) capital costs and slightly lower (92–95%) operating costs than conventional multistage comminution circuits.

large steel balls) mills are employed. If the mining operation generates excessively large rocks, a primary crusher has to precede the grinding stage. Screening and classification—by stages, in cyclones—complete the closed grinding circuit (see Figure 4-3). The final overflow of cycloning should contain particles finer than the maximum size prescribed for complete liberation and very high gold extraction.

The comminution of ores containing a large proportion of clay is affected severely in wet or winter climates. The wet clays tend to stick on conveyors and to accumulate within crushers, blocking the free flow of solids. In some cases, washing the fine clays on a screen after primary crushing may be required during the wet seasons.

Autogenous grinding or semi-autogenous grinding—as part of a simple comminution circuit—are recommended for high-clay ores treated in winter or in wet climates. The autogenous mill may constitute the first grinding step and be followed by closed-circuit ball milling (see Figure 4-3). The controlled fine-slurry overflow from the cyclones of the closed grinding circuit is the feed to the cyanidation plant.

Autogenous Grinding-Ball Milling-Crushing (ABC)

In some instances increases in the hardness of the ore or demand for increased mill capacity cannot be handled by an existing SAG system. Keeping up the grinding rate, in spite of ores harder than anticipated by the mill design, or satisfying the demand for increased capacity can be achieved with the installation of a secondary cone-crusher in the comminution circuit, and the change of the semi-autogenous mill to an autogenous one. This conversion to an *A*utogenous grinding-*B*all milling-*C*rushing closed circuit (an ABC circuit) can significantly increase the ore throughput and reduce the grinding cost per ton of mill feed (Figures 4-4 and 4-5). Hence, the use of a secondary crusher along with a combination of autogenous grinding and ball mill is recommended for hard ores (Freeport McMoRan Gold Co., 1986).

Tumbling grinding mills with steel media (balls or rods) should be preceded by elaborate multistage crushing circuits (see Figures 4-3 and 4-4). The multistage crushing supplies a controlled, economically acceptable size of feed (– ⅝″ to – ½″) to conventional grinding circuits.

Rod mills are high-throughput, open-circuit, primary grinders with cylindrical steel rods as the grinding media. The coarse mill output has to be further ground in secondary closed-circuit ball mills. Ball mills have a lower throughput per volume unit than rod mills. The ball-mill loading (expressed as a percentage of the mill volume) and the maximum allowable speed (in rpm) determine the mill's maximum grinding efficiency.

All tumbling mills are internally lined with mill liners; the mill liners—

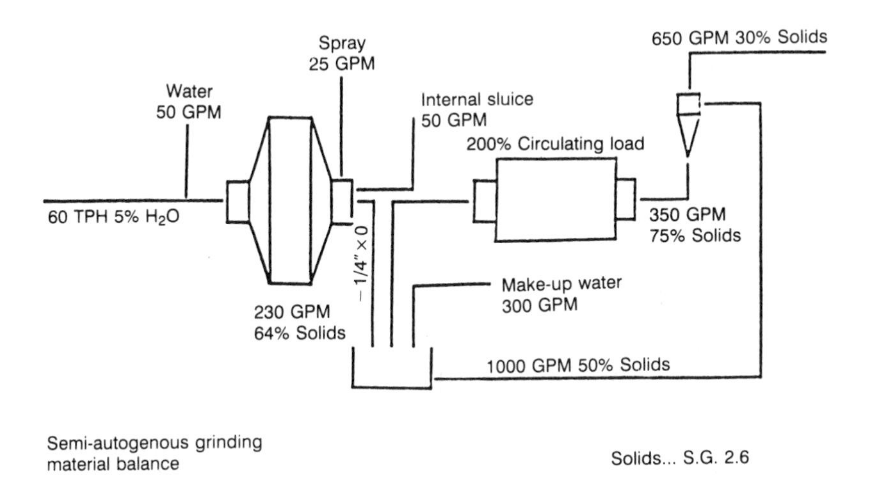

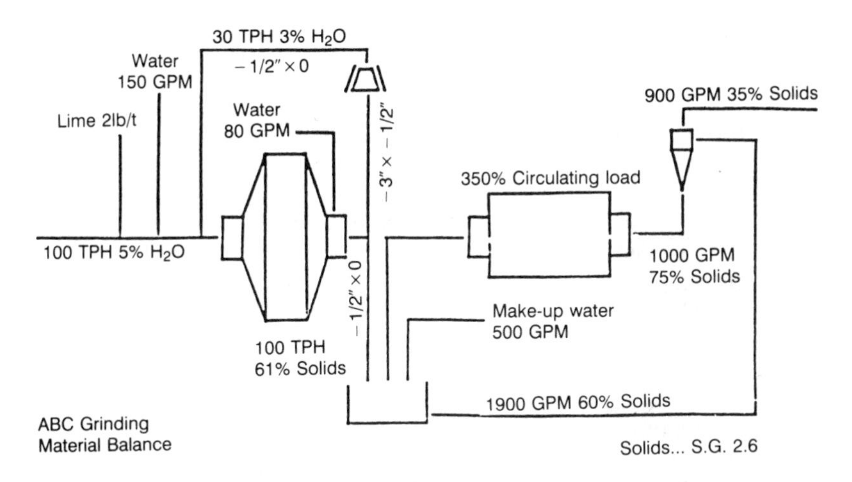

FIGURE 4-4. Material balances of SAG and ABC grinding, with the same size of basic equipment and the same ore.

Reprinted by permission. From Gold Grinding Circuit. Freeport-McMoRan Gold Company. Paper presented at the 1986 SME-AIME Annual Meeting, at New Orleans, LA. Freeport-McMoRan Gold Co.—Jerrit Canyon Joint Venture Elko, NV.

in addition to protecting the mill shell from wear—reduce the slip between the shell and the grinding media. Liners wear off and are replaceable. The materials commonly used for liners are cast steel and rubber. Although the rate of liner wear is lower than the rate of wear on the steel grinding media, the effect of the liners on the operating cost is more significant, due to the

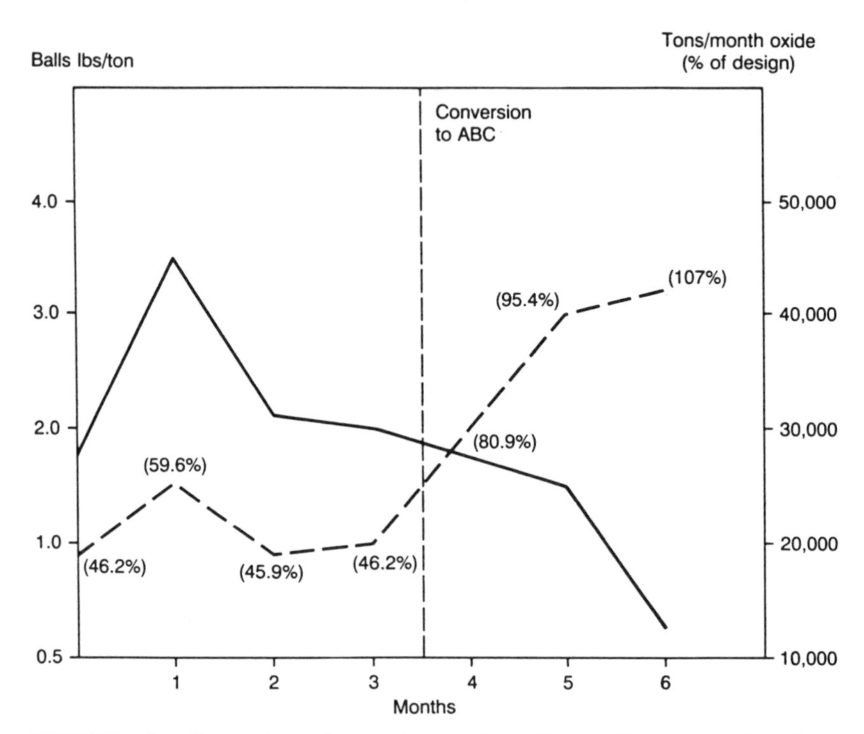

FIGURE 4-5. Comparison of throughput and grinding media consumption of an ABC circuit (see Figure 4-4).

Reprinted by permission. From Gold Grinding Circuit. Freeport-McMoRan Gold Company. Paper presented at the 1986 SME-AIME Annual Meeting, at New Orleans, LA. Freeport-McMoRan Gold Co.—Jerrit Canyon Joint Venture, Elko, NV.

loss of production caused when relining the mill. Rubber liners significantly reduce the noise in the plant and are easier and safer to replace than steel liners. In mills with smaller ball sizes, and in grinding hard ores, rubber liners have longer lives than steel liners.

Ball mills always operate in closed circuit with classifiers and/or cyclones; the coarse underflow of the cyclones/classifiers can be fed to a gravity-concentration circuit (if fully liberated recoverable particles of gold exist) or fed directly to the cyanidation mill (see Figures 4-2 and 4-4).

The following design features are indispensable for the efficient operation of the grinding circuit:

- Ball hopper
- Mill-liner handler
- Automatic regulation of the feed rate
- Sloping floors for easy cleanup
- Open areas over pumps for easy accessibility

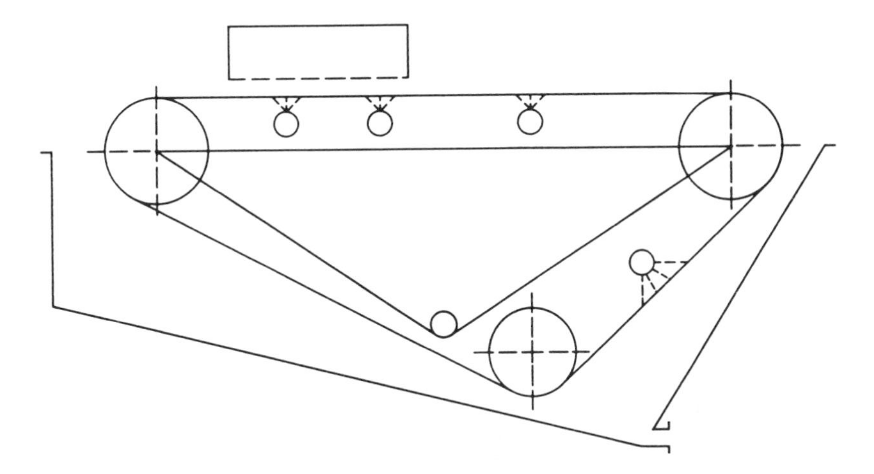

FIGURE 4-6. The linear screen.

Reprinted by permission, Canadian Institute of Mining and Metallurgy (Montreal). J. A. Tumilty, M. J. Wilkinson, and M. P. Collins, in *Gold Metallurgy*, edited by R. S. Salter et al. Pergamon Press (New York), 1987.

The Linear Screen (Delcor Screen) as a classifier in a milling circuit was patented by the Anglo-American Corporation in the early 1980s. It consists essentially of an endless screen cloth revolving on rollers (Figure 4-6). Pulp is distributed evenly from the feed box across the width of the belt. The majority of the undersize material will pass through the screen. Water jets across the width of the screen and disturbs the bed, allowing more fines to pass through the screen.

The oversize material is carried over the end of the drive roller and is washed into the oversize collection box. Text work has indicated that replacing a hydrocyclone in a milling circuit with a Delcor screen results in less overgrinding of the minerals and leads to significant increases of mill capacity (Tumilty et al., 1987).

Partial Direct Recovery of Gold

Gravity Concentration

Although leaching gold with cyanide solution is a very efficient process, the recovery of as much gold as possible by gravity concentration (before the cyanide leaching) has significant economic advantages. Liberated, sizable grains of gold (1 mm to 0.03 mm) should be removed from the underflow of the cyclones/classifiers by gravity concentration. Gold is malleable; once liberated, it cannot be further ground and should be recovered as soon as possible to prevent it from settling and locking up behind the mill liners.

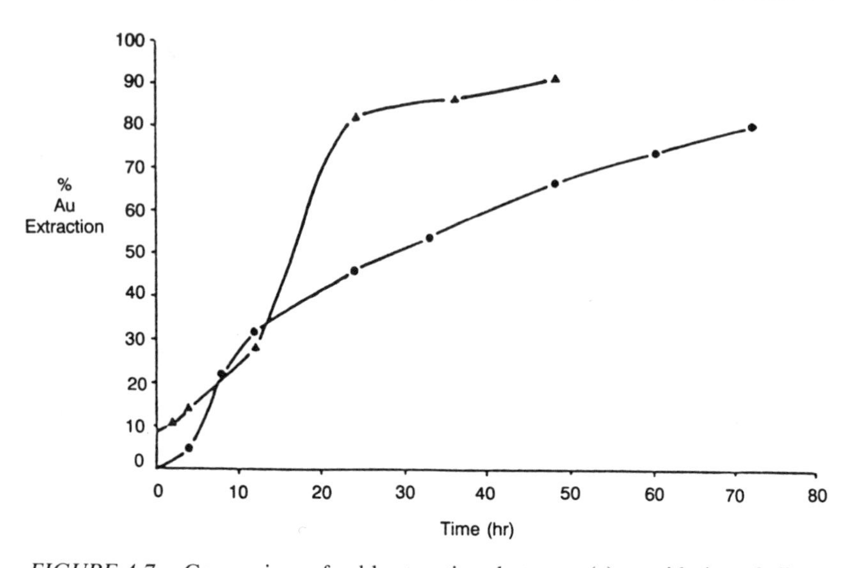

FIGURE 4-7. Comparison of gold extractions between (•) cyanidation of all ore *vs.* (▲) recovery of coarse gold by gravity separation followed by cyanidation.
Reprinted by permission. D. E. Spiller, *Mineral & Metallurgical Processing*, Society for Mining, Metallurgy and Exploration (Littleton, Co), August 1984: 118–38.

Gold is as good as money, and the sooner it reaches the market the more profitable it is (Figure 4-7). The early recovery of gold by gravity concentration—when feasible—has some very significant additional advantages.

- Reduced size of the cyanidation plant, since the plant avoids the excessive dissolution time required for large gold grains
- Recovery of coated gold grains that may resist or decelerate dissolution by cyanide
- Savings in reagent consumption
- No significant locking of gold behind the mill liners

Taking into account that gravity concentration recovers the coarse gold, the reduction of the size of the cyanidation plant is bound to lead to major savings in capital cost. Over 50% of the gold mines in South Africa employ gravity concentration ahead of the cyanidation circuit. Some of the South African mills also recover, by gravity, fine particles of osmiridium that occur in the gold reefs.

Many devices are used for gravity concentration of the cyclone underflow (Figures 4-8 and 4-9). Stationary devices include plane tables with riffles, corduroy blankets, Reichert Cones, and spirals. Moving devices include jigs, shaking tables.[2] The gravity concentrate has to be treated

---

[2]See description of gravity-concentration equipment in Chapter 3.

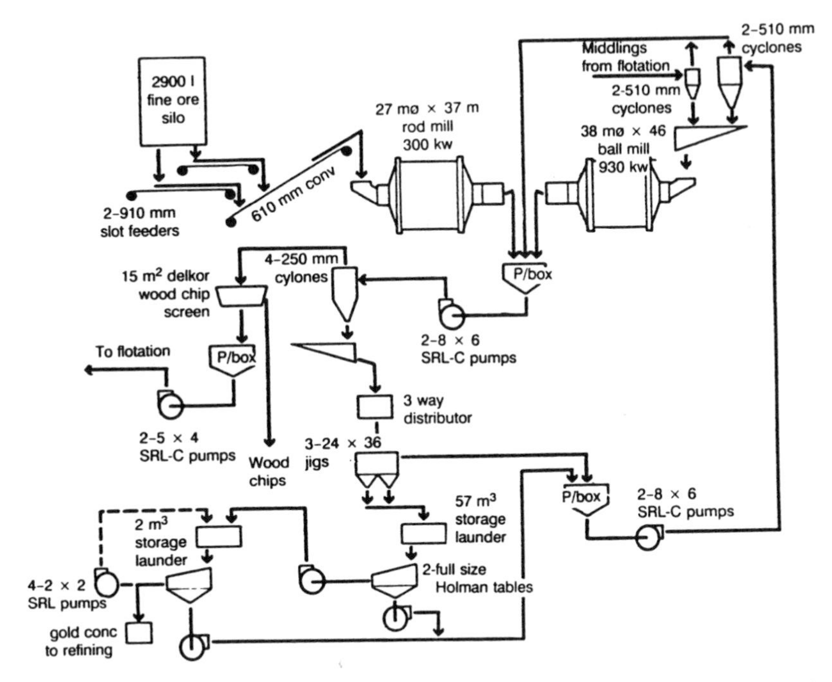

FIGURE 4-8. Grinding and gravity concentration flowchart of Campbell Red Lake Mines, Ltd.

Reprinted by permission, Canadian Institute of Mining and Metallurgy (Montreal). S. Roberts and J. Starkey, in *Gold Metallurgy*, edited by R. S. Salter et al. Pergamon Press (New York), 1987: 189.

further, as a rule, by either amalgamation or intensive cyanidation (see Chapter 8).

Amalgamation is a very old method of recovering gold from an ore slurry by dissolving it in mercury and forming a liquid mercury-gold alloy. Amalgamation has been used to recover fine alluvial gold, free gold from a ground ore, and free gold from the concentrate of a gravity-separation circuit. The use of mercury for amalgamation of gold was mentioned by Theophrastus (about 300 B.C.) and by Vitruvius (13 B.C.). The date of the first use of mercury for amalgamation is not known, but an educated guess has been ventured that "it has been used for this purpose for the last 2000 years" (Rose & Newman, 1937).

Stamp milling and amalgamation made up the universal method of recovering lode gold before the introduction of cyanide leaching on an industrial scale. Amalgamation of stamp-mill slurry was retained even after the introduction of cyanidation in 1889. Direct amalgamation of the whole ore stream is now obsolete. Modern plants employ gravity concentration, and may subject the high-grade concentrate to amalgamation.

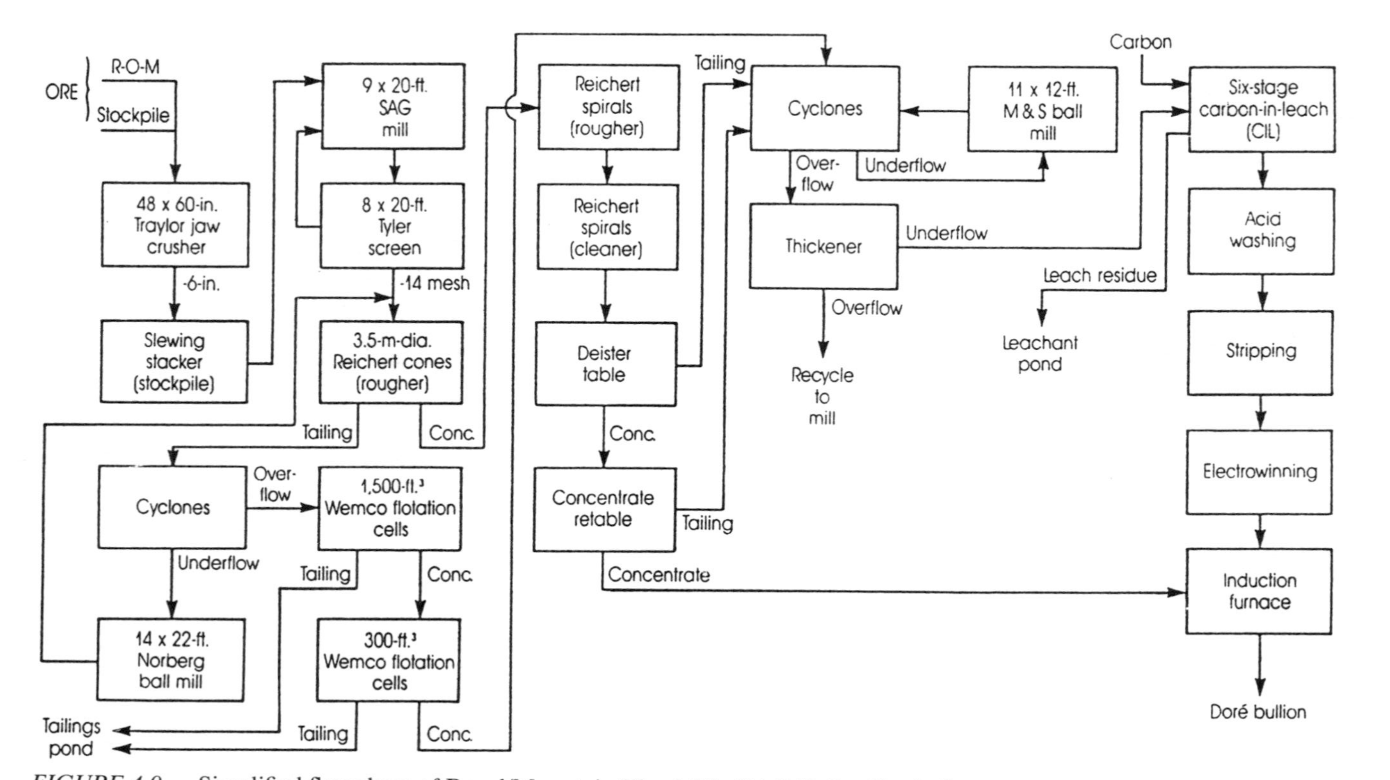

FIGURE 4-9. Simplified flow sheet of Royal/Mountain King Mill. (SAG-Ball mill grinding—gravity concentration—carbon in leach cyanidation.)

Reprinted by permission. G. O. Argall, Jr., in *Eng. Min. J.,* October, 1988.

Amalgamation can recover most of the coarse gold quickly at an early stage of the milling process, and may thus facilitate the cyanide leaching treatment.

Amalgamation has been used in mills treating ores with relatively coarse gold, in order to remove such gold from the circulating load of the grinding mill(s). The underflow slurry from the cyclone(s) is passed over the surface of mercury-treated (amalgamated) copper plates in a thin stream.

A method of preparing the copper plates is to coat them with electrode-posited silver and then have the mercury applied to the thin silver deposit. As the ore slurry flows over the plates, the gold particles adhere and form amalgam. This amalgam of gold has to be scraped off the plates from time to time. A duplicate line of copper plates allows the continuous operation of the grinding and amalgamation circuits. Small amounts of mercury can be added to the rod and ball mills (0.45–0.50 oz. of mercury per ton of mill feed) to enhance the amalgamation of free coarse gold.

The excess of mercury and water in the scraped amalgam are filtered through a press. The "dry" amalgam is retorted in iron retorts, where sponge gold is recovered and the mercury vapors are strictly collected and condensed. The mercury condensate is recycled to the mercury-feed tank of the amalgamation circuit.

The use of amalgamation in gold mills has declined significantly in recent years to prevent costly losses of mercury, theft of amalgam, and, especially, serious environmental problems. In the few plants that use amalgamation, strict precautions are taken for the collection of mercury vapors and fumes, the prevention of mercury spills, and the safety of the amalgam.

Leaching of Pulverized Gold Ores

The gold cyanidation process has become the universal gold-extraction practice since J. S. MacArthur and the brothers R. and W. Forrest patented it in 1889. The optimum concentration of the cyanide solution depends on the characteristics of the ore and should be determined experimentally. The levels of cyanide commonly used are 0.05–0.20% NaCN (1–4 lb. NaCN per ton). Cyanide is added in the grinding mill and in the first leaching tanks. Lime has to be added to the solution to keep its pH between 10.5 and 11.5. A large excess of lime has been found to increase the cyanide consumption.

Low concentrations of cyanide for gold leaching are strongly recommended. They offer economic (lower cost) and technical (less impurity dissolution) advantages. Minimum cyanide strength can be assured by employing a continuous cyanide analyzer.

The cyanidation process replaced the rather awkward chlorination method. The world's gold production doubled in 20 years after its first industrial application.

$$4Au + 8NaCN + O_2 + 2H_2O \rightarrow 4NaAu(CN)_2 + 4NaOH$$

The major factor affecting the dissolution of gold by cyanide is the oxygen concentration in solution. The solubility of oxygen in the cyanide solution depends on the partial pressure of oxygen (and hence on the altitude of the site), the temperature, the ionic strength of solution, and the intensity of the agitation. Other major factors affecting the gold-cyanidation process are the pH and redox potential of the slurry, the concentration of cyanide, and the presence of cyanicides (other than gold ions reacting with cyanide).

The pH level of the slurry and a minute addition of lead nitrate are important factors in minimizing side reactions in the leaching circuit. A very low concentration of lead ions significantly accelerates the dissolution of gold.

Calcium hydroxide tends to slow down gold dissolution in cyanide significantly more than sodium hydroxide (see Figure 8-8), but it is used almost universally in practice because of its lower cost. The deceleration of gold dissolution in the presence of calcium hydroxide becomes more pronounced with increasing pH in the range of 10.5 to 12.0. The retarding effect of calcium ions on the rate of gold cyanidation is minimal at a pH of approximately 10.5. Lime addition and the appropriate regulation of pH prevent the loss of cyanide by hydrolysis or by reactions with acidic substances (CO_2 in the air or bicarbonates and acidic contaminations in the mill water and in the ore).

Gold Milling Flow Sheets

In addition to the comminution plant, gold mills contain the following (see Figure 4-8):

- The cyanide leach circuit
- The solids-liquid separation area and/or a system of gold recovery from solution
- The gold melting and refining area
- The tailings disposal area

Preaeration

Preaeration of the slurry (Figures 4-10 and 4-11) may be required if the ore contains inordinate proportions of oxygen consumers. The cyclone overflow should be thickened from 35% solids to 50% solids, and the thickened pulp must be vigorously aerated at pH 10.5–11.0 in Pachucas[3] or in agitated tanks.

[3]The Pachuca tank, also known as the Brown tank, is a bubble-column slurry reactor used as an agitated leaching reactor in extractive hydrometallurgy. The Pachuca is a large, cylindrical tank with a conical bottom (up to 16 m total height and 10 m diameter). Compressed air is injected from the tank bottom to agitate the ore slurry and keep it in suspension.

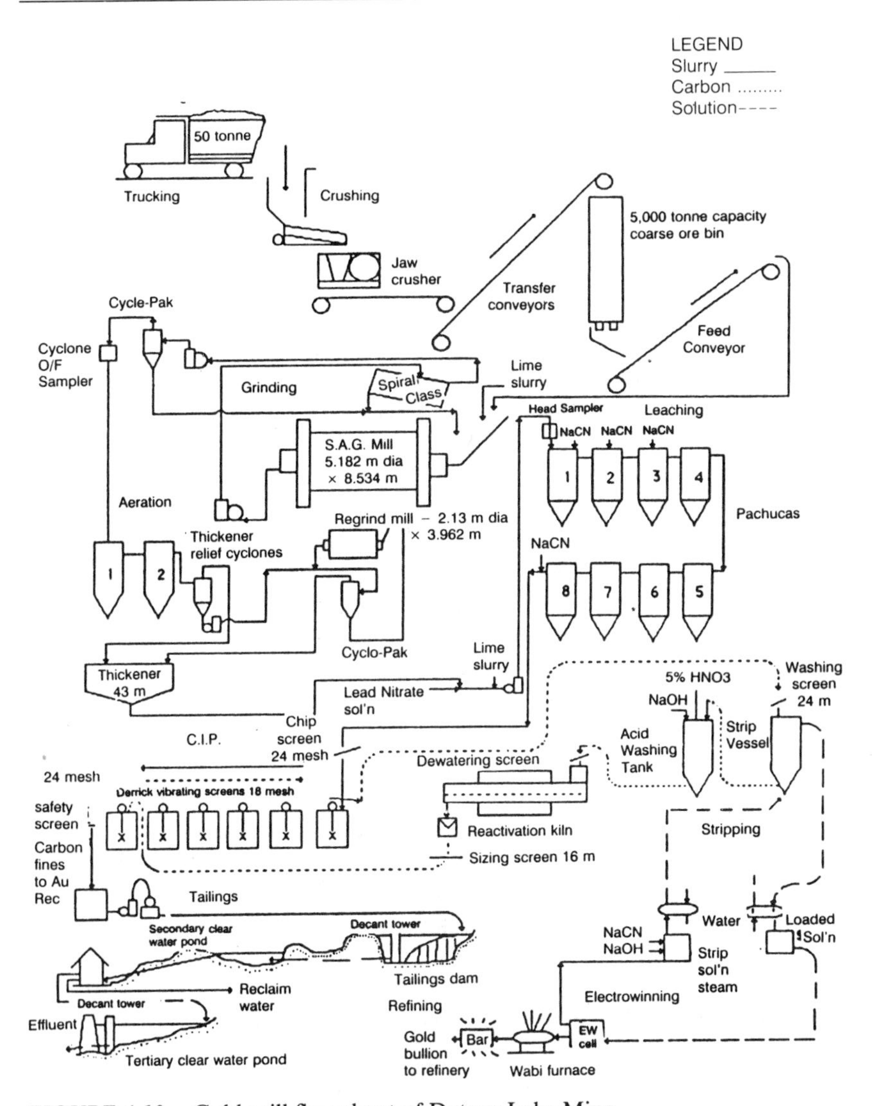

FIGURE 4-10. Gold-mill flow sheet of Detour Lake Mine.

Reprinted by permission, Canadian Institute of Mining and Metallurgy (Montreal). D. Rollwagen et al., in *Gold Metallurgy*, edited by R. S. Salter et al. Pergamon Press (New York), 1987: 43.

The aerated pulp is filtered—if cyanicides have been generated—and the filter cake, repulped with barren solution (cyanide solution from which dissolved gold has been recovered) is fed to the leaching circuit. Preaeration supplies oxygen to the thickened slurry, saturates the liquid phase with oxygen in advance of cyanidation, and promotes the oxidation of any oxygen consumers.

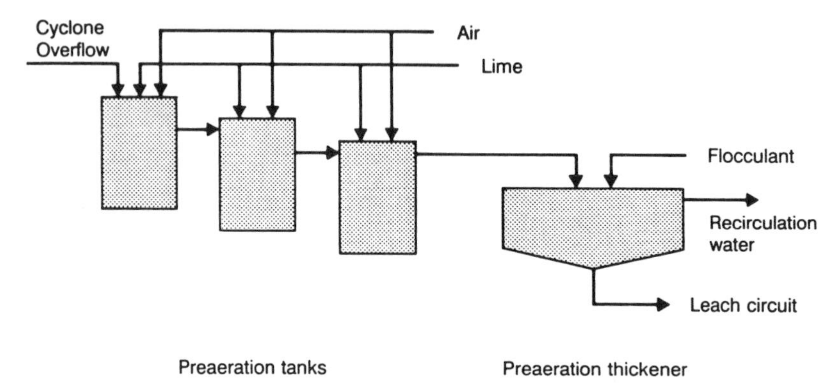

FIGURE 4-11. Preaeration of the slurry before cyanidation.

The Leaching Circuit

A surge tank or, preferably, a thickener smooths out fluctuations in the grinding rate and in the density of the ground-ore pulp. A pump, with a controlled flow rate to maintain the content of solids at the range of 44–48%, feeds the slurry into a cascade of leaching tanks. The thickener overflow is recycled into the grinding circuit.

Lime slurry must be added to the thickener underflow to maintain a pH of 10.5 to 10.6 during the cyanide leaching. Pachucas, where the air flow provides agitation and oxygen for the cyanidation reaction, are often used in gold mills (see Figure 4-10). A constant stream of bubbling air is maintained under the propellers of the agitated leaching tanks.

As a rule, long reaction times of 24 to 28 hours are allowed for cyanidation. Slurry samples taken from a tank or Pachuca before the last third of the circuit are analyzed to check the progress of the leaching circuit.

Recovery of Gold from Solution

Solids-Liquid Separation

When the dissolution of gold is complete, the pregnant solution must be separated from the leached solids if gold is to be recovered by zinc precipitation. The leached solids must be washed thoroughly to minimize any loss of dissolved gold with the solids.

Counter-current decantation (CCD) in a series of thickeners is designed to provide solids-liquid separation with thorough washing of the solids (Figure 4-12). With a series of thickeners, the solids are washed as they flow from the first to the last thickener and to the tailings disposal. The liquid washings overflow and are pumped from the last to the first thickener, counter to the movement of the solids.

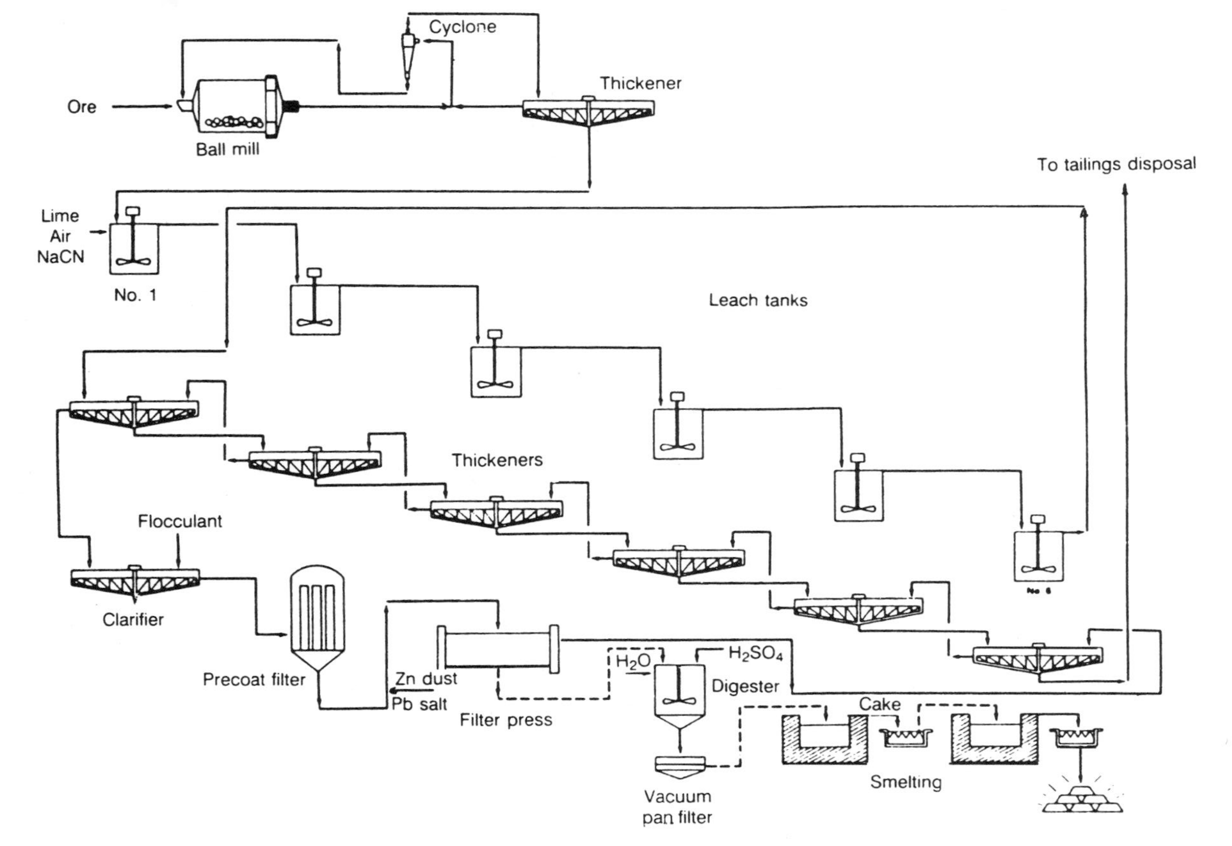

FIGURE 4-12. Gold-mill flow sheet with counter-current decantation (CCD) and Merrill-Crowe zinc cementation process.

In a CCD system, the following design features (Buttler & Collins, 1986) must be provided.

- Bypass piping to remove any one thickener from the circuit, if necessary
- Mixing boxes feeding each thickener
- Stand-by pumps to all thickeners
- Multiple points of flocculant addition in thickener feed launders (fitted with mixing baffles)
- Steeply sloping floors for easy cleanup

The Merrill-Crowe zinc cementation process consists of four basic steps.

1. Clarification of the pregnant cyanide solution
2. Deaeration
3. Addition of zinc powder and of lead salts
4. Recovery of zinc-gold precipitate

$$Zn + NaAu(CN)_2 + H_2O + 2NaCN \rightarrow Au + Na_2Zn(CN)_4 + NaOH + 0.5H_2$$

The pregnant solution is pumped through Stellar Candle filters that have been precoated with diatomaceous earth for thorough removal of any suspended minute solids. The clarified solution is splashed through the plates of the under-vacuum Crowe tower to remove any dissolved oxygen. (Even traces of oxygen in solution will passivate zinc surfaces and inhibit gold precipitation.) Zinc dust is added to the de-oxygenated solution as it flows to the precipitation filters. Pressure Candle filters—precoated with diatomaceous earth and a secondary coating of zinc dust—are used. These filters are totally enclosed to safeguard the precious precipitate.

Activated-Carbon Systems [4]

Activated carbon is a highly porous material with distinct adsorptive properties. Gold complexes with either chloride or cyanide are strongly adsorbed by activated carbon. Gold recovery from solution by granular, activated carbon is widely used in gold mills, and consists of the following distinct operations.

1. Loading the carbon
2. Elution or stripping of the carbon

2a. Carbon regeneration

3. Gold production by electrowinning or cementation from the eluate solution

---

[4] See also Chapter 10.

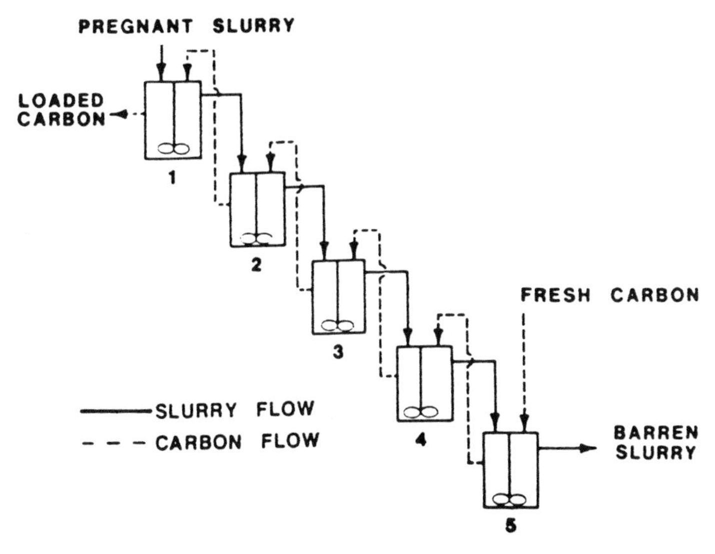

FIGURE 4-13. Flow diagram of a five-stage carbon-in-pulp (CIP) circuit.
Reprinted by permission. K. C. Garner, N. J. Peperdy, and C. N. Moreton, in *Mineral and Metallurgical Processing*. Society for Mining, Metallurgy and Exploration (Littleton, CO), February 1986.

The activated-carbon systems used in gold mills are:

- *Carbon in pulp (CIP)* The activated carbon is mixed with the leached slurry and adsorbs the gold from solution (see Figures 4-10 and 4-13).
- *Carbon in leach (CIL)* The activated carbon is added to the ore slurry in the leaching tanks and adsorbs the gold from solution as cyanidation of the ore proceeds (see Figures 4-9 and 4-14).
- *Carbon in columns* The granular activated carbon is packed in columns and adsorbs gold from clarified solutions as they percolate through.

Electrowinning of Gold

Electrowinning of gold[5] from dilute pregnant solution was developed by Zadra (1950), who proposed and tested a cathode compartment packed with steel wool. The extensive surface of the cathode allowed an acceptable rate of gold electrowinning from dilute pregnant solution. A modern EW cell is rectangular, and has pervious cathodes packed with steel wool and stainless steel mesh anodes (Figure 4-15). The electrodes are sized to

[5]See also Chapter 10.

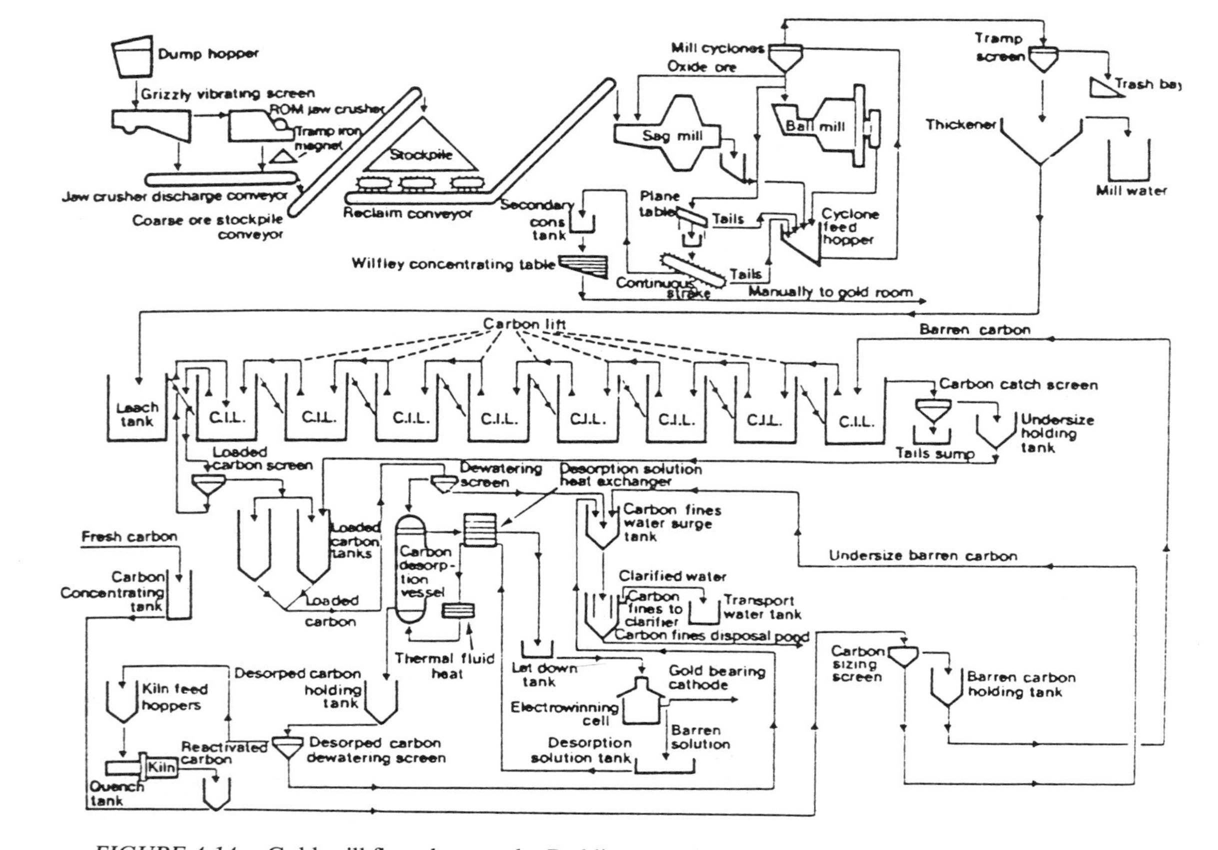

FIGURE 4-14. Gold-mill flow sheet at the Paddington Mine.

Reprinted by permission. *The Mining Journal Ltd.* (London), November 1985: 385.

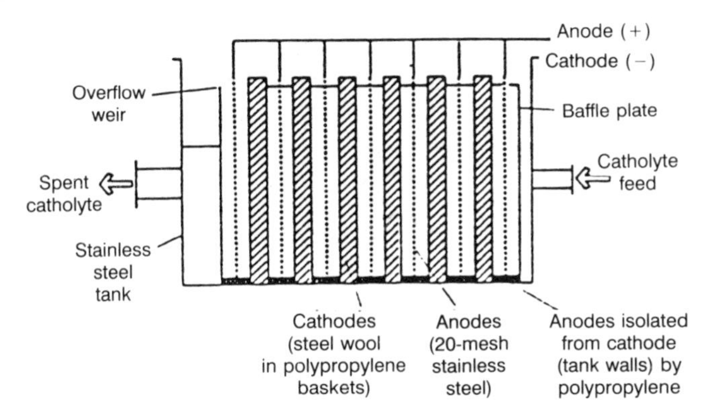

FIGURE 4-15. The MINTEC electrowinning cell with steel-wool cathodes. Reprinted by permission. A. P. W. Briggs, in *J. S. Afr. I.M.M.*, October 1983: 251.

fit snugly down the sides and along the bottom of the cell, so that the electrolyte has to flow through the electrodes.

Suppressing Mercury Extraction

Several low-grade gold ores contain a small amount of mercury (usually less than 15 ppm). During cyanidation, up to 30% of the mercury is extracted along with the gold and silver. Mercury is also adsorbed by activated carbon in CIP circuits, and is later stripped from the carbon (along with gold and silver).

$$Hg^{++} + 4CN^- \rightarrow Hg(CN)_4^{2-}$$
$$2Hg + 8CN^- + O_2 + 2H_2O \rightarrow 2Hg(CN)_4^{2-} + 4OH^-$$

During the electrowinning of gold onto steel-wool cathodes, mercury is deposited along with gold (Laxen et al., 1979). The cathodes have to be retorted under vacuum at 650–700° C to remove mercury prior to the pyrorefining of gold.

The U.S. Bureau of Mines developed a method of suppressing mercury extraction with the addition of calcium sulfide in the ore (0.5–2.9 Kg CaS per kg of mercury contained in the ore). After treatment with CaS, gold-to-mercury ratios from 100:1 to 400:1 were obtained in the leach solution of mercury-containing gold ores (Staker et al., 1984).

Selective Electrowinning of Mercury

The U.S. Bureau of Mines also developed a method of selective electrowinning of mercury from cyanide solutions by using mercury-coated copper wool as the cathode. With applied potentials of 1.0 to 1.5 v, 96% of the mercury and 98% of the silver have been selectively electrowon from mill carbon-strip solutions that contained 1.9 to 20 ppm Hg, 0.35 to 39.5 ppm Ag, and 20 to 345 ppm Au. Gold electrolytic extractions ranged between zero and 20% (Sheya et al., 1989).

Refining of Gold [6]

The laden gold cathodes may contain 70–75% Au and 10–15% Ag, with the remainder being copper and steel wool. To fire-refine the electrowon gold, a pan is filled with 30% sodium nitrate, 40% borax, and 30% silica (all in % of cathode weight).

Sodium nitrate ($NaNO_3$) oxidizes base metals and combines with silica (SiO_2) to lower the melting temperature of the mix. Borax ($Na_2B_4O_7$) is an acid flux that dissolves metal oxides and also lowers the melting temperature of the charge. Silica combines with borax and contributes to the dissolution of metal oxides.

The charge (1:1 cathode/flux) is placed in a silicon-carbide crucible in a tilting gas-fired furnace and heated to 1,093° C. After melting, it remains here for about 1.5 hours and separates into two molten phases, slag and bullion. After the slag is skimmed off, the crude bullion is further refined in clay crucibles.

References

Note: Sources with an asterisk (*) are not cited in text but are recommended for further reading.

Argall, Jr., G. O. 1989. Royal/Mountain King Mine brings new technology to the mother lode. *Eng. Min. J.*, October: 40–42.

Butler, L. J., and D. J. Collins. 1986. FMC's Paradise Peak Mill—a gold-silver cyanide and CCD leach mill. Paper presented at the AMC International Mining Show, 5–8 October, at Las Vegas, NV.

Freeport McMoRan Gold Co. 1986. Freeport-McMoRan gold grinding circuit. Paper presented at the 1986 SME-AIME Annual Meeting, New Orleans, LA.

Garner, K. C., N. J. Peperdy, and C. N. Moreton. 1986. Process and process control design using dynamic flowsheet simulation. *Min. & Metal. Proc.* February: 41–44.

*Janish, R. P. 1986. Gold in South Africa. *J. S. Afr. I.M.M.*, (86): 273–316.

*Kelly, E. G., and D. Spottiswood. 1982. *Introduction to Mineral Processing.* New York: J. Wiley & Sons, 127–236.

[6]See also Chapter 11.

Laxen, P. A., G. S. M. Becker, and R. Rubin. 1979. Developments in the application of carbon-in-pulp to the recovery of gold from South African Ores. *J. S. Afr. I. M. M.,* 79(11): 315–26.

Nordberg Inc. 1989. Nordberg introduces high-efficiency waterflush crushing. *Eng. Min. J.*, January: 106–107.

Rose, T. K., and W. A. C. Newman. 1937. *The Metallurgy of Gold*. Philadelphia: J. B. Lippincott Co., 168–87.

Sheya, S. A. N., J. H. Maysilles, and R. G. Sandberg. 1989. Selective electrowinning of mercury from gold cyanide solutions. Paper presented at the 1989 TMS Annual Meeting, 27 February–2 March, at Las Vegas, NV.

Spiller, D. E. 1984. Applications of gravity beneficiation in gold hydrometallurgical systems. *Min. & Metal. Proc.*, August: 118–28.

Staker, W. L., W. W. Simpson, and R. G. Sandberg. 1984. Mercury removal from gold cyanide leach solution. *Min. & Metal. Proc.*, May: 56–61.

Tumilty, J. A., M. J. Wilkinson, and M. P. Collins. 1987. The linear screen as classifier in a milling circuit. In *Gold Metallurgy*, edited by R. S. Salter, D. M. Wyslouzil, and G. W. McDonald. New York: Pergamon Press, 97–104.

Wills, B. A. 1988. *Mineral Processing Technology*, 4th ed. New York: Pergamon Press, 292–94.

Zadra, J. B. 1950. A Process for the Recovery of Gold from Activated Carbon by Leaching and Electrolysis. U.S.B.M. RI4672.

CHAPTER 5

Treatment of Refractory Gold Ores

Problem ores that yield low (say less than 80%) gold recovery by cyanidation when "normally ground" are defined as *refractory*. The term, as well as the technical methods used to treat such ores, are relatively new. Rose and Newman's very comprehensive book, *The Metallurgy of Gold*, does not even contain the term *refractory* in its latest, fourth edition (1937).[1] Dorr and Bosqui's *Cyanidation and Concentration of Gold and Silver* (2d edition, 1950) has a brief chapter (XIII) on "Cyanicides and Refractory Ores," with only four pages devoted to refractory ores. *Gold Metallurgy in South Africa*, edited by R. J. Adamson and published by the Chamber of Mines of South Africa (1952), contains about six pages with meaningful contents on "Refractory Properties of the Ores and their Treatment."

Mineralogy of Refractory Gold Ores

Gold is found in nature mainly as native gold. A number of gold alloys (the most important being gold-silver alloy, known as electrum) and tellurides constitute the main gold minerals (Table 5-1). Electrum is a variety of native gold containing 20% or more silver. There is no important mineralogical difference between native gold and electrum; their optical properties and crystal structures are similar. The other alloys and tellurides have different optical properties and structures than electrum has, and

[1]However, Rose and Newman have chapters on "Roasting Gold Ores" and "The Cyanide Process—Special Methods" that describe treatments of ores not responding fully to cyanidation.

TABLE 5-1. Gold-bearing minerals.

Native gold	*Au*
Electrum	(Au,Ag)
Cuproauride	(Au,Cu)
Porpezite	(Au,Pd)
Rhodite	(Au,Rh)
Iridic Gold	(Au,Ir)
Platinum	(Au,Pd)
Bismuthian Gold	(Au,Bi)
Amalgam	Au_2Hg_3(?)
Maldonite	Au_2Bi
Auricupride	$AuCu_3$
Rozhkovite	$(Cu,Pd)_3Au_2$
Calaverite	$AuTe_2$
Krennerite	$(Au,Ag)Te_2$
Montbrayite	$(Au,Sb)_2Te_3$
Petzite	Ag_3AuTe_2
Muthamannite	(Ag,Au)Te
Sylvanite	$(Au,Ag)Te_4$
Kostovite	$AuCute_4$
Nagyagite	$Pb_5Au(Te,Sb)_4S_{5-8}$
Gold Tellurate	(?)
Uytenbogaardtite	Ag_3AuSb_2
Aurostibnite	$AuSb_2$
Fishchesserite	Ag_3AuSe_2

Reprinted by permission. G. Casparini, *CIM Bull.*, March 1983: 145. Canadian Institute of Metallurgy (Montreal). p. 145.

they are scarce. (Antimonides of gold are rare. Gold selenides—and some tellurides—hosted in copper-sulfide minerals end as slimes during the electrolytic refining of copper.)

To the above three types of gold-bearing minerals (native gold/gold alloys/gold compounds) a fourth class must be added: "invisible" gold dispersed in the form of solid solution or submicron inclusions in sulfide minerals and sulfosalts (Gasparini, 1983; Chryssoulis et al., 1987). Gold, in minor to trace amounts, occurs in a number of minerals such as tellurium, atokite (Pt_3Pd_3Sn) (Kuhnel et al., 1980), atheneite (Pd_3Hg_3As), palladoarsenide (Pd_2As), and svyagintsevite (Pd_3Pb) (Cabri, 1976; Gasparini, 1983).

Refractory gold ores are unoxidized (or partially oxidized) ores containing carbonaceous materials, sulfides, and/or tellurides. Gold associated with arsenic, and sometimes with bismuth, is often only partially soluble in cyanide solution. Gold is found in veins, sizable (visible) grains, submicron particles, and finer particles not visible even at 15,000 × SEM resolutions (Casparini, 1983; Hausen, 1981).

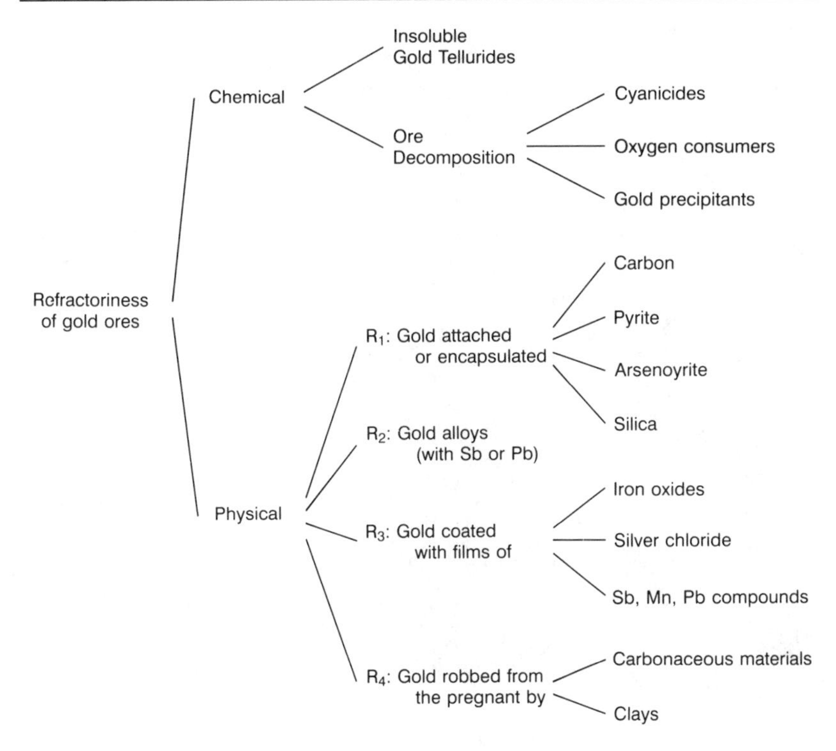

FIGURE 5-1. Classification of refractory gold ores.

The refractoriness of gold ores can be broadly distinguished as being of a chemical physical nature. The *chemical refractoriness* is relatively rare and confined to three conditions:

- Insoluble gold tellurides
- Ingredients of the ore that might decompose and react with cyanide (cyanicides)
- Ingredients of the ore that consume oxygen

The last two conditions will cause precipitation of gold from solution (Figure 5-1). Casparini (1983) pointed out that a very wide range of tellurides exhibit different solubilities in cyanide solution, with some being insoluble.

The preponderant refractoriness of gold ores is of a physical nature, and should be specified by microscopic mineralogy. Figure 5-1 indicates the possibility of distinguishing physically refractory ores of five types.

- R_1 contains fine gold encapsulated or attached to host matter
- R_2 contains gold alloys
- R_3 contains gold coated with films (mineral films or films created during the treatment of the ore)

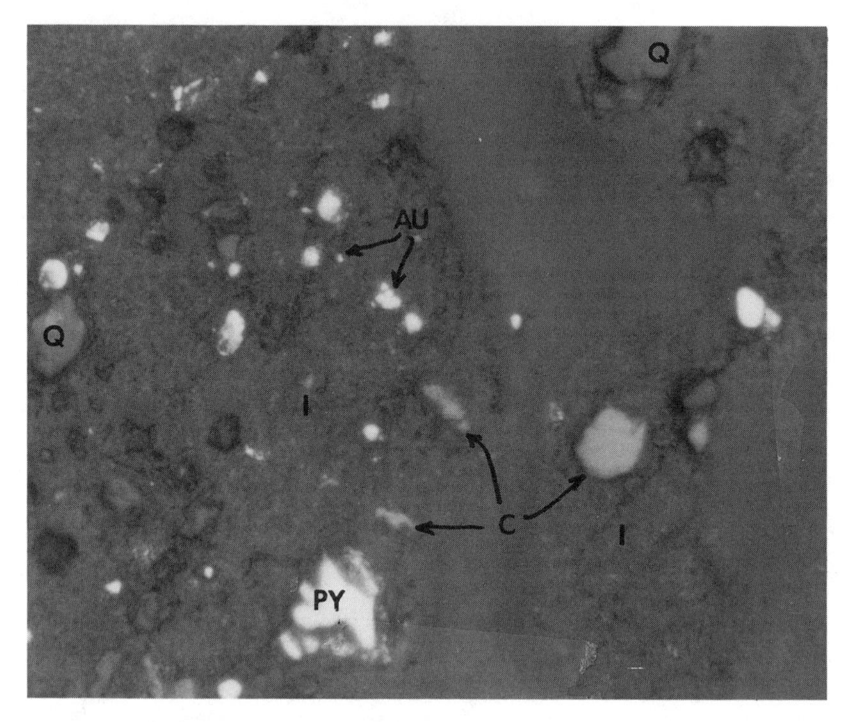

FIGURE 5-2. Particles of pyrite (Py), organic carbon (C), quartz (Q), and free gold (Au) in a matrix of illitic clays (I); Carbon and recleaner concentrate, light sands (<2.4 sp.gr.) fraction; Incident light, X666.

- R_4 contains adsorbent materials (carbonaceous and/or clays) that rob dissolved gold from the pregnant solution (preg robbing[2] materials)
- R_5 contains decompositions of host minerals (sulfides can decompose and form cyanicides; sulfides, thiosulfites, arsenites, and ferrous ions can consume oxygen; humic acid can interact with gold complexes)

It was recognized early (in 1968) that three types of refractory ores occurred in the Carlin (Nevada) deposit (Figures 5-2 and 5-3). *Type R1* contained refractory gold locked with carbon and pyrite that would not adsorb additional gold from the pulp. *Type R4* contained an active form of carbon that would adsorb additional gold from the mill circuit (Hausen & Bucknam, 1985). *Type R5* contained an organic acid similar to "humic acid," with functional groups that would interact with gold complexes (Guay, 1981b).

---

[2]*Preg robbing* is the active adsorption of gold from cyanide-pregnant solutions by components of ore (Hausen & Bucknam, 1985).

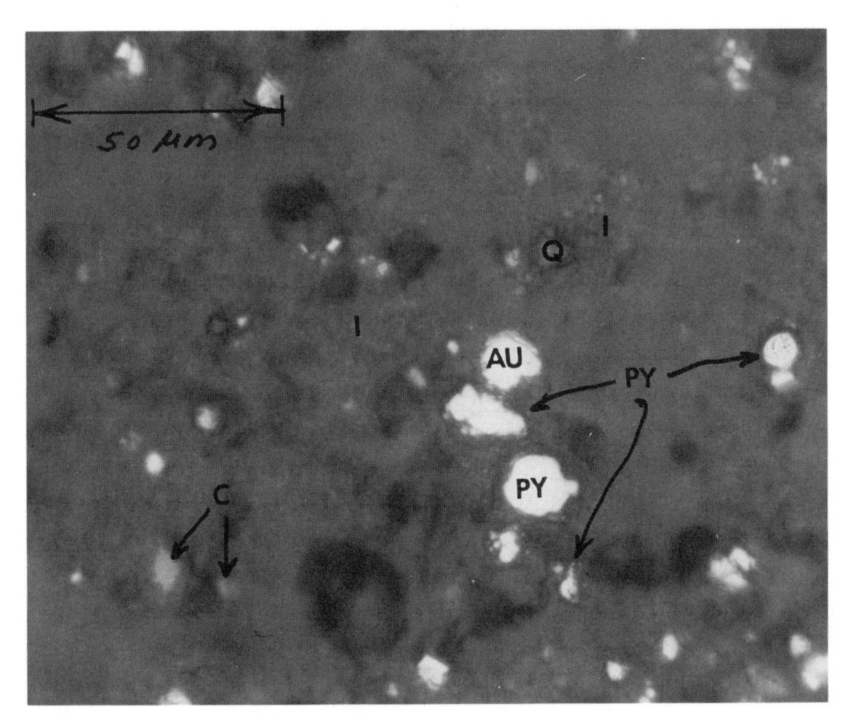

FIGURE 5-3. Relatively coarse (8–10 μm) particle of gold (Au) in illitic clay matrix (I) with embedded grains of pyrite (Py), organic carbon (C), and heavy sands (> 2.4 sp.gr.), Incident light, X666.

Most of the Nevada "carbonaceous" ores contain from 0.4 to about 1% organic carbon, with occasional samples having as much as 4%. The term *carbonaceous* cannot be rigorously defined. Guay (1981b) proposed an arbitrary classification of the Carlin ores by fire assay, by atomic adsorption (AA) on the raw ore and on ore roasted at 500° C, and by the preg-robbing value of the ground ore, which is equated to its affinity to adsorb gold from pregnant solutions over specified time periods.[3]

The AA assay on raw or roasted ore samples is a hot cyanide leach of a small sample of pulverized ore followed by an AA assay on the pregnant solution. If the AA assay on raw ore gives more than 80% gold extraction, the ore is classified as "amenable to cyanidation" (nonrefractory) by Guay (1981b).

[3]The preg-robbing value is defined by an adsorption process that is influenced by the concentration of gold in the pregnant solution, temperature, pH, Eh, and the concentrations of other (competing) ions. All those conditions have to be standardized in any particular laboratory.

TABLE 5-2. Host materials for gold.

Pyrite	*Most common host*
Arsenopyrite	Relatively common
Sulfides	
chalcocite	
covellite	
pyrotite	
galena	Relatively common
Chalcopyrite	Not very common
Nickel sulfides	
arsenides	
sulfo-arsenides	Relatively common
Iron oxides	Relatively common
Uranium minerals	Common in South Africa
Silicates and carbonates	Relatively common, gold enclosed or distributed along grain boundaries
Carbonaceous material	Common host

Reprinted by permission. G. Casparini, *CIM Bull.*, March 1983: 145. Canadian Institute of Metallurgy (Montreal).

Hausen and Bucknam (1985) established through lab experiments that "... the rate of adsorption by carbonaceous ore from the Gold Quarry and Carlin orebodies (Nevada) was approximately four times faster than the activated carbon sample studied ..." (Westates carbon CC321 with "medium" activity and loading capacity). In other words, Hausen and Bucknam proved experimentally that the carbonaceous material naturally occurring in those ores is a very potent adsorbent of gold and it prevents gold recovery by simple cyanidation.

The most common refractoriness occurs by encapsulation of fine gold. Minerals that act as hosts to the enclosed gold grains are listed in Table 5-2.

Some arsenic compounds are readily soluble in cyanide solution but, during the gold-precipitation step with either zinc or aluminum, there is the hazard of forming poisonous arsine (AsH_3) with the hydrogen evolved. When arsenopyrite ($FeAsS$) occurs in a gold ore, a proportion of the gold is frequently in intimate association with the sulfide lattice. Arsenopyrite in an ore can yield alkaline arsenites when the finely ground ore is agitated with a strong concentration of lime, or can yield other alkalis in the presence of air.

$$2FeAsS + 2Ca(OH)_2 + 5.5O_2 \rightleftharpoons 2FeSO_4 + 2CaHAsO_3 + H_2O$$

This reaction is a strong oxygen consumer. It might affect the rate of cyanidation if arsenopyrite is not decomposed ahead of the leaching operation.

When stibnite (Sb_2S_3) occurs in an ore, it acts as a weak acid. With lime or alkalis it forms thio-antimonites and thio-antimonates, and thus affects the alkalinity of the cyanidation pulp. Sufficient lime is required to maintain a permanent alkaline protection; otherwise, any latent acidity of the ore will decompose the thio-antimonates to antimony pentasulfide and hydrogen sulfide, which are potent cyanicides.

Hausen (1981) has shown that the term *auriferous pyrite* relates to an association of gold with pyrite in a variety of textural forms that affect metallurgical treatment. Differences in morphological features may affect the rate and efficiency of oxidative pretreatments. Following native gold, gold associated with sulfides is the most important source of the precious yellow metal (see Table 5-2).

The manner of occurrence of gold in the host mineral varies. Casparini (1983) summarized the potential mineral associations:

- Gold distributed in fractures or at the border *between grains of the same mineral*
- Gold distributed along the border *between grains of two different minerals* (e.g., two different sulfides, a sulfide and a silicate, or a sulfide and an oxide)
- Gold totally *enclosed in the host mineral*

Gold in ores containing sulfides may be distributed as free leachable gold, fine gold attached to the sulfide grains that can be exposed for leaching by fine grinding, and gold finely disseminated in the sulfide mineral.

Since the 1960s extensive mineralogical work has attempted to define the exact nature of refractoriness in recalcitrant gold ores and concentrates, and to assist in the development of pretreatments that contribute to maximum gold recoveries from such ores.[4] Microscopic examination of polished sections of ore particles can lead to a meaningful assessment of the gold occurrence and the type of refractoriness, if any (see Figures 5-2 and 5-3).

Observations on the basic mineralogy of refractory gold ores provide valuable information in the early stages of process evaluation. In general, ores are partially refractory. As a measure of refractoriness, Guay (1981b) and Hausen (1984) proposed the ratio of "percentage extracted" by cyanide to the total gold contained (established by fire assay). The lower this ratio, the higher the refractoriness of the ore.

---

[4]Hausen and Kerr, 1968; Radtke and Scheiner, 1970; Scheiner et al., 1971; Radtke et al., 1972; Henley, 1975; Hausen, 1981; Robinson, 1983; Hausen and Bucknam, 1985.

Flotation of Some Refractory Ores

Gravity preconcentration[5] and/or flotation often precede the treatment of a refractory gold ore. Very common types of refractoriness are due to fine gold attached or encapsulated in pyrite, arsenopyrite, or other sulfides (see Table 5-1). "Invisible" gold, as a solid solution in sulfide minerals, is known to exist (Chryssoulis et al., 1987).

When gold is locked in or associated with sulfides, the concentration of sulfides by flotation must be seriously considered. After removing coarse sulfides from the underflow of the cyclones by gravity concentration, fine grinding is required, as a general rule, for the complete liberation of the gold-containing sulfide minerals. Sonora Mining Corporation has reported 90% −325 mesh in the cyclone overflow, which is feed to flotation (Argall, 1987).

Conditioning agents are commonly used, especially with ores that are partially oxidized. The alkalinity of the pulp is regulated with soda ash, since lime acts as a depressor of free gold and inhibits pyrite flotation. Copper sulfate can be helpful in accelerating the flotation of pyrite and arsenopyrite. Sodium sulfide is needed for the flotation of partially oxidized ores, but it should be used with caution because it depresses free gold. The rougher concentrate often requires regrinding before cleaning. The flotation treatment of refractory gold ores must be designed to achieve high recovery of gold, even with a relatively low grade of concentrate, if necessary.

Extensive laboratory and pilot testing of flotation, along with a rigorous economic analysis of the experimental results, must precede any decision that is made on the concentration of refractory gold ores. Flotation recoveries range between 90% and 95%, with some gold remaining in the flotation tails. Gold in nonsulfide tails may be amenable to recovery by cyanide leaching, and it is advisable to test and estimate the costs of such (secondary) recovery.

In one North American operation, ore with 1.5% pyrite (containing virtually all the gold and the silver) is treated by flotation. Cleaner concentrates average 25% FeS, with 99% pyrite recovery. Flotation is practiced at natural pH (6.5–7.5), with sodium isopropyl xanthate as the primary collector, Aerofroth 76 as the frother, and Aerofloat 208 as the secondary collector. Rougher-scavenger concentrates are combined and cleaned in a single bank of 60-cu. ft. cells. Cleaner tails are recirculated back to the beginning of the rougher flotation.

Most of the causes of refractoriness (see Figure 5-1) can be eliminated by an oxidation treatment. In practice, refractory ores are submitted, as a rule, to some form of oxidation treatment before cyanidation (Figure 5-4).

[5]See also Chapter 3.

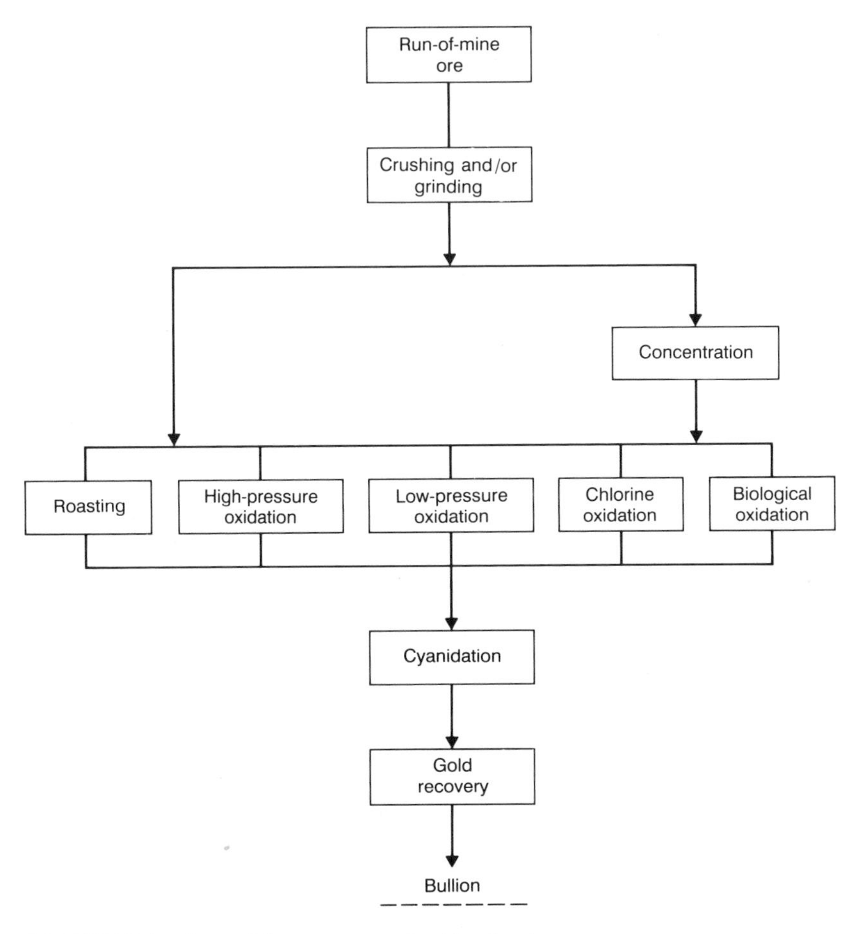

FIGURE 5-4. Potential treatments of gold refractory ores.

High-Temperature Oxidation: Roasting

Roasting pretreatment of refractory gold ores or concentrates has the objective of liberating any gold particles encapsulated or attached in sulfides, arsenopyrites, or carbon, and of destroying carbonaceous material and any potential cyanicides. Hence, a porous, fluffy calcine—which neither encapsulates nor coats the gold particles—is the desirable product of roasting.

Roasting, or "calcination," as a pretreatment of refractory gold ores containing sulfur and/or arsenic, antimony, or tellurium is a very old operation. It probably is as old as extracting gold from complex ores, and it might have been used by alchemists. Roasting was industrially applied on a large scale even when chlorination was the established leaching method for gold ores (before cyanidation). Roasting is, in effect, the most widely applied oxidative pretreatment process for gold refractory ores and concentrates (see Figure 5-4). Roasting has the objective of removing any sulfur, arsenic, antimony, and other volatile post-oxidation substances, and of oxidizing any tellurides and base metals contained in the ore.

The Chemistry of Roasting Refractory Gold Ores and Concentrates

A multitude of chemical reactions are possible in roasting, depending on the host mineral(s) causing the refractoriness, the excess of air, and the prevailing temperature during roasting. The main reactions of roasting—oxidations of pyritic and arsenopyritic minerals—are strongly exothermic, and the gold-containing particles are sites of heat generation. The temperature of the reacting minerals has to be controlled to avoid "flash" roasting, the fusion of the calcines and re-lockup of the gold micrograins.

In the absence, or deficiency, of air, decompositions of sulfides and arsenides are possible.

$$FeS_2 \rightarrow FeS + S \quad (5.1)$$
$$4FeAs_2 \rightarrow Fe_4As + 7As \quad (5.2)$$
$$FeAsS \rightarrow FeS + As \quad (5.3)$$
$$2FeAsS + 2FeS_2 \rightarrow 4FeS + As_2S_2 \quad (5.4)$$

Depending on the excess of air and the prevailing temperature, the following oxidation reactions can be expected, among others.

$$4FeS_2 + 11O_2 \rightarrow 2Fe_2O_3 + 8SO_2 \quad (5.5)$$
$$4FeS + 7O_2 \rightarrow 2Fe_2O_3 + 4SO_2 \quad (5.6)$$
$$3FeS_2 + 8O_2 \rightarrow Fe_3O_4 + 6SO_2 \quad (5.7)$$
$$3FeS + 5O_2 \rightarrow Fe_3O_4 + 3SO_2 \quad (5.8)$$
$$FeS_2 + O_2 \rightarrow FeS + SO_2 \quad (5.9)$$
$$FeS_2 + 3O_2 \rightarrow FeSO_4 + SO_2 \quad (5.10)$$
$$FeAsS + 3O_2 \rightarrow FeAsO_4 + SO_2 \quad (5.11)$$
$$2SO_2 + O_2 \rightarrow 2SO_3 \quad (5.12)$$
$$4As + 3O_2 \rightarrow 2As_2O_3 \quad (5.13)$$
$$2As_2S_2 + 7O_2 \rightarrow 2As_2O_3 + 4SO_2 \quad (5.14)$$
$$As_2O_3 + O_2 \rightarrow As_2O_5 \quad (5.15)$$
$$2FeAsO_4 \rightarrow Fe_2O_3 + As_2O_5 \quad (5.16)$$
$$FeO + SO_3 \rightarrow FeSO_4 \quad (5.17)$$
$$4Fe_3O_4 + O_2 \rightarrow 6Fe_2O_3 \quad (5.18)$$

The main roasting reaction of pyrite concentrates, or of ores with pyritic refractoriness, is the highly exothermic reaction (5.5) producing hematite (Figure 5-5). Hematite (red oxide) is more amenable to cyanidation than

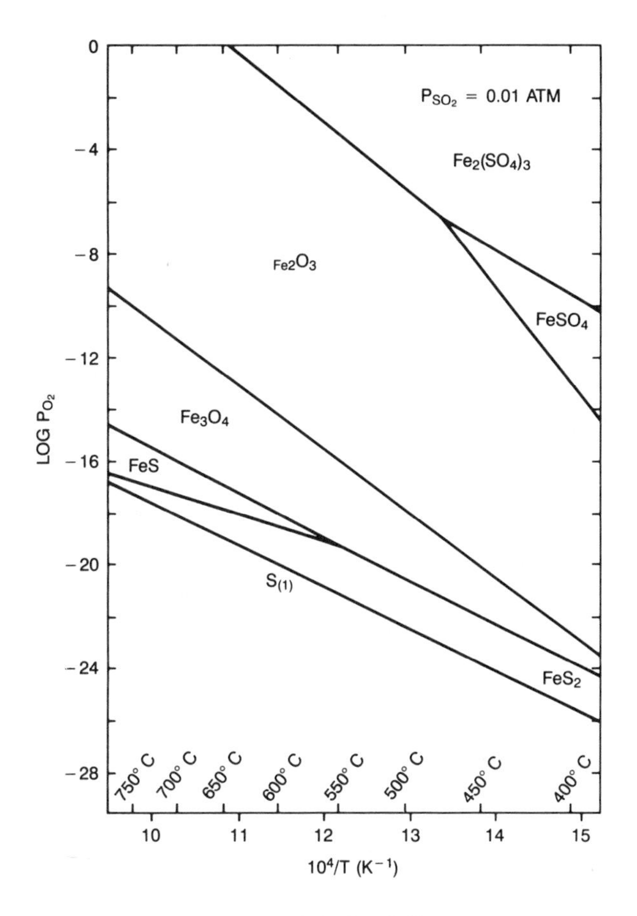

FIGURE 5-5. Stability diagram of the Fe-O-S system.

Reprinted by permission. V. Kudryk, D. A. Corrigan, and W. W. Liang, eds., *Precious Metals: Mining, Extraction, and Processing*. The Metallurgical Society (Warrendale, PA), 1984.

magnetite, since it constitutes a finer-grained product, in which gold particles are more freely exposed than in magnetite (black oxide). During the roasting of refractory ores, any contained carbonaceous material will be burned, and the calcine is not expected to be preg robbing.

In salt (or chlorination) roasting, the following reactions can take place (among others), producing volatile chlorides.

$$2Au(s) + Cl_2(g) \rightleftharpoons Au_2Cl_2 \quad \text{(sublimation at 650° C)} \tag{5.19}$$

$$2Au(s) + 3Cl_2(g) \rightleftharpoons Au_2Cl_6 \quad \text{(sublimation 200°–350° C)} \tag{5.20}$$

$$2Au(s) + Fe_2Cl_6(g) + 3Cl_2(g) \rightleftharpoons 2AuFeCl_6 \quad (5.21)$$
(enhanced volatility)

In the presence of chlorine, auric chloride (Au_2Cl_6) volatilizes at 200–350° C, and aurous chloride (Au_2Cl_2) has high vapor pressure above 650° C (James & Hager, 1978). The U.S. Bureau of Mines has experimentally proven that the volatility of gold chloride can be enhanced by complexing it with ferric chloride.[6]

$$2Au(s) + Fe_2Cl_6(g) + 3Cl_2(g) \rightarrow 2AuFeCl_6(g)$$

The above reaction can be reversed at temperatures higher than 450° C, and gold can be recovered by dissociation of the complex auric-ferric-chloride, whereas chlorine and ferric chloride are recycled.

$$2AuFeCl_6(g) \xrightarrow{475^\circ C} 2Au(s) + 3Cl_2(g) + Fe_2Cl_6(g)$$

Dunn (1982, 1983) has proposed a detailed flow sheet for treating refractory gold ores by chlorination and volatilization. The ore is first roasted with air to oxidize any sulfide minerals, and then it is chlorinated at 350° C. Palmer et al. (1986) advocated "rapid extraction of gold . . .", by high-temperature chlorination (and volatilization), ". . . without the generation of noxious waste streams" (as if chlorine and volatile chlorides were not noxious streams). The fact is that chlorine and high temperature make a very nasty combination that should be avoided in industrial practice, if possible. Chloride roasting is not popular in practice, mainly due to the corrosive and polluting nature of the operation.

Conditions Affecting the Roasting of Refractory Gold Ores and Concentrates

The temperature of roasting pyritic and arsenopyritic refractory gold ores must be controlled in order to produce the optimum quality of calcine. Studies of the solid phase changes that take place during the roasting of pyritic and arsenopyritic ores and concentrates suggest the phase diagram presented in Figure 5-5. The final reactions in the proposed mechanisms are strongly exothermic, and hence the temperatures of the reacting particles cannot be easily controlled. The particle temperature during normal roasting may be several hundred degrees higher than the furnace atmosphere; fusion of particles can occur, associated with a low-melting (950° C) eutectic mixture of FeO and FeS (Carter & Samis, 1952).

[6]Dunn, 1982; Eisele and Heinen, 1974; Heinen et al., 1974; Eisele et al., 1970.

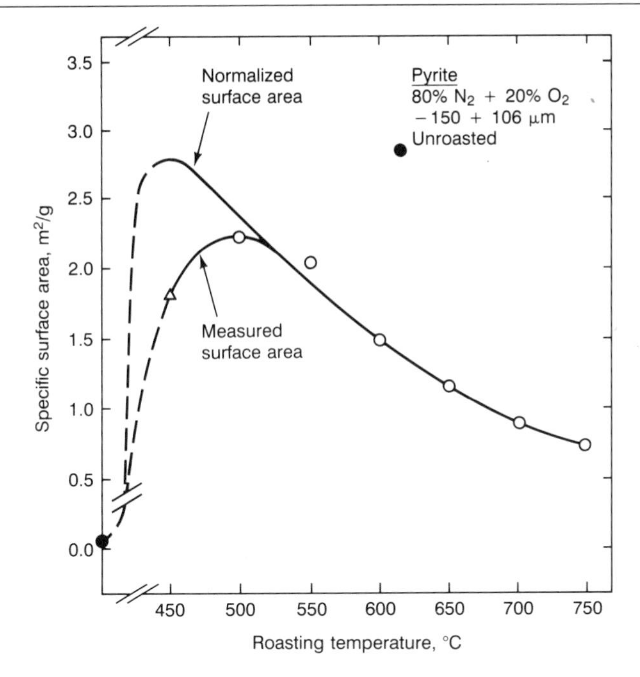

FIGURE 5-6. Effect of roasting temperature on the surface area of pyrite calcines. Reprinted by permission. V. Kudryk, D. A. Corrigan, and W. W. Liang, eds., *Precious Metals: Mining, Extraction, and Processing*. The Metallurgical Society (Warrendale, PA), 1984.

The importance of temperature control during the roasting of pyritic- and arsenopyritic-gold refractory ores has been emphasized by many authors.[7] The optimum roasting temperature to produce "fluffy" hematite varies from ore to ore (depending on the ore's composition and mineralogy), and has to be established experimentally. In general, however, it is less than 650° C (Figure 5-6).

Arriagada and Osseo-Assare (1984) proved experimentally that the calcine acquires the maximum surface area per unit of mass (and hence the maximum rate of leaching) at the optimum roasting temperature. They concluded that: "The marked sensitivity of the structural characteristics of the roast products to the temperature emphasizes the need for extremely close control of the roasting operations, where parameters such as surface area, porosity and pore size distribution of the calcines must be incorporated as controlled variables in the process and considered in any optimization efforts."

[7]Arriagada and Osseo-Assare, 1984; Osmanson, 1987; Swash and Ellis, 1986; Smith et al., 1985; Mason et al., 1984; Hansen and Laschinger, 1967; Carter and Samis, 1952; Djingeluzian, 1952.

Since the most important reactions in roasting gold refractory ores (reactions 5.5–5.8 and 5.11) are highly exothermic, the sulfide and/or arsenide particles are sites of heat generation. In order to control the temperature of the reacting particles and prevent their fusion, the rate of heat transfer from the reacting solids to the gas phase must equal the rate of heat generation. This is a rather delicate control, difficult to maintain in industrial operations. The temperature of the gaseous atmosphere of a roaster—which can be controlled—is not necessarily identical to the temperature of all reacting particles. The more efficient the control of the solid-gas heat-transfer rate during roasting, the less "flash" roasting occurs, and the fewer overheated, fused, solid particles produced.

Close temperature control—to prevent fusion of overheated particles—may result in *calcines with some residual sulfur and/or arsenic* (such as sulfide, sulfite and sulfate, and arsenite and arsenate) *that act as cyanicides* during the subsequent cyanidation. To overcome those ill effects, elaborate water-washing circuits are included prior to the cyanidation of calcines in some roasting plants (Jha & Kramer, 1984). Aeration of the calcine slurry in the presence of lime, to remove alkali and oxygen-consuming substances prior to cyanidation, may be advantageous in some roasting operations. It is advisable to wash calcines as rapidly as possible, because it appears that prolonged contact of gold with the dissolved components of the calcine can make coatings form on the gold. Such coatings are not rapidly soluble in dilute cyanide solutions, although they might be soluble in concentrated cyanide (Mrkusic & Laschinger, 1967).

The temperature control in an industrial roaster cannot be perfect, and some overheating (and fusion) of reacted solids may occur. Fine grinding of gold-bearing calcines prior to cyanidation has resulted, as a rule, in increased gold extractions (as compared to extractions from the same calcines after direct cyanidation) (Filmer, 1982; Dunne & Liddell, 1984). The decision about spending additional capital for a calcine-mill has to be based on the achievable increase of gold extraction by grinding, if any.

Control of the Gaseous Phase During Roasting. The composition and flow rate of the gaseous phase during roasting have to be controlled in order to sustain the oxidation reactions at their optimum rate, and to contribute to the control of the solid-gas heat transfer rate within the roaster. An excess of air is required to oxidize pyrite and arsenopyrite to hematite (Fe_2O_3) rather than magnetite (Fe_3O_4).

Rabbling or Fluidization Mixing. Mechanical rabbling, or fluidization of fine sulfide solids, is required to keep exposing fresh ore surfaces to the oxidizing gas phase and sustain acceptable high rates of roasting. Continuous mixing increases also the heat transfer rate from reacting solids to the gas and contributes to the temperature control of the roasting operation.

Roasting with Additives. *Chloridizing roasting* of pyritic and arsenopyritic refractory gold ores converts the contained iron to ferrous chloride and fixes the sulfur and the arsenic as sulfate and arsenate.

$$\mathrm{FeS + 4NaCl + 40 \rightarrow FeCl + 2Na\ SO + Cl}$$
$$\mathrm{FeAsS + 4NaCl + 40 \rightarrow FeCl + Na\ SO + Na\ AsO + Cl}$$

According to Palmer et al. (1986), "the arsenic in arsenopyrite can be removed with chlorine at 130° C." Iron chlorides react with oxygen at temperatures above 200° C, and generate gaseous chlorine.

$$\mathrm{6FeCl + 30 \rightarrow 2FeCl + 2Fe\ O + 3Cl}$$
$$\mathrm{2FeCl + 1.50 \rightarrow Fe\ O + 3Cl}$$

An excess of chlorine during chloridizing roasting may form volatile gold chlorides (Palmer et al., 1986).

Gold Loss by Volatilization. Very high volatilization of gold—more than 90%—could be obtained from a gold ore by salt roasting. According to Hansen and Laschinger (1967), "gold is volatilized, but the form of the gold in the vapor (phase) has never been determined." As reported above, some auric chlorides can dissociate at moderately high temperatures and form very fine metallic colloidal gold in the gas phase. This contributes to gold losses.

The number of chloridizing roasting plants for gold refractory ores has been very limited. The industry has always considered chloridizing roasting as the last resort in the treatment of refractory gold ores. In addition to the serious risk of gold loss by volatilization, the chlorine environment, at relatively high temperatures, is very corrosive to construction materials.

Roasting with Sodium Carbonate. This process has been tested in the absence of air to remove arsenic from arsenical refractory ores, at the laboratory of the U.S. Bureau of Mines (Hansen & Laschinger, 1967; Archibald, 1949). Those laboratory experiments were successful in removing arsenic and achieving high gold extraction by cyanidation of the calcine, but the development of an industrial roaster operating in the absence of air has been considered impractical.

The U.S. Bureau of Mines at Reno, Nevada, has tested a method for roasting, in the absence of air, gold-containing sulfide concentrates mixed with trona ($Na_2CO_3/NaHCO_3$) and briquetted. The sulfide concentrate was mixed with 80–120% of the estimated stoichiometric requirements of trona for oxidation. The briquettes were roasted at 525° C in a rotary kiln. The sulfide was oxidized to sulfate, and the contained heavy metals and silver formed carbonates. The roasted briquettes were leached with sulfuric acid containing some sodium chloride, to keep silver in the precipitate. After filtration, the residue—which contained the gold and silver chloride—was leached with cyanide.

If the sulfuric acid solution contains substantial copper, it can be treated with a caustic agent to precipitate Fe, As, and Sb as jarosites and leave copper in solution (to be recovered by electrowinning). Although laboratory roasting experiments with sodium carbonate were successful in oxidizing the sulfides, removing arsenic, and yielding high gold extraction by cyanidation, the development of a roaster that operates in the absence of air has been considered practically impossible.

Industrial Furnaces for Roasting Gold Ores and Concentrates. Four kinds of furnaces have been used for roasting refractory gold ores and concentrates. The *rotary kiln* is used very rarely for special reasons. *The Edwards roaster* was developed in Australia and has been used in several Canadian gold mines to roast arsenical and/or pyrite concentrates (Djingheuzian, 1952). It is a single-hearth roaster, long and narrow, with the charge moving along the hearth by gravity. Rabbling is usually done with a double row of revolving, air-cooled rabble arms. The rabbling causes considerable mixing of the feed and prevents local overheating and sintering of the calcine. The roasting temperature is also controlled by

- The sulfur content of the concentrate (maintained at 26–27% with the addition of plant residues)
- The moisture of the concentrate filter cake
- The draught of the furnace (as a last resort)

The vertical multi-hearth furnace (known as a Wedge or Hereshoff roaster; Archibald, 1949) is cylindrical and tall. The solid feed is charged at the top hearth and the hot gas is introduced from the bottom of the furnace (Figure 5-7). The roaster is well insulated and is equipped with a central revolving shaft that supports two rabble arms per hearth. The roaster shaft and rabble arms are internally air-cooled. The rabble arms are slow in stirring the solid charge and dropping it from hearth to hearth.

The prevailing-temperature, draught, and air-inlet controls are interdependent during roasting of sulfide concentrates. The roaster gas exits from the top and, through a balloon flue, is brought to an electrostatic precipitator (Cottrell) for the collection of fine entrained solids.

The fluosolids roaster is a relatively "new" roasting furnace (1947) in which fine solids are suspended (fluidized) by and within the oxidizing gas (Matthews, 1949; Wright, 1961). The roaster consists of a cylindrical shell built of steel plate and lined with insulating brick and refractory firebrick (Figures 5-8 and 5-9). There are two outlets in the dome top of the reactor. One is a gas discharge to the cyclone dust collectors, and the other is an auxiliary stack opening that is normally used when the reactor is preheated at the starting of a campaign. Both outlets are lined with refractory materials. Fine feed (like flotation concentrates) can only be charged in a fluosolids roaster. The best results are obtained when sulfide concentrates

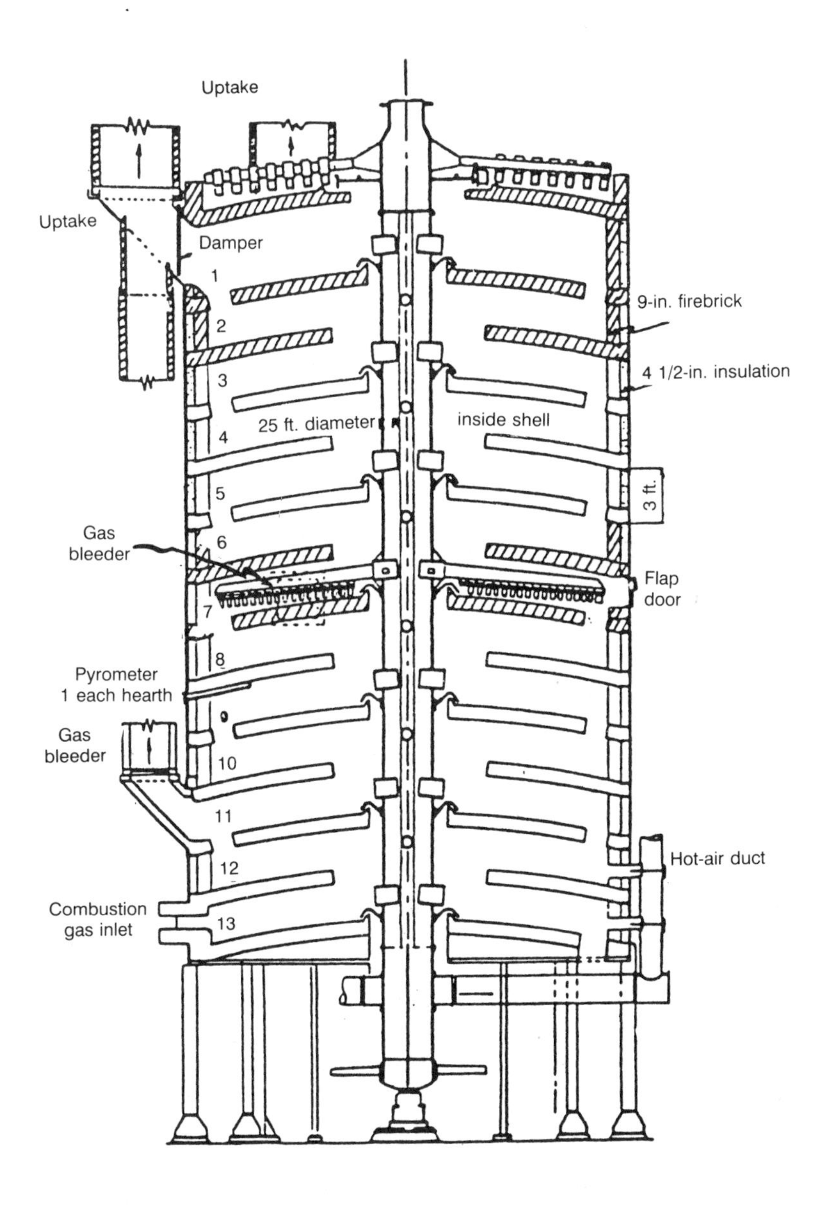

FIGURE 5-7. Diagramatic section of multi-hearth roaster

Reprinted by permission. F. R. Archibald, Roasting arsenical gold ores and concentrates, *Canadian Mining & Metallurgy Bulletin*, Vol. 42 (443): 129–39. The Canadian Institute of Mining and Metallurgy (Montreal), 1949.

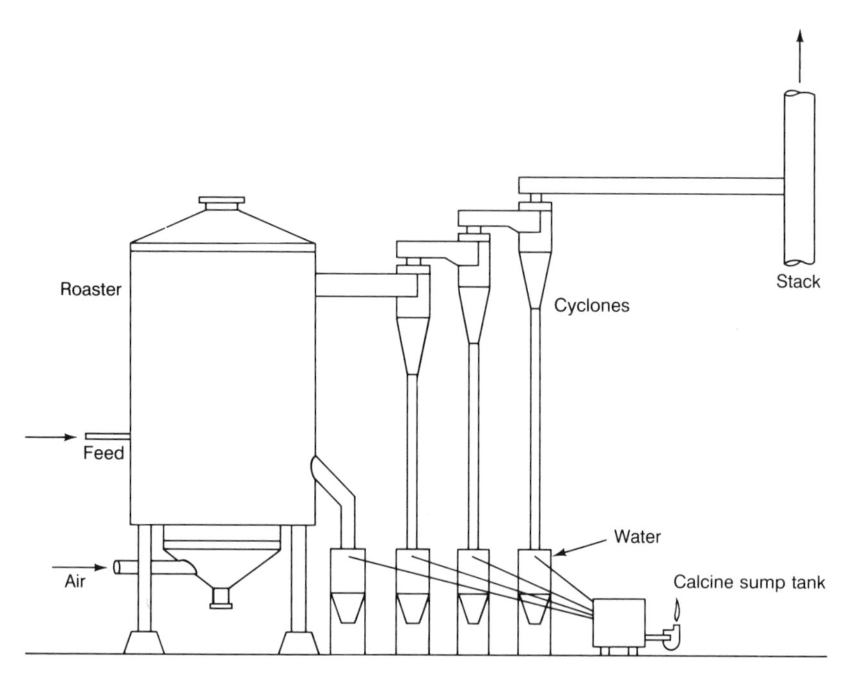

FIGURE 5-8. Fluosolids roaster.

Reprinted by permission. K. R. N. Hansen and J. E. Laschinger, The Roasting of Refractory Gold Ores and Concentrates: A Literature Survey. Report No. 85: South African Institute of Mining and Metallurgy (Johannesburg), 1967.

are introduced with the right proportion of air, so that there is only a trace of oxygen in the exit gas. Excess or insufficiency of air adversely affects the production of porous calcine.

Thermocouples are inserted at the lower and upper parts of the fluidized bed, at the top of the furnace chamber, and at the bottom of the stack. Temperature control can be achieved by injecting water into the bed. In most operations, temperature control around 595° C leads to complete desulfurization and fluffy porous calcine. The optimum temperature of roasting has to be established experimentally for any particular concentrate. It has been shown in the past that maximum recovery of gold, from concentrates containing arsenic and/or antimony, can be obtained after a low-temperature (450° C) roasting.

A significant fraction of the calcine is carried by the product gas, which has to be vented through cyclones for adequate solid-gas separation. The cyclone overflow is treated in a wet scrubber to collect fine particles of the calcine.

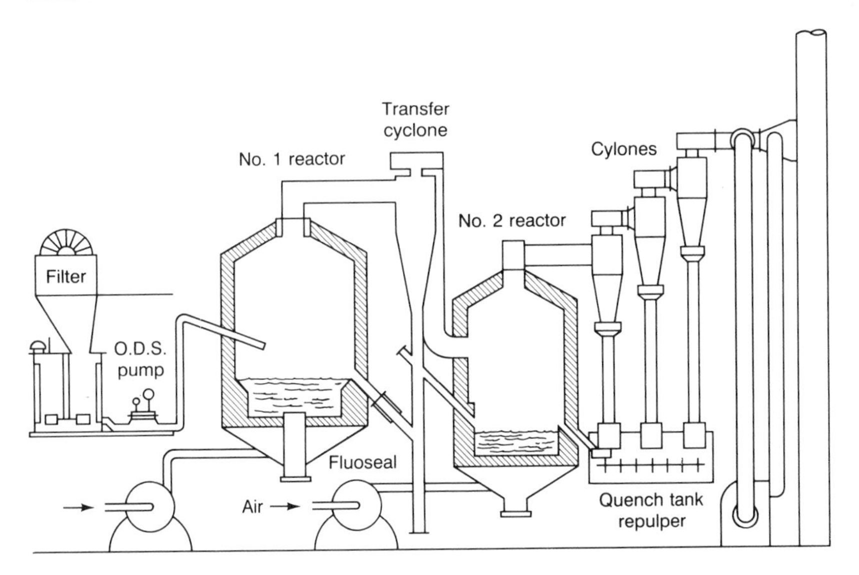

FIGURE 5-9. Diagrammatic section of a two-stage fluosolids roaster.

Reprinted by permission. K. R. N. Hansen and J. E. Laschinger, The Roasting of Refractory Gold Ores and Concentrates: A Literature Survey. Report No. 85: South African Institute of Mining and Metallurgy (Johannesburg), 1967.

Problems in Roasting Gold Ores and Concentrates

Pollution of the Atmosphere. Refractory gold ores and concentrates contain, as a rule, sulfides and/or tellurides, arsenides, and mercury compounds. In the case of concentrate roasting, low to high, contents of sulfur and arsenic oxides, tellurium, and mercury contaminations—along with fine particulates are expected in the roaster gas. Therefore, *expensive gas-treatment installations are required to control the loss of particulates and polluting emissions.*

Fusion of Calcines. A number of highly exothermic reactions occur during the roasting of gold pyritic ores, as discussed already. Particles with a high content of sulfides and/or arsenides are combustible, and localized temperature extremes may occur during roasting that cause fusion to eutectic mixtures of iron oxides and iron sulfides. Fused particles may occlude gold, which aggravates the losses as not accessible to cyanidation. A *calcine-grinding installation*, often required to maximize gold extraction by cyanidation, *significantly increases the cost of the roasting plant.*

Gold Lockup in Recrystallized Hematite. Close temperature control and adequate air excess during roasting are required for the production of porous hematite without gold lockup in its crystal structure. Closing of the pores by recrystallization of hematite, and solid solution of gold in hematite, were suggested by Norwood (1939) as causes of gold lockup.

Washing of Calcines. The presence of significant arsenites, sulfites, and soluble sulfides in the calcine may increase cyanide and lime consumption during cyanidation. Some gold mills find it advantageous to wash the calcine prior to cyanidation (Jha & Kramer, 1984). The tailings dam of the plant must be able to accommodate such washings.

Calcite Content of the Ore. Calcium carbonate in the roaster feed increases the energy requirements due to the exothermic decomposition $CaCO_3 \rightarrow CaO + CO_2$. It also changes the composition of the gas phase in the roaster (by the evolution of CO_2), and may form calcium sulfide and/or sulfate. Calcium sulfide is a cyanicide and a reducing agent; hence, it is deleterious to cyanidation. Calcium sulfate may plug the pores of the calcine, and it also may be deposited as gypsum in the pumps and conduits of the mill. Loss of gold extraction has been attributed to high calcitic gangue in calcines of refractory gold ores (Jha & Kramer, 1984).

High-Pressure Oxidation

The application of high-pressure oxidation to refractory gold ores and concentrates (in order to enhance their amenability to cyanide leaching) was first demonstrated and patented by American Cyanamid Company (Hedley & Tabachnik, 1957). The gold extraction from ores and concentrates with gold locked up in arsenopyrite, pyrite, stibnite, realgar, orpiment, and carbonaceous material was improved from the 18–86% range to the 92–99.5% range, after pressure oxidation over a 225–245° C temperature range.

The process chemistry of the oxidation of arsenopyrite and pyrite in an aqueous medium at elevated temperature was summarized by Berezowsky and Weir (1984) as follows.

$$4FeAsS + 11O_2 + 2H_2O \rightarrow 4HAsO_2 + 4FeSO_4$$
$$2FeS_2 + 7O_2 + 2H_2O \rightarrow 2H_2SO_4 + 2FeSO_4$$
$$4FeSO_4 + O_2 + 2H_2SO_4 \rightarrow 2Fe_2(SO_4)_3 + 2H_2O$$
$$2HAsO_2 + O_2 + 2H_2O \rightarrow 2H_3AsO_4$$

Under milder oxidizing conditions (100–160° C), and in the presence of excessive proportions of sulfuric acid and ferric sulfate, elemental sulfur may be produced as the preferred, or intermediate, product of arsenopyrite or pyrite oxidation.

$$4FeAsS + 5O_2 + 4H_2SO_4 \rightarrow 4HAsO_2 + 4FeSO_4 + 4S^\circ + 2H_2O$$
$$2FeAsS + 7Fe_2(SO_4)_3 + 8H_2O \rightarrow 2H_3AsO_4 + 16FeSO_4 + 5H_2SO_4 + 2S^\circ$$

The oxidation of pyrite with simultaneous production of sulfur is rare, but feasible, in the absence of oxygen excess.

$$FeS_2 + 2O_2 \rightarrow FeSO_4 + S^\circ$$
$$FeS_2 + Fe_2(SO_4)_3 \rightarrow 3FeSO_4 + 2S^\circ$$

The formation of elemental sulfur is favored, at the initial stages of pyrite oxidation, at relatively low temperatures (90–120° C) and high acidities. Such generation of elemental sulfur during the oxidation of auriferous ores and concentrates should be avoided in order to prevent the following.

- Occlusion of unreacted sulfides by sulfur, hindering the completion of oxidation
- Occlusion of gold particles, penalizing the extraction by cyanidation
- Increased consumption of cyanide and oxygen, during cyanidation

Therefore, the aqueous oxidation of sulfides should be conducted at pressures allowing temperatures above 160° C to promote the complete oxidation of sulfides, and of any elemental sulfur, to sulfates.

$$2S^\circ + 3O_2 + 2H_2O \rightarrow 2H_2SO_4$$

Iron and arsenic may precipitate as ferric arsenate under the pressure-oxidation conditions.

$$Fe_2(SO_4)_3 + 2H_3AsO_4 \rightarrow 2FeAsO_4 + 3H_2SO_4$$

Hydrolysis of ferric sulfate may precipitate hematite, basic ferric sulfate, and/or jarosites.

$$Fe_2(SO_4)_3 + 3H_2O \rightarrow Fe_2O_3 + 3H_2SO_4$$
$$Fe_2(SO_4)_3 + 2H_2O \rightarrow 2Fe(OH)SO_4 + H_2SO_4$$
$$3Fe_2(SO_4)_3 + 14H_2O \rightarrow \underset{\text{hydronium jarosite}}{2H_3OFe_3(SO_4)_2(OH)_6} + 5H_2SO_4$$
$$3Fe_2(SO_4)_3 + K_2SO_4 + 12H_2O \rightarrow \underset{\text{potassium jarosite}}{2KFe_3(SO_4)_2(OH)_6} + 6H_2SO_4$$

Any other metal ions in solution (e.g., silver, mercury, lead) may precipitate as jarosites, either substituting for the potassium, sodium, or hydronium ions or being occluded in the jarosite precipitate.

Oxidation conditions are critical to the chemistry within the autoclave and can significantly affect gold extraction (Figures 5-10 and 5-11). The oxidation of pyritic and arsenopyritic refractory gold ores, under oxygen pressure, has to be tested and optimized in a laboratory autoclave and confirmed in a pilot pressure vessel. Important parameters to optimize are the fineness of grinding; the pressure/temperature, density, and pH of the slurry in the autoclave; and the total required retention time.

Regrinding of flotation concentrate before pressure oxidation may be economically advantageous. Prolonged leaching under pressure would

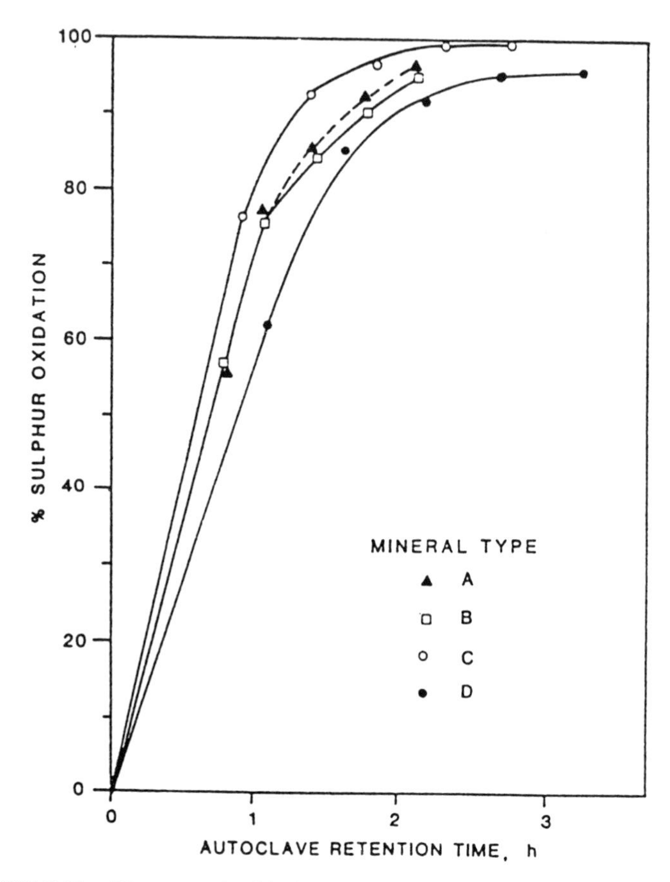

FIGURE 5-10. The rate of oxidation in an autoclave is affected by the mineral type. (Mineral types are pyrite concentrates from four mines.)

Reprinted by permission. D. R. Weir, J. A. King, and R. C. Robinson, in *Minerals and Metallurgical Processing*, Society for Mining Metallurgy and Exploration Inc. (Littleton, CO), November 1986.

increase both capital and operating costs by far more than the cost of a reasonable expansion of grinding capacity.

Autoclave Oxidation Technology

Pressure leaching is conducted, on an industrial scale, to extract alumina from bauxite ores, copper from copper concentrates in acidic or alkaline media, and nickel from laterites with sulfuric acid. Oxidation of pyritic or arsenopyritic ores and concentrates containing gold results in highly acidic slurries. The combination of abrasion, low pH, and high pressure and temperature restrict the selection of autoclave materials to titanium,

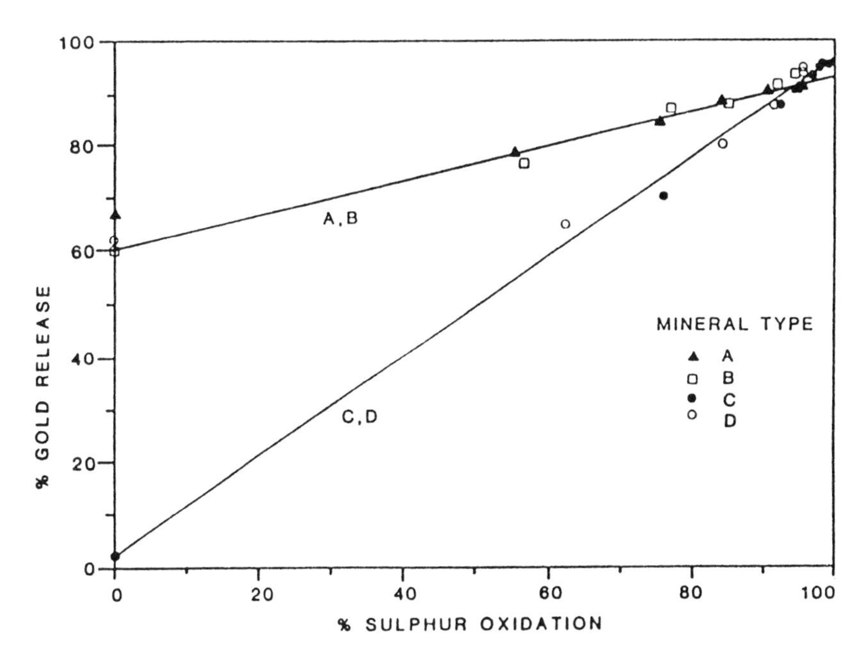

FIGURE 5-11. Effect of sulfur oxidation on subsequent gold extraction. (Mineral types are pyrite concentrates from four mines.)

Reprinted by permission. D. R. Weir, J. A. King, and R. C. Robinson, in *Minerals and Metallurgical Processing*, Society for Mining, Metallurgy and Exploration (Littleton, CO), November 1986.

titanium-clad carbon steel, or acid-resistant brick-lined carbon steel. Reliable high-pressure slurry-pumping systems are required. An efficient system of transferring heat from the autoclave's discharge slurry to the incoming feed is the concentric-pipe, slurry-to-slurry heat exchanger developed for the uranium industry in the 1950s.

A Modern Gold Plant with Pressure Oxidation

Pretreatment. Mineralogical studies of the McLaughlin deposit in California confirmed the existence of a complicated orebody with fine gold disseminated in sulfide minerals. Gold typically occurs as electrum (18% to 26% silver), primarily associated with sulfosalts and also locked in the gangue. The major sulfide mineral is pyrite, with small contents of chalcopyrite, sphalerite, and cinnabar. Among the sulfosalts contained in the ore are miargyrite ($AgSbS_2$), polybasite ($[Ag,Cu]_{16}\ Sb_2S_{11}$), and pyrargyrite (Ag_3SbS_3). In addition, carbonaceous materials and clays in some ore types present further complications by adsorbing gold during leaching (they are preg-robbing materials).

Batch and continuous test work at the McLaughlin project (of Homestake Mining Company) established the following conditions for continuous autoclave treatment.

Temperature in	160° to 180° C
Temperature out	180° to 190° C
Pressure	320 lb./sq. in.
Pulp density	40% to 45% solids
Final acidity	15 to 25 g/l H_2SO_4
Final EMF	420 mv
Oxygen consumption	80 to 100 lb./sh. tn.

The oxidized slurry is washed in two counter-current decantation thickeners. After adding lime, the washed pulp is submitted to cyanidation. Gold is recovered from the pregnant solution by electrowinning.

Pyrolysis and pressure leaching has been proposed for the treatment of pyritic and arsenopyritic refractory gold ores (Dry & Coetzee, 1986). When pyrite is heated to 600° C or higher in the absence of oxygen, it decomposes to pyrrhotite and gaseous elemental sulfur.

$$FeS_{2(S)} \rightarrow FeS_{1+X(S)} + ((1-X)/2)S_{2(g)}$$

Similarly, arsenopyrite, heated in the absence of air, decomposes to pyrrhotite and elemental arsenic.

$$FeAsS_{(S)} \rightarrow FeS_{(S)} + 0.25As_{4(g)}$$

Pyrite reacts with arsenopyrite, when heated in the absence of air, as follows.

$$2FeAsS_{(s)} + 2FeS_{2(s)} \rightarrow 4FeS_{(s)} + As_2S_{2(g)}$$

Mild oxidative leaching of pyrrhotite, under oxygen pressure, promotes the following reactions.

$$FeS + 2Fe^{+++} \rightarrow 3Fe^{++} + S^{\circ}$$
$$FeS_{1+x} + 2Fe^{+++} \rightarrow 3Fe^{++} + {}^{(1+x)}S^{\circ}$$
$$0.5O_2 + 2H^+ + 2Fe^{++} \rightarrow 2Fe^{+++} + H_2O$$

The ferrous ions are re-oxidized to ferric ions by the dissolved oxygen. The re-oxidation step consumes acid, thus enhancing the iron precipitation.

$$Fe^{+++} + 2H_2O \rightarrow FeO(OH)_s + 3H^+$$

Although gold extraction can be obtained by alkaline cyanidation of the residue, roasting in the absence of air is expensive and impractical.

High-pressure low-alkalinity cyanidation has been tried in the treatment of refractory gold ores. According to Dry and Coetzee (1986), 82% to 87% gold extractions were obtained by high-pressure cyanidation of a stibnite concentrate (as compared to less than 1% extraction by conventional cyanidation). The same authors report that, in general, high-pressure low-alkalinity cyanidation achieves much faster gold dissolution than does conventional cyanidation, but it does not yield higher gold extraction.

Pietsch et al. (1984) proposed cyanidation under pressure in a pipe reactor. The effect of increasing temperature from 20° C to 120° C, under oxygen pressure of 50 bars, was investigated. A slight increase in cyanidation rate, with increasing temperature, was recorded, but with significant decomposition of cyanide (which obliterated the advantage of the minor increase in the reaction rate).

Muir and Hendricks (1985) patented the cyanidation of refractory sulfide gold ores containing arsenic and antimony in a pipe reactor at pressures between 2 and 10 MPa, a temperature lower than 60° C, and a pH of 10 or less. The application of oxygen "overpressure" resulted in a "marked" increase of the gold extraction rate, especially from a stibnite concentrate.

Acid-pressure oxidation of sulfide concentrates has been considered as an alternative to smelting, without practical success (Figure 5-12). Pressure leaching under acidic conditions is an exothermic and extremely corrosive process that requires lining the autoclave with lead and using titanium for the agitator and internal piping. Although autoclaving of auriferous sulfide ores offers the advantage of a treatment without air pollution, the large capital investment required precludes industrial applications.

Arsenic Fixation in Residues. Two proprietary pressure-oxidation processes claim successful treatment of gold-containing arsenopyrite and arsenical materials in recovering gold and fixing the arsenic in highly insoluble (and safely rejectable) precipitates. *The Arseno process*, developed by Arseno Processing Ltd., proposes three major stages:

- Acidic leaching of concentrate in a stainless-steel autoclave, under 200 psig oxygen pressure, followed by filtration and recovery of the gold-bearing residue
- Partial neutralization of the solution (which contains dissolved arsenic, sulfur, and iron)
- Precipitation of ferric arsenate, hematite, and gypsum in an autoclave (100–200° C) to form a safely rejectable residue

After the removal of the ferric arsenate residue, the filtrate is recycled to the leaching autoclave (Anonymous, 1983[a], Anonymous, 1983[b]).

The Cashman process proposes the aeration, under pressure of 40–50 lb./sq. in. (approximately 120° C), of finely ground arsenic-bearing ore, concentrate or flue dust, with ground limestone and/or lime, and/or calcium chloride. The reaction takes place in a titanium reactor and requires 15 minutes to 2 hours, depending on the complexity of the charge. Arsenic is fixed in the form of highly insoluble precipitates (Anonymous, 1987).

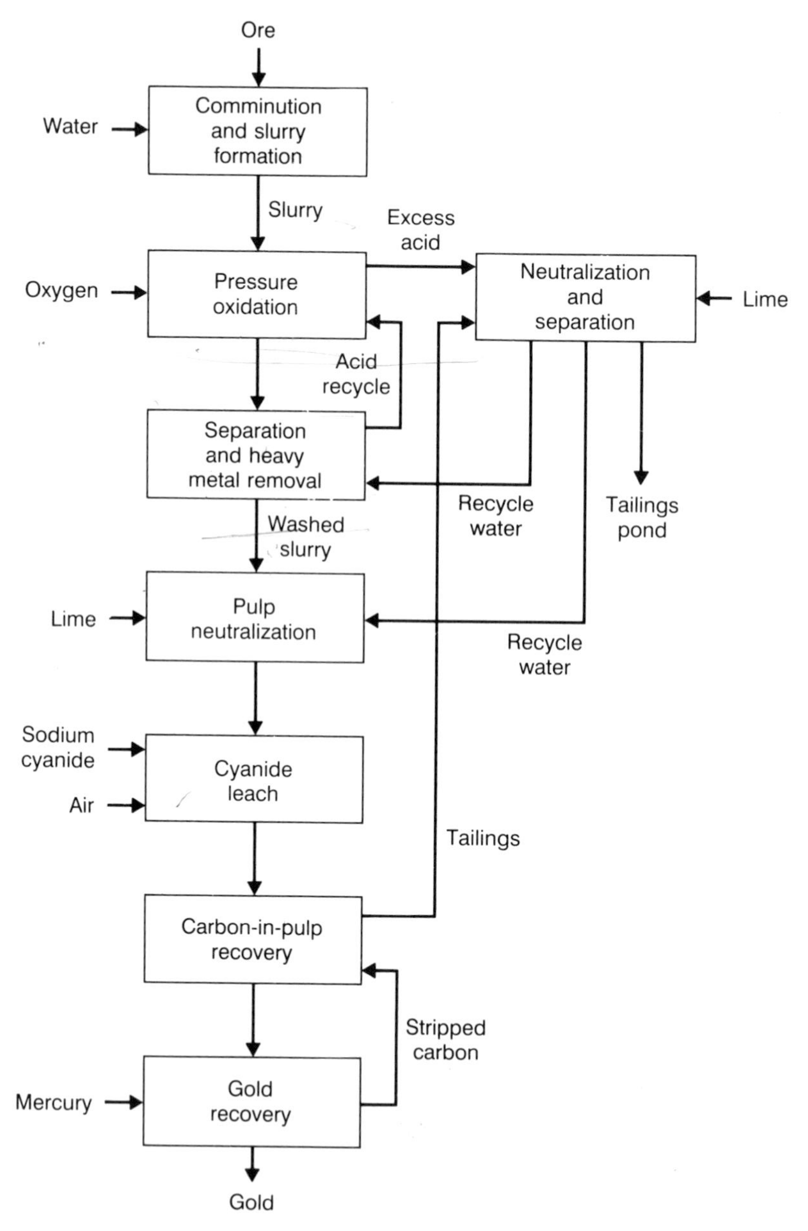

FIGURE 5-12. Pressure oxidation of sulfide refractory ore in acidic conditions with subsequent neutralization of the pulp.

From R. S. Kunter, J. R. Turney, and R. D. Lear, "McLaughlin Metallurgical Development: The Project, the Problems, the Process." International Metal Symposium (Los Angeles), February 1984.

Biological Oxidation

Sulfide (pyritic or arsenopyritic) minerals are very often hosts to gold particles and causes of refractoriness for gold ores and concentrates. Oxidation of sulfides by high-temperature and/or high-pressure methods has been discussed in the previous pages.

An alternative to these methods is microbial pretreatment, by which iron-sulfide minerals are degraded and the encapsulated gold grains are liberated and leachable by cyanide. Bioleaching of refractory gold ores is a relatively new concept. It appears to have originated from bacterial studies at the Pasteur Institute, in Paris, in the mid-1960s (Pares, 1964a, 1964b). Its first practical applications occurred in the USSR in the late 1970s. Comparative tests have demonstrated that bioleaching can achieve the same or better results than roasting or pressure oxidation, in terms of gold recovery and process economics.[8]

The leaching microorganism used has been *Thiobacillus ferrooxidans*, which can be adapted to oxidize iron and (sulfide) sulfur under acidic conditions. This microorganism can be grown easily on pyritic and arsenopyritic ores. It grows and multiplies at moderate temperatures, but it is inactivated at temperatures higher than 40° C. Hence, bioleaching ore with a high sulfur content presents a problem, since the reactions enhanced by bacteria are exothermic. Therefore, a microbial oxidation system for sulfide ores and concentrates should be equipped with a cooling system, which increases the operating cost.

The chemistry of oxidations enhanced by bacteria is depicted by the following equations:

$$FeS_2 + 3.5O_2 + H_2O \rightarrow FeSO_4 + H_2SO_4$$
$$FeS_2 + H_2SO_4 + 0.5O_2 \rightarrow FeSO_4 + H_2O + 2S$$
$$2FeAsS + 5.5O_2 + 3H_2O \rightarrow 2FeSO_4 + 2H_3AsO_3$$
$$2FeSO_4 + 0.5O_2 + H_2SO_4 \rightarrow Fe_2(SO_4)_3 + H_2O$$
$$H_3AsO_3 + 0.5O_2 \rightarrow H_3AsO_4$$
$$2S + 3O_2 + 2H_2O \rightarrow 2H_2SO_4$$

Neutralization reactions may occur and cause precipitations.

$$2H_3AsO_4 + Fe_2(SO_4)_3 \rightarrow \underline{2FeAsO_4} + 3H_2SO_4$$
$$3Fe_2(SO_4)_3 + 14H_2O \rightarrow \underline{2H_3OFe_3(SO_4)_2(OH)_6} + 5H_2SO_4$$
$$H_2SO_4 + CaCO_3 + H_2O \rightarrow \underline{CaSO_4 \cdot 2H_2O} + CO_2$$
$$H_2SO_4 + CaO + H_2O \rightarrow \underline{CaSO_4 \cdot 2H_2O}$$

[8]Gilbert et al., 1986; Hackl et al., 1985; Karaivko et al., 1977; Fridman and Savari, 1983; Livasy-Goldblatt et al., 1983.

Pooley (1987) described *T. ferrooxidans* and *T. thiooxidans*—the most active bacteria in sulfide oxidation—as "aerobic, acidophilic, chemolitho-autotrophic rod-shaped bacteria," 1-2μ long. He listed their growth requirements:

- pH of 1 to 5 (usual optimum pH < 1.8)
- Temperature of 35° C
- Energy sources include oxidation of ferrous iron, sulfides, and sulfur
- Nutrients include nitrogen, phosphate, and trace amounts of calcium, magnesium, and potassium; oxygen and carbon dioxide (for organic synthesis) from the air

The thiobacilli are very common in mining sites with sulfide minerals; acidic mine waters are evidence of their presence. Enriched culture of bacteria should be obtained from the mine or site, where the minerals to be leached originate, and be used for pilot studies and ultimately in the planned leaching operation.

There are other groups of bacteria—in addition to *T. ferrooxidans* and *T. thiooxidans*—that degrade iron-sulfide minerals, including some that grow at high temperatures (thermophiles). Brierley et al. (1980) described two groups of thermophiles adaptable to bioleaching operations.

- The moderately thermophilic bacteria that thrive and exhibit optimum growth in the temperature range of 45° C to 55° C.
- The extremely thermophilic *Sulfolobus* species, which grow in the temperature range of 50° C to 80° C.

The testing of bacterial oxidation has the objective of optimizing four parameters.

1. *The optimum particle size* for bacterial leaching is not standard for all ores and concentrates, and has to be established experimentally. Current practice seems to favor grinding to less than 35μ in particle size.
2. *The solids concentration* of the pulp on which bacterial oxidation is performed has been reported as a critical parameter, more influential than even the particle-size distribution of the feed. Maximum bacterial leaching rates of sulfide minerals have been reported to occur with 15% to 20% solids concentrations (Pooley, 1987). Excessive sulfide material concentrations may cause a very significant reduction in the growth of bacteria and in the leaching rate, even to the extent of completely stopping oxidation of the mineral. Furthermore, the concentration of sulfide minerals affects the chemical conditions in the system. The microorganisms present in the leach may be inadequate to maintain optimum leaching conditions at certain sulfide concentrations.

3. *The optimization of nutrient addition* should be established during testing. Small amounts of potassium, phosphorus, and nitrogen are required in solution to maintain bacterial action and growth. Additions of potassium hydrogen phosphate and ammonium sulfate should be optimized by testing.
4. *Aeration* at relatively low rates (0.5–1.0 l/min, per liter of slurry) has been reported as adequate to maintain the oxidation rate and bacterial growth. A temperature of around 35° C has to be maintained for optimum bacterial activity and growth, as already mentioned. (In large-scale operations, the bacterial oxidation reactors have to be equipped with cooling systems in order to maintain the operating temperature.)

Bacterial leaching flow sheets are shown in Figure 5-13. After adequate grinding, the feed is transferred to mixing vessels, along with ferric sulfate solution containing bacteria and nutrients.

A bio-heap leach project, in which a refractory gold ore is biologically oxidized on a heap, was reported by Peng Yu-mei et al. (1988). After biological oxidation, the heap has to be neutralized with lime milk and subsequently leached with cyanide.

Chemical Oxidation

In the late 1960s, the U.S. Bureau of Mines at Reno, Nevada, demonstrated experimentally that the deleterious effects of the carbonaceous materials in Carlin-type gold ores could be overcome by oxidation of aqueous ore slurries. *Chlorine and sodium hypochlorites*, generated *in situ* by electrolysis of an ore pulp containing brine, were chosen as oxidation agents for the Carlin carbonaceous ores. The electrolytic method required high capital and low operating costs, whereas chlorination, with purchased chlorine gas, generated low capital costs and high operating costs (mainly the cost of chlorine). The use of chlorine at Carlin proved to be very efficient, yielding 83%–90% gold extractions from refractory ores. Initially, the Carlin plant treated ores that required 15 kg of chlorine per ton of ore, but later, more recalcitrant refractory ore reserves required over 50 kg per ton of ore.

Double Oxidation

A double-oxidation process, consisting of aeration followed by chlorination, was developed in the late 1970s by W. J. Guay (1980b). Air is blown into agitated tanks containing a slurry of finely ground ore in a solution of sodium carbonate. The oxygen and the soda react with pyrite and form soluble sulfates and iron oxides. The dissolution of sulfates enhances the

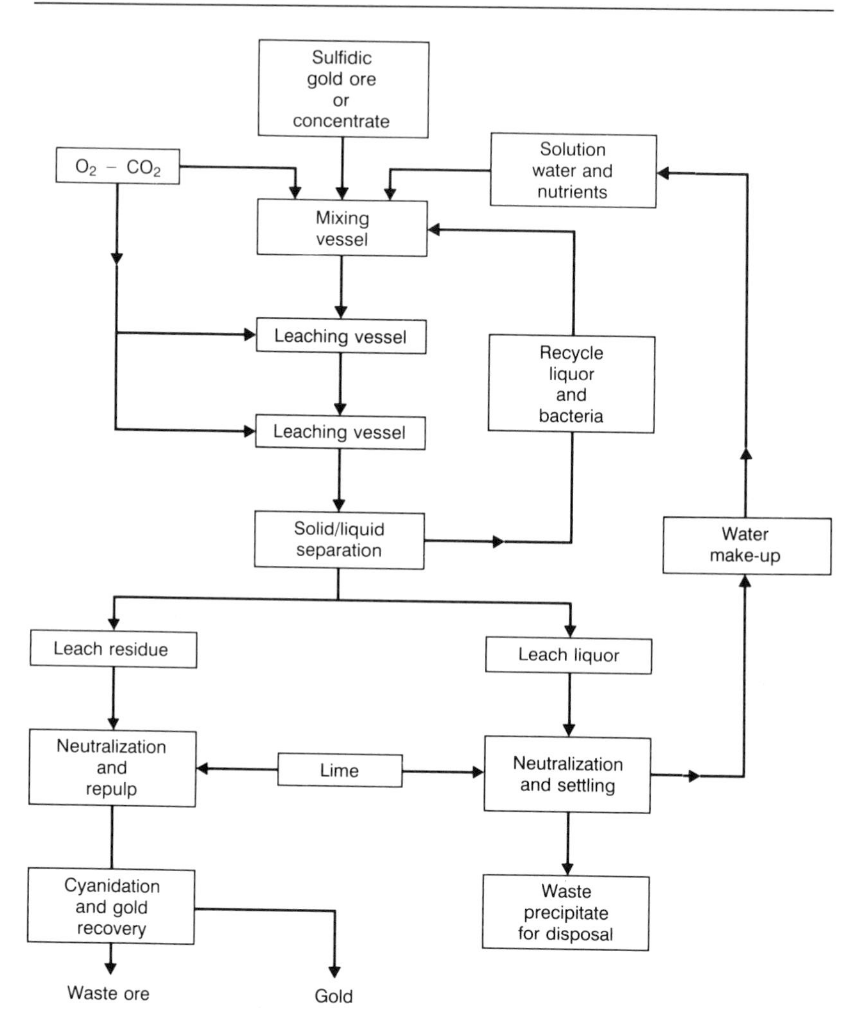

FIGURE 5-13. Bacterial treatment of gold-bearing sulfide ore/concentrate.

exposure of new surfaces to oxidation. During the aeration, some carbonaceous material is partially oxidized, but most remains unaffected. The objective of the aeration pretreatment is to reduce chlorine consumption during the following chlorination treatment.

The ore slurry, preheated through a heat exchanger, flows into the first aeration tank at approximately 50° C, where it is further heated to 82° C by stainless steel panel-type heat exchangers mounted inside the tank. The pulp, at a pH of around 6, flows by gravity to three more aeration tanks. There it is maintained at 82° C by sparging live steam to counterbalance the heat lost in the exit air. The pulp is pumped from the least aeration tank onto a falling-film panel-type heat exchanger, through which water (to be used in the grinding circuit) is preheated to an average of 38° C, and the pulp is cooled to approximately 50° C.

The ore pulp (40–45% solids) is fed to a series of chlorination tanks, where it is agitated by propellers and reacts with chlorine gas sparged into the first three tanks (with most or all of the chlorine added to the first tank). Chlorine reacts with the limestone in the ore to form calcium hypochlorite. Lime is added to the last chlorination tank to bring the slurry to a pH of 11. The reaction of dissolved chlorine (hypochlorite) with carbonaceous material and pyrite is very rapid, and the rate-controlling step is the dissolution of chlorine gas in the liquid.

There should be no excess hypochlorite going with the pulp into the cyanidation circuit, in order to prevent its reaction with cyanide (conservation of cyanide). The rate of chlorine addition is controlled by maintaining a certan low concentration of hypochlorite (approximately 0.100%) in the discharge of the third chlorination tank. Small amounts of sodium bisulfite may be added to the discharge of the last chlorination tank to ascertain the complete consumption of hypochlorite.

Soluble chlorides in the cyanide pregnant solution increase the consumption of zinc, during the gold precipitation, and contribute to low grade precipitates with high impurity levels. The quality of the precipitates and the reduced consumption of chlorine are strong incentives for the adoption of the "Double Oxidation" process in the treatment of carbonaceous and/or pyritic refractory gold ores.

Flash Chlorination

The reaction of dissolved chlorine with a carbonaceous ore can be very rapid, and the rate-controlling step is that of chlorine dissolution in the liquid. In order to accelerate the chlorine dissolution, the gas-liquid interface must be increased with very intensive and vigorous agitation of the gas-liquid-solids system. The intensive agitation accelerates the rate of chlorine diffusion in the liquid phase and thus increases the rate of chlorination of the ore. Thus, intensive (flash) chlorination achieves a very efficient use of the chlorine gas and significantly reduces the required reactor size.

The Nitrox Process

The nitrox process proposes the use of nitric acid to oxidize sulfide and arsenosulfide minerals.

$$3FeAsS + 14HNO_3 + 2H_2O \rightarrow 3FeAsO_4 \cdot 2H_2O + 3H_2SO_4 + 14NO$$

$$3FeS_2 + 18HNO_3 \rightarrow Fe_2(SO_4)_3 + Fe(NO_3)_3 + 3H_2SO_4 + 6H_2O + 15NO$$

For economic and environmental reasons, the nitrogen monoxide (NO) has to be collected and re-oxidized in a nitric acid plant. The Nitrox process is a neat concept, but it is questionable if a chemical plant making nitric acid from nitrous oxide—with all its corrosion nightmares—can be operated at mining sites.

Silica-Locked Gold

Fine gold encapsulated in silica is found in certain Nevada ores. Silica-locked gold, by definition, cannot be recovered by cyanidation of "normally" ground silica ore (say 60% to 70% −200 mesh).

Silica is chemically resistant to all acids except hydrofluoric acid (HF). Silica dissolves in hot, concentrated alkali, and more rapidly in fused alkali carbonates and silicates. Therefore, there is no affordable chemical pretreatment of ores with silica-locked gold.

Concentration of auriferous silica by flotation, if feasible, may be advisable. The concentrate is shipped as flux material to a nearby copper smelter, where gold will be recovered along with the copper. Chloridizing roasting of auriferous silica and/or silicate ores may yield satisfactory recovery of volatile gold chlorides. All the options of treating silica-locked gold ores should be tested and evaluated, including very fine grinding and cyanidation of the finely ground pulp.

References

Note: Sources with an asterisk (*) are recommended for further reading.

Mineralogy, Flotation, and General Papers

Adamson, R. J. (ed.) 1952. *Gold Metallurgy in South Africa*. Johannesburg: Chamber of Mines of S. Afr., 414–20.

Argall, Jr., G. O. 1987. Sonora gold makes flotation concentrates at Jamestown. *Eng. Min. J.*, December: 38–39.

Arriagada, F. S., and K. Osseo-Assare. 1984. Gold extraction from refractory ores—roasting of pyrite and arsenopyrite. In *Precious Metals: Mining, Extraction and Processing*, edited by V. Kudryk et al. Warrendale, PA: TMS-AIME.

Boyle, R. W. 1979. The geochemistry of gold and its deposits. *Canada Geolog. Survey Bull. 210*.

Cabri, L. J. 1976. Glossary of platinum group munerals. *Econ. Geol.* 71: 1476–80.

*Casparini, C. 1983. The mineralogy of gold and its significance in metal extraction. *C.I.M. Bull.* March: 145–53.

Chryssoulis, S. L, L. J. Cabri, and R. S. Salter. 1987. Direct determination of invisible gold in refractory sulphide ore. In *Gold Metallurgy*, edited by R. S. Salter et al. New York: Pergamon Press, 235–44.

Dorr, J. V. N. and F. L. Bosqui. 1950. *Cyanidation and Concentration of Gold and Silver Ores*. New York: McGraw-Hill, 238–52.

Dry, M. J., and C. F. B. Coetzee. 1986. The recovery of gold from refractory ores. In *Gold 100: Proceedings of the Internat. Conference on Gold*, Vol. 2. Johannesburg: S. Afr. I.M.M., 259–74.

Hausen, D. M., and P. F. Kerr. 1968. Fine gold occurrence at Carlin, Nevada. In *Ore Deposits of the United States, 1933–1967,* ed. by J. D. Ridge. New York: AIME, 908–40.

Hausen, D. M. 1981. Process mineralogy of auriferous pyritic ores at Carlin, Nevada. In *Process Mineralogy, Metal Extraction, Mineral Exploration and Energy Materials.* New York: AIME, 271–89.

Hausen, D. M. 1984. Process mineralogy of select refractory Carlin-type gold ores. Paper presented at the 23rd C.I.M. Conference of Metallurgists, August 16–19, Quebec City, Que., Canada.

*Hausen, D.M., and C. H. Bucknam. 1985. Study of preg-robbing in the cyanidation of carbonaceous gold ores from Carlin, Nev. *Proceedings Internat. Conference of Applied Mineralogy (ICAM) 84.* New York: TMS-AIME, 833–57.

Henley, K. J. 1975. Gold ore mineralogy and its relation to metallurgical treatment. *Minerals Sci. Eng.* 7(4): 289–312.

Kuhnel, R. A., J. J. Prins, and H. J. Rooda. 1980. *The Delft System for Mineralogical Identification: I. Opaque Minerals.* Delft, Netherlands: Delft University Press.

Pooley, F. D. 1987. Use of bacteria to enhance recovery of gold from refractory ores. In *Minprep 1987: Internat. Symposium on Innovative Plant and Processes for Minerals Engineering*, March 31–April 2. Doncaster, UK: IMM, 22–29.

Rose, T. K., and W. A. C. Newman. 1937. *The Metallurgy of Gold* (7th ed.). Philadelphia: J. B. Lippincott.

Roasting

Adamson, R. J. (ed.) 1952. *Gold Metallurgy in South Africa*. Johannesburg: Chamber of Mines of S. Afr., 150–51.

Archibald, F. R. 1949. Roasting arsenical gold ores and concentrates. *Can. Min. Metal. Bull.* vol. 42: 129–39.

Carter, R., and C. S. Samis. 1952. The influence of roasting temperature upon gold extraction by cyanidation from refractory gold ores. *Can. Min. Metal. Bull.*, vol. 45: 160–66.

Djingeluzian, L. E. 1952. Theory and practice of roasting sulphide concentrates. *Can. Min. Metal. Bull.* vol. 45: 352–61.

Dunn, W. E. 1982. Chlorine extraction of gold. U.S. Patent 4,353,740.

Dunn, W. E. 1983. Chlorine extraction of gold. In *Gold, Silver, Uranium and Coal*, edited by M. C. Fuerstenau and B. R. Palmer. New York: AIME, 174–88.

Dunne, R. C., and S. J. Liddell. 1984. Improved recoveries of gold from auriferous calcines and pyrites by fine milling. *MINTEC Rpt. No. M161.*

Eisele, J. A., and H. J. Heinen. 1974. Recovery of gold. U.S. Patent 3,834,896.

Eisele, J. A., D.D. Fisher, H. J. Heinen, and D. G. Kesterbe. 1970. Gold transport by complex chloride vapors. *U.S. Bureau of Mines RI 7289.*

Filmer, A. O. 1982. The dissolution of gold from roasted pyrite concentrates. J. S. Afr. I.M.M., March: 90–94.

Hansen, K. R. N., and J. E. Laschinger. 1967. The roasting of refractory gold ores and concentrates: A literature survey. *Nat. Inst. of Metallurgy, Report No. 85*, Johannesburg.

Heinen, H. J., J. A. Eisele, and D. D. Fisher. 1974. Recovery of gold from ores. U. S. Patent 3,825,651.

James, S. E., and J. P. Hager. 1978. High temperature vaporization chemistry in the gold-chlorine system including formation of vapor complex species of gold and silver with copper and iron. *Metal. Trans.* December: 501–508.

Jha, M. C., and M. J. Kramer. 1984. Recovery of gold from arsenical ores. In *Precious Metals: Mining, Extraction and Processing*, edited by V. Kudryk et al. Warrendale, PA: TMS-AIME, 337–65.

Mason, P. G., R. Pendreigh, F. D. Wicks, and L. D. Kornze. 1984. Selection of the process flowsheet for the Mercur gold plant. In *Precious Metals: Mining, Extraction, and Processing*, edited by V. Kudryk et al. Warrendale, PA: TMS-AIME, 435–45.

Matthews, O. 1949. Fluo-solids roasting of arsenopyrite concentrates at Cochenour Willans. *Can. Min. Metal. Bull.*, April: 178–87.

Mrkusic, P. G., and J. E. Laschinger. 1967. Refractory gold ores—report on investigations into roasting of gold concentrates in the Edwards roaster of the Fairview and the New Consort reduction plants. *Nat. Inst. of Metallurgy, Report No. 211,* Johannesburg.

Norwood, A. F. B. 1939. Roasting treatment of auriferous flotation concentrates. *Proceedings Austr. I.M.M.* 116: 391–412.

Osmanson, R. D. 1987. Gold occurrence and the effect of roasting temperature on cyanidation recovery from the Gold Acres, Nev., deposit. Paper presented at the 1987 TMS Annual Meeting, 23–27 February, at Denver, CO.

Palmer, B. R., D. D. Garda, A. K. Temple, and W. E. Dunn. 1986. Recovery of gold by high-temperature chlorination. In *The Reinhardt Schuhman Internat. Symposium*, edited by D. R. Gadkel et al. Warrendale, PA: TMS-AIME.

Radtke, A. S., and B. J. Scheiner. 1970. Studies of hydrothermal gold deposition. Carlin gold deposit, Nevada: The role of carbonaceous material in gold deposition. *Econ. Geol.* 67: 87–102.

Radtke, A. S., C. Heropoulos, B. P. Fabbi, B. J. Scheiner, and M. Essington. 1972. Data on major and minor elements in host rocks and ores, Carlin gold deposit, Nevada. *Econ. Geol.* 69: 975–78.

Robinson, P. G. 1983. Mineralogy and treatment of refractory gold from the Porgera deposit, Papua, New Guinea. *Trans. I.M.M.*, June: 926–86.

Scheiner, B. J., R. E. Lindstrom, and T. A. Henrie. 1971. Oxidation process for improving gold recovery from carbon-bearing gold ores. *U.S. Bureau of Mines RI 7573.*

Smith, E. H., J. W. Forest, Ph. Minet, and Ph. Gauve. 1985. Selective roasting to de-arsenify enargite/pyrite concentrate from St. Joe's El Indio mine: From pilot plant to commercial operation. In *Complex Sulfides—Processing of Ores, Concentrates, and By-products*, edited by A. D. Zunkel et al. New York: TMS-AIME, 421–40.

Swash, P. M., and P. Ellis. 1986. The roasting of arsenical gold ores—a mineralogical perspective. In *Gold 100, Vol. 2: Extractive Metallurgy of Gold*. Johannesburg: S. Afr. I.M.M. 235–50.

Wright, K. P. 1981. Fluid bed roasting practice in the Red Lake Camp. *Can. Min. Metal. Bull.*, August: 595–600.

High-Pressure Oxidation

Anonymous. 1987. Cashman process may offer key to treating arsenical ores. *Eng. Min. J.,* November: 55.

Anonymous. 1983a. Gold recovery from arsenopyrite by the Arseno process. *Western Miner*, March: 21.

Anonymous. 1983b. Process for gold recovery from arsenopyrite developed. *Eng. Min. J.*, April: 39.

*Berezowsky, R. M. G. S., and D. R. Weir. 1984. Pressure oxidation pretreatment of refractory gold. *Miner. & Metal. Proc.* May: 1–4.

Hedley, N., and H. Tabachnik. 1957. Process for recovering precious metals from refractory source materials. U.S. Patent 2,777,764.

Kunter, R. S., J. R. Turney, and R. D. Lear. 1984. McLaughlin Metallurgical Development: The Project, the Problems, the Process. Paper presented at the AIME International Metal Symposium (Los Angeles), February 1984.

Muir, C. W. A, and L.P. Hendricks. 1985. Leaching refractory gold ores, U.S. Patent 4,559,209.

Pares, M. Y., 1964a. Action d' Agrobacterium Tumefacien dans la mise en solution de l'or. *Annales de L'Institut Pasteur,* 107(1): 141.

Pares, M. Y. 1964b. Action de quelques bacteries heterotrophes banales dans le cycle de l'or. *Ibid.* 107(4): 573.

Peng Yu-mei, et al. 1988. Innovative technology for gold and silver recovery from Au-Ag bearing slags and tailings. Paper presented at the Randol Gold Forum 88, 23-24 January, at Scottsdale, AZ.

Pietsch, H. B., W. M. Turke, and G. H. Rathje. 1983. Research of pressure leaching of ores containing precious metals. *Erzmetall.* 36(6): 261-65.

Pietsch, H., W. Turke, E. Bareuther, F. Kampf and H. Bings. 1984. Method of extracting gold and silver from an ore. U.S. Patent 4,438,076.

Robinson, P. C. 1983. Mineralogy and treatment of refractory gold from the Porgera deposit, Papua, New Guinea. *Trans. I.M.M.* June: C83-C89.

Weir, D. R., J. A. King and P. C. Robinson. 1986. Preconcentration and pressure oxidation of Porgera refractory gold ore. *Miner. & Metal. Proc.*, November: 201-8.

Biological Oxidation

Brierly, C. C., J. A. Brierly, P. R. Norris, and D. P. Kelly. 1980. Metal tolerance of microorganisms in hot acid environments. In *Microbial Growth and Survival in Extremes of Environment*, edited by G. W. Gould and J. E. Corry. London: Academic Press.

Karaivko, G. I., S. I. Suynetsov, and A. I. Golnoiyik. 1977. *The Bacterial Leaching of Metals from Ores* (trans. by W. Burns). Stonehouse, England: Technicopy Ltd.

Chemical Oxidation

Guay, W. J. 1981a. Recovery of gold from sedimentary gold-bearing ores. U.S. Patent 4,259,107.

*Guay, W. J. 1981b. The treatment of refractory gold ores containing carbonaceous material and sulfides. In *Gold and Silver Leaching: Recovery and Economics*, edited by W. J. Schlitt, W. C. Larson, and J. B. Hiskey. New York: SME-AIME, 17-22.

Guay, W. J. 1980a. How Carlin treats gold ores by double oxidation *World Min.*, May: 17-49.

Guay, W. J. 1980b. Recovery of gold from carbonaceous gold-bearing ores. U.S. Patent 4,188,208.

CHAPTER 6

Leaching Low-Grade Gold Ores

A dilute cyanide solution is an efficient solvent of gold. However, as in any other leaching process, the solvent has to come in contact with the solid gold particle. Hence, extensive crushing and meticulous two-stage, closed-circuit grinding are needed to liberate the gold particles in the ore. Multistage agitation leaching followed by elaborate solids-liquid separation is required to recover most of the gold (up to approximately 95%) in solution, in a reasonably short time (see Chapter 4). Large tonnage of a relatively high-gold-content ore is a prerequisite for the substantial capital and operating costs required for milling.

Small tonnage, and/or low-grade gold deposits have to be addressed with a process of low capital and operating costs, if possible. What if one could skip fine grinding—thus circumventing agitation leaching and solids-liquid separation—and still recover a substantial fraction of the gold contained in low-grade deposits? Oxidized copper ores have been leached in heaps or in vats, without any grinding or elaborate preparation, since the mid-16th century. However, cyanide heap leaching of gold and silver ores is a recent development. Discoveries of numerous low-grade gold deposits, mainly in the Western United States, along with a hefty price of the precious metal, stimulated the development of low-cost cyanide leaching in massive scale in the late 1960s. Carlin Gold Mining Company started a relatively small heap-leaching operation in Northern Nevada at that time (Figure 6-1).

The concept is simple. Crushed ore is stacked on an impervious pad and sprinkled with cyanide solution to dissolve gold and silver. For the heap leaching to be successful, the gold-bearing ore must be porous and contain fine-sized clean gold particles. The ore, after having been crushed and

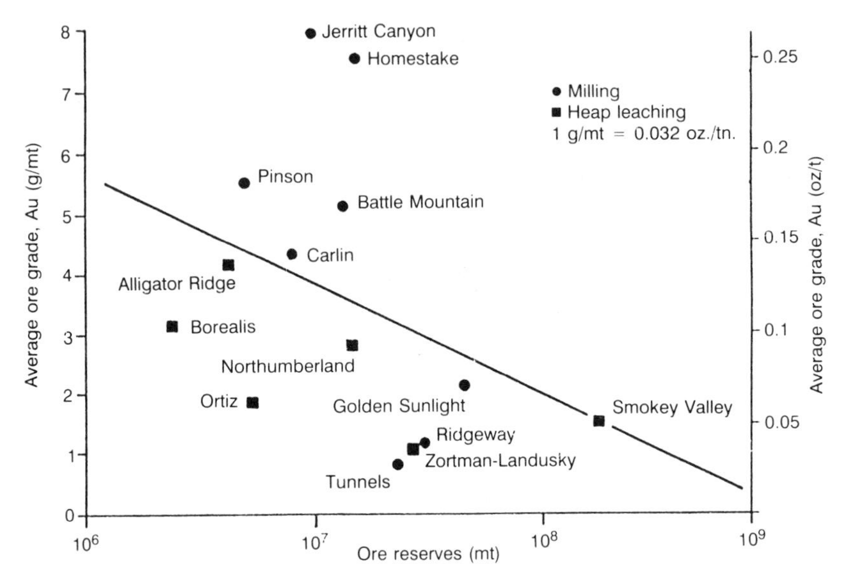

FIGURE 6-1. Ore reserves plotted as a function of ore grade.

Reprinted by permission. R. Dorey et al., in D. J. A. van Zyl, I. P. G. Hutchinson, and J. E. Kiel, eds., *Introduction to Evaluation, Design and Operation of Precious Metal Heap Leaching Projects.* Society for Mining, Metallurgy, and Exploration, Inc. (Littleton, CO), 1988.

stacked, must have good permeability, allowing uniform distribution of the solution through the heap, with affordable consumption of cyanide. Specific characteristics required for an ore to be amenable to heap leaching include the following (Dorey et al., 1988):

- Content of extremely small or flattened gold particles
- Porous and permeable host rock
- Absence of refractory, carbonaceous, or preg-robbing materials (which handicap the gold leaching)
- Absence of cyanicides (materials consuming cyanide)
- Absence of fines and/or clays that impede uniform cyanide solution percolation (agglomeration is required with excessive fines and clays)
- Absence of acid-forming constituents (which cause high consumption of cyanide or lime)

Ore Testing

Heap leaching is not a panacea for all low-grade gold deposits. The gold has to be amenable to cyanide leaching and have most of the above listed characteristics. The extent of the required ore crushing—to achieve good permeability, uniform distribution of the cyanide leach solution, and satisfactory gold extraction—has a paramount effect on the economics of the projected heap leaching. An excessive proportion of clays in the ore, or of fines generated by crushing, may slow the percolation flow rate of the leach solution, cause channeling, and produce dormant bulks of material within the heap. Agglomeration of the crushed ore may be required to get a permeable and uniform feed to the heaps. Crushing circuits and agglomeration systems are capital intensive and should be used in a commercial operation only if and when their capital and operating costs can be more than justified by the overall economics of the operation.

The importance of meticulous testing to ascertain the feasibility and establish the optimum conditions of the projected heap leaching cannot be overemphasized (Potter, 1981). The ultimate objective is to establish the design and operating conditions leading to the lowest cost per ounce of gold extracted. This goal can only be reached if the optimum combination of low operating costs and high gold recoveries is achieved. Those two major factors can often be in conflict. For example, the higher a heap, the lower is the cost per ton for pads, ponds, pipes, and operating labor; however, the increased height of a heap might adversely affect gold extraction. Once the crushed (and agglomerated, if needed) ore is stacked on the impermeable pad and the solution lines and sprinklers are installed, maximizing the gold recovery is the only way to boost the profitability per ton of ore and per ounce of gold recovered.

Climatic conditions at the location of the projected heaps should be taken under consideration during the testing and design of a particular heap-leaching operation. The rainfall rate and cycle, and the temperature levels throughout the year, may have a significant effect on the design and economics of a heap-leaching operation.

Sampling for Complete Evaluation of the Orebody

A close cooperation among the field geologists, the mineralogists, and the test metallurgists has to determine several things:

- The salient geological characteristics of the ore and their effect on the projected leaching treatment
- The requirements and availability of core samples for testing deep-level ore
- The availability of near-surface and deep-level ore for large-scale testing of bulk samples.

Gold and silver values in ore samples are generally determined by fire assaying. However, samples should also be assayed by Atomic Absorption (AA), which permits the assessment of the fraction of nonrefractory gold leachable by cyanide (see Chapter 5). Cyanide-solubility tests should be run on four- or five-foot intervals of the drill-hole samples, with a cut of the same samples fire-assayed in order to establish potential gold extraction and refractoriness of the orebody.

Ore Mineralogy

A poor understanding of the ore mineralogy may lead to problem-laden heap leaching. A detailed mineralogical study is indispensable during the early stages of planning a heap-leaching operation. The representative samples of the deposit-to-be-leached have to be explained mineralogically in order to establish four parameters:

- The host mineral bearing the precious metal
- The precious-metal grain size range
- The distribution and preferred location of the precious-metal mineral in the host mineral
- The existence of minerals causing refractoriness (if any)

Head samples for column tests (described below) and their respective tailings should be screened into size fractions to be assayed and mineralogically examined in order to determine the effect of crushed size on gold extraction and the potential causes of gold losses in the tailings.

Bottle Tests

Bottle tests are normally conducted with a 400-g ground ore sample in 600 ml of cyanide solution. Lime is added to the pulp to adjust its pH to 11. The pulped ore is agitated by rolling the bottle continuously, on the proper set of rollers, for 72 hours. The leach (pregnant) solution and the washed tailings are then assayed for gold (and silver, if any) values. This test can be used to compute the reagent requirements of the ore sample.

Column Leach Tests

Percolation leach tests should be performed in lab-size (transparent plastic) columns (6–12 in. dia. by 4–8 ft. high) to establish, on a preliminary basis, essential design parameters for heap leaching. The cyanide solution is sprayed on the top of the ore column and allowed to trickle down. The solution that has percolated through the column of ore is measured daily, and determinations of gold (and silver) content, cyanide concentration, and pH are made (Figure 6-2). The essential design parameters that should be collected with column tests are listed below.

- The effect of crushing size on the rate of gold recovery and the maximum gold extraction from the optimum ore size achieved in 60 days

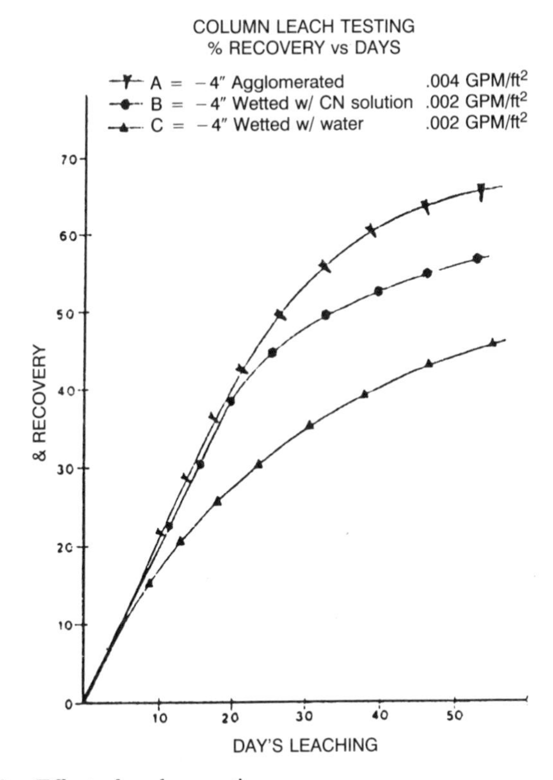

FIGURE 6-2. Effect of agglomeration on recovery.

Reprinted by permission. R. E. Brewer, The Barrick-Mercur Leach Project. The Minerals, Metals & Materials Society (Warrendale, PA), 1986: A36–86.

- The effect of cyanide concentration on the rate of gold recovery and on the consumption of cyanide
- The optimum pH and the required consumption of alkalis (lime and/or NaOH)
- The volume of solution required to saturate the ore column
- The volume of solution that drains from saturated ore
- The draining and washing times
- The effects of agglomeration on the rate of gold recovery, and on the volume parameters

Pilot-Scale Leach Tests

One to three pilot tests with small-size heaps (2,000 to 3,000 tons) are required to confirm and optimize the results collected with column experiments. Such field tests—especially in remote locations without any support facilities—are very expensive (in the hundreds of thousands of dollars). If there are no support facilities close to the prospective mine, testing in huge concrete columns (e.g., 10-ft. I.D. by 25-ft. high) should be considered.

Feasibility Study

The objective of meticulous testing of heap leaching, including alternative operating schemes, is to produce the maximum quantity of gold at the lowest cost from a given low-grade gold deposit. As metallurgical information becomes available, various cash-flow and return-on-investment scenarios must be reviewed to determine the optimum operating conditions, design, and rate of treatment. The mining plan has to be meshed with the optimum metallurgical operation for the highest possible return on investment.

Heap leaching is simple in principle, but is still sensitive to operating pitfalls. Its profitability depends on a rather narrow margin; hence the leachability of the ore and the optimum parameters of the projected heap leaching have to be thoroughly investigated, from bottle tests to large heap-pilot tests.

Heap Leaching and Dump Leaching

Heap leaching means leaching ores that have been mined, crushed, and transported on impervious pads for leaching by sprinkling or ponding, and percolation of the solution through the stack of the ore (Figure 6-3). *Dump leaching* means leaching dumps or accumulations of very low-grade ore or overburden, often without the use of prepared pads under them. The choice of whether to use a dump- or heap-leaching process depends on the grade and tonnage of the available "ore," although its permeability is the critical factor. Without adequate permeability, there will be no percolation, no dissolution, and no economically acceptable gold extraction. Poor permeability should exclude even dump leaching—in spite of its lower cost compared to heap leaching.

Heap-Leaching Methods

Although variations exist, three distinct heap construction methods can be identified. First, the *reusable-pad method* involves the construction of a series of stable, durable pads onto which the prepared ore is loaded, leached, washed, and unloaded for disposal (Figure 6-4). The main requirements for this method are flat, firm ground; a durable pad liner; ore with consistent leachability; and an arid climate without temperature extremes. Significant drawbacks of the method are the double handling of

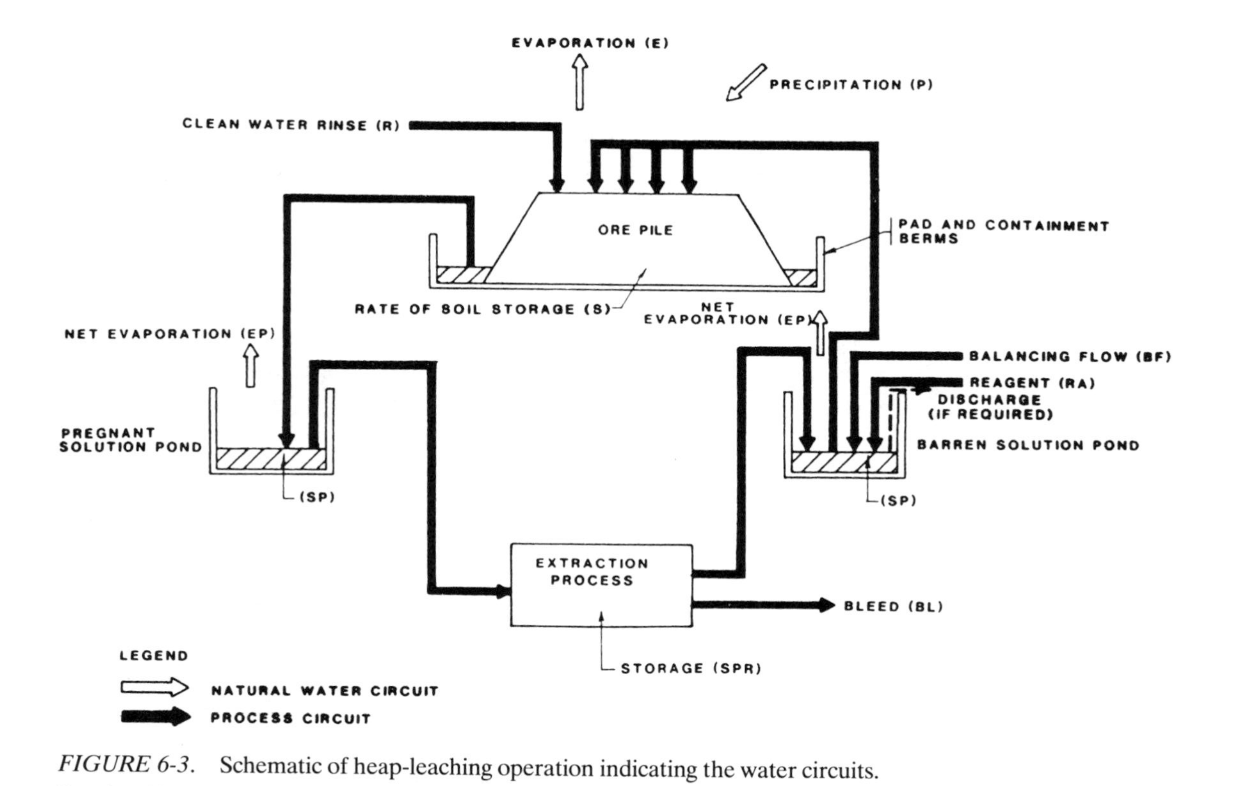

FIGURE 6-3. Schematic of heap-leaching operation indicating the water circuits.

Reprinted by permission. I. Hutchinson, in D. J. A. van Zyl, I. P. G. Hutchinson, and J. E. Kiel, eds., *Introduction to Evaluation, Design and Operation of Precious Metal Heap Leaching Projects*. Society for Mining, Metallurgy, and Exploration (Littleton, CO), 1988.

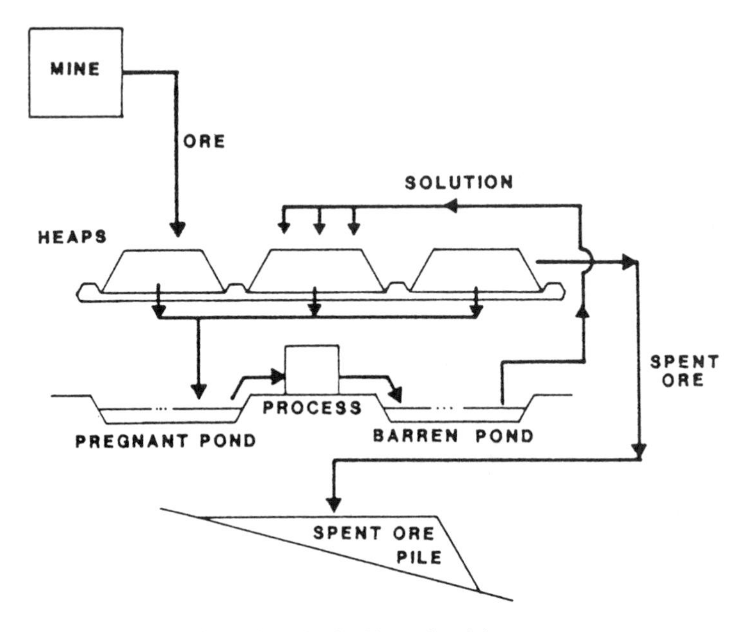

FIGURE 6-4. Reusable-pad method of heap leaching.

Reprinted by permission. R. Dorey et al., in D. J. A. van Zyl, I. P. G. Hutchinson, and J. E. Kiel, eds., *Introduction to Evaluation, Design and Operation of Precious Metal Heap Leaching Projects*. Society for Mining, Metallurgy, and Exploration, Inc. (Littleton, CO), 1988.

the ore, potentially lower gold recovery (since the heap is not left to "age" on the pad), and lack of flexibility in the leach time. The main advantages of the reusable-pad method are the low per-ton costs of pad and ponds construction.

Second, the *expanding-pad method* involves ore preparation, placement of the ore on a leaching pad and the abandonment of the leached residue in place (Figure 6-5). Subsequent releaching and/or washing are feasible, if required. Additional lifts can be added to the "exhausted" heaps. The main requirements for this method are a large area available for leaching pads and large-capacity ponds for rains and storms. The main drawbacks of the expanding method is the spoiling of an extensive area, the relatively high cost of the pad per ton of ore, and the potential multiplication of "polluting" discharges. The significant advantage of the method is high gold recovery, since the ore can age on the pad.

Third, the *valley-leach method* involves placing the prepared ore behind a retaining structure in a valley. Subsequent lifts of ore progress up with a slope (Figure 6-6). This method requires a strong retaining structure, durable lumpy ore, and a very-high-integrity liner due to the hydraulic head. The ore remains in contact with the leaching solution during the life of the operation, and high gold recovery can be achieved.

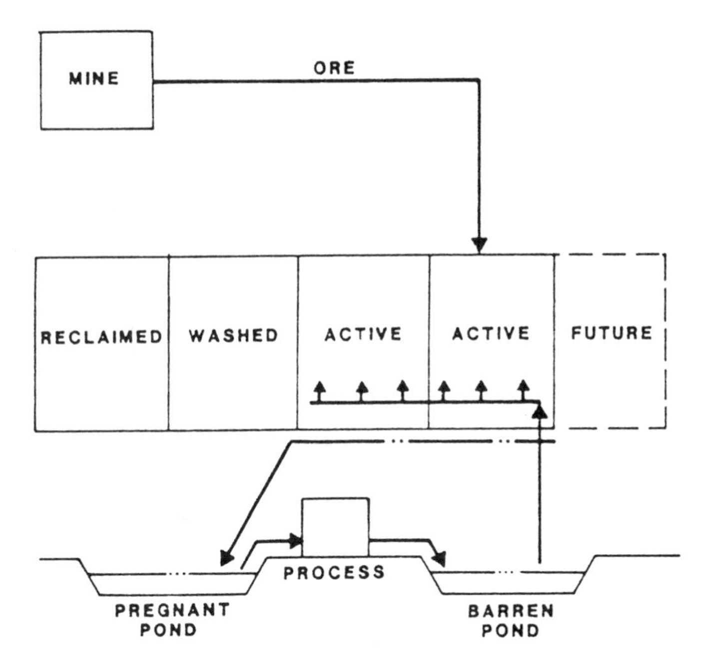

FIGURE 6-5. Expanding-pad method of heap leaching.

Reprinted by permission. D. J. A. van Zyl, I. P. G. Hutchinson, and J. E. Kiel, eds., *Introduction to Evaluation, Design and Operation of Precious Metal Heap Leaching Projects*. Society for Mining, Metallurgy, and Exploration (Littleton, CO), 1988.

Geotechnical Design

The pad on which the ore mass has to be stacked must be formed on firm ground, be able to withstand the heavy load of the ore, and have three ponds adjacent to one of its sides:

- The pregnant-solution pond
- The barren-solution pond
- The overflow pond

The topsoil and vegetation have to be scrubbed, and a uniformly sloping ground configuration—with 3–4% along the length of the pad—has to be prepared by grading and filling as required. The pad area has to be graded and compacted as much as possible.

The three solution ponds have to be prepared, at the drainage side of the pad, with adequate capacity (usually each pond is 20 ft. by 40 ft. by 10 ft. deep). Since the pregnant-solution pond collects the drainage from the heap, it should be designed with a capacity large enough to accommodate the maximum run-off from the expected rainfall at the specific site.

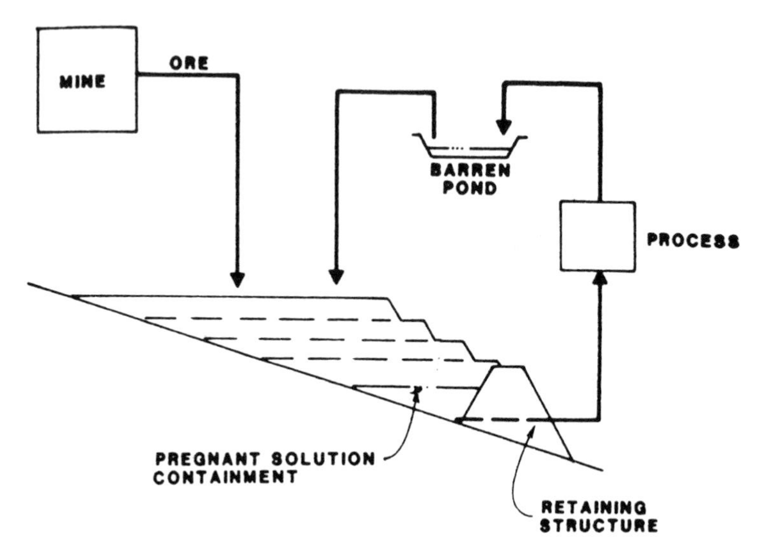

FIGURE 6-6. Dump leaching in a valley.

Reprinted by permission. D. J. A. van Zyl, I. P. G. Hutchinson, and J. E. Kiel, eds., *Introduction to Evaluation, Design and Operation of Precious Metal Heap Leaching Projects.* Society for Mining, Metallurgy, and Exploration (Littleton, CO), 1988.

Leaching Pad Preparation

The leaching pad must be the strong base for a (usually) gigantic heap of ore. When geomembrane[1] liners are used, the pad is the firm foundation upon which the liner is placed. When a clay-impermeable surface is collecting the pregnant solution, it is the upper surface of the clay pad (Figure 6-7).

Heap-leaching pads are built on flat terrain, adequately graded and compacted, with a 3–4% slope along their length and 1% toward one side. All flows run toward the collecting corner where the pregnant solution discharges, through a lined ditch, and to the pregnant-solution pond. The pad has to be effectively impermeable to prevent seepage and loss of precious pregnant solution, and also to prevent contamination of the ground water with cyanide solution.

Compacted clay is a very effective sealant and, as a rule, forms the first—and in some instances, the only—line of defense against seepage (Dayton, 1986). Other means of constructing impermeable pads include the following three methods.

---

[1]Geomembrane is, by definition (Giroud & Frobel, 1983), an impermeable manufactured liner (other than compacted clay).

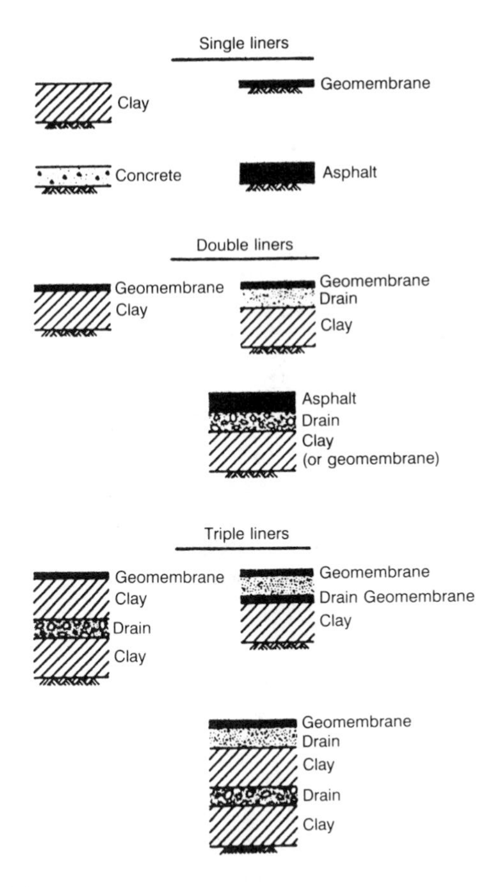

FIGURE 6-7. Liner systems.

Reprinted by permission. C. Strachan and D. van Zyl, in D. J. A. van Zyl, I. P. G. Hutchinson, and J. E. Kiel, eds., *Introduction to Evaluation, Design and Operation of Precious Metal Heap Leaching Projects*. Society for Mining, Metallurgy, and Exploration, Inc. (Littleton, CO), 1988.

- Mixing small amounts of bentonite, cement, or lime into existing layers of sand or soil with a KHD-Koering soil stabilizer
- Using an asphalt pad with a flexible (rubber or plastic) geomembrane impermeable layer
- Using compacted clay with a geomembrane between sand layers (compacted clay with a sand geomembrane sandwich)

Figure 6-8 depicts a multilayer leaching pad on top of an initial compacted-clay layer in a U.S. state with strict environmental regulations. The bottoms and sides of the three solution ponds must be sealed also. Usually, they are sealed with a layer of well-packed clay covered with a geomembrane (Harper et al., 1987).

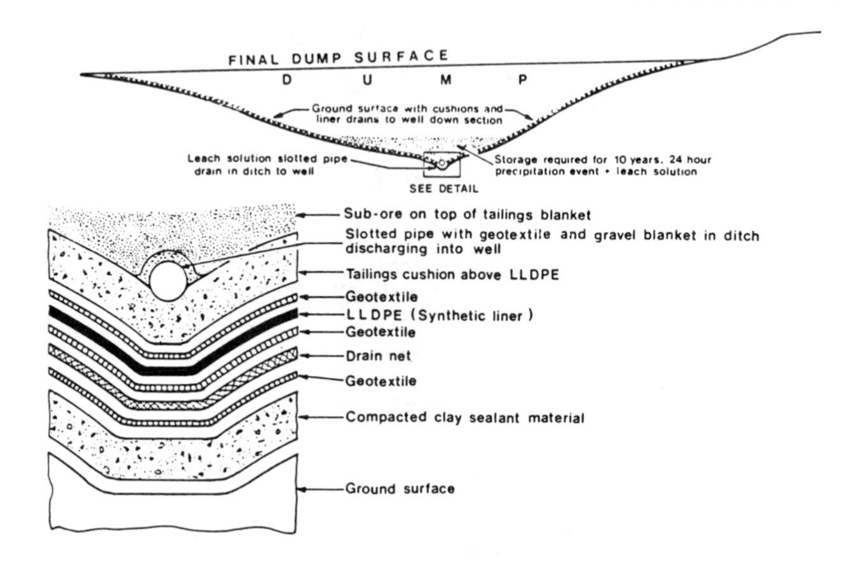

FIGURE 6-8. Cross-section of the pad of a dump-leaching operation.
Reprinted by permission. R. E. Brewer, The Barrick-Mercur Leach Project. The Minerals, Metals & Materials Society (Warrendale, PA), 1986: A36–86.

Ore Preparation

Most ores require some preparation in order to be heap leached. The mineralogy and testing of the ore must establish, in advance, the required degree of crushing, if any; the effects of ore sizing and agglomeration on the permeability of the heap and the rate of gold extraction; and the proportion of lime to be added for pH control. Automated crushed ore sampling and "weight" of the ore on the pad (usually by count of the number of trucks or by a weightometer on a belt) are used for the estimates of gold recoveries and gold balances in heap leaching.

Run-of-mine ore may be used to extract gold from noncrushed material of very low grade (sub-mill or sub-heap-leach grade). Run-of-mine ore is usually generated by blasting, and ranges in size from fines to boulders. A dump with run-of-mine ore can be operated for years, and should be shut down when the effluent deteriorates to its economic limit.

Crushing the ore to the optimum size distribution must be done with the appropriate crushing circuit (see Chapter 4). Lime should be mixed with the ore during crushing and added to the ore before loading onto the leach pad. The rate of lime addition depends on the mineralogy of the ore; it is usually 3–4 pounds per ton of ore (Argall, 1985).

Agglomeration of crushed ore to produce a porous and more uniform feed material for heap leaching has been proven to be a beneficial pretreatment for gold ores with a high content of clays and/or fines generated during crushing.

The need for agglomeration of a particular crushed ore should be determined by testing, before committing capital for a commercial operation. The presence of clays and fines in the heap slows the percolation flow of the leach solution, which may result in channeling and dormant bulks of ore within the heap. This may result in very long leaching times and poor gold extraction. In extreme cases, the clays and/or slimes may completely seal the heap, forcing the leach solution to run off the sides of the heap instead of percolating through the stacked ore.

Agglomeration of gold ores is different than pelletizing. In pelletizing, all particles are fines and form balls by aggregation and sintering. In agglomeration of gold ores, the clays and the fines are being attached to larger particles.

The gold ores that have to be agglomerated for heap leaching should be crushed to the optimum size for maximum profitability. Fine crushing liberates gold values and improves recovery. In the past, fine crushing was in conflict with the requirement of a uniformly high permeability of the heap. Agglomeration now makes fine crushing and high permeability of heaps compatible. Three important agglomeration parameters were determined from the U.S. Bureau of Mines work for successful pretreatments of crushed ores with poor percolation characteristics McLelland et al., 1983.

- The quantity of binder (Portland cement) added to the crushed ore
- The amount of moisture added to the binder/ore mixture
- The curing period required to form calcium silicate binding

Crushed ores can be agglomerated by mixing 5–10 pounds of Portland cement or lime per ton of dry solids, adding 8–16% moisture as either strong cyanide solution or water, mechanically tumbling the wetted mixture, and curing the agglomerated feed for at least 8 hours—and preferably for 24 to 48 hours—before heap leaching. The data about the effect of using cyanide solution during agglomeration on the performance of heap leaching (as reported in the literature) are conflicting. Therefore, the advisability of using a cyanide solution for agglomeration, and its optimum concentration, should be established by column testing during the evaluation of each specific orebody.

Although lime is not as persistent a binder as cement, it does promote fine-particle bonding that improves permeability. Lime has, of course, the additional advantage of providing protective alkalinity for the leach solution.

TABLE 6-1. Ore agglomeration systems and binders.

Type of Material to be Agglomerated	*Description*	*System Used*	*Binder*
Gold or silver ore	Coarse (−0.5 in. or similar). No clays.	Belt system as simple as possible (belt discharging to the ore pile).	Lime solution
Gold or silver ore	Fine (−0.5 in. or finer). No clays.	Belt plus inclined vibrating chute. Very simple three belt system.	Lime
Gold or silver ore	Coarse (−0.5 in. similar). Clays present.	Belt cascade system (three belts at least plus mixing bars).	Cement Lime Cement & lime
Gold or silver mill tailings	−65 mesh or finer.	Rotary drum. Inclined rotating plate.	Cement Lime Cement & lime

Reprinted by permission. M. R. Lastra and C. K. Chase, Permeability, solution delivery and solution recovery. *Min. Eng.*, November 1984.

Good agglomeration is a prerequisite for a uniformly permeable heap of ores containing clays and fines (Table 6-1). The agglomerator must provide the blending, agitation and compaction needed to produce agglomerates with the necessary green strength. The major types of agglomerators in use are:

- Cascade-belt systems
- A combination of belts with an inclined vibrating chute
- A mixing drum (like a concrete-mixing drum)
- Revolving-pan agglomerators

The cascade-belt system uses conveyors for tumbling and compaction. The belts may be steeply inclined to induce rolling, or may have multiple transfer points at which agitation and compaction occur. The belt system may discharge onto stockpiles of ore or directly onto a heap. Tumbling agitation and compaction of the agglomerates also occur in drums. The pan-type agglomerator uses a flat inclined disc (as in pelletizing) to produce, through tumbling and compaction, uniform agglomerates with excellent green strength.

Desliming should be studied as a potential alternative to agglomeration when testing gold ores containing clays and/or fines after crushing. Most operations have reported satisfactory results with heap leaching of agglomerated gold ores. Nevertheless, there have been heaps that encountered problems with ponding, channeling, and blinded areas due to migration of fines and compaction of the heaps (Herkenhoff and Dean, 1987).

Since fines and slimes may cause troubles, desliming of the crushed ore and separate treatment of the fines (say, −100 mesh) with agitation leaching—and of the deslimed coarse ore with heap leaching—should be considered. The option of desliming is even more attractive when the crushed fines are higher in gold content than the coarse fractions. This is due to the higher gold extraction expected from agitation leaching than from percolation of the solution through a heap. Desliming is bound to have higher cost per ton of ore than agglomeration. Hence its use can only be justified by a gold extraction significantly higher than the one expected from heap leaching of the same agglomerated ore.

In the treatment of low-grade gold ores with a high content of clays and fines, agglomeration is definitely more in use than desliming. Only one Australian and one American operation are reported to include desliming in the treatment of low-grade gold ores.

Heap Construction and Operation

High gold recoveries from heap leaching operations depend mainly on three critical factors.

- The pretreatment of the ore (crushing and agglomeration if needed)
- The placement of the ore on the leach pad (heap-building techniques)
- The wetting of the ore with cyanide (solution-application techniques)

Dump trucks haul crushed ore to the heap, as a rule, and bulldozers, scrapers, front-end loaders, or conveyors spread the ore on the heaps. Although bulldozers are the most common spreading equipment, gantry cranes, conveyors, and stackers are often used to keep heavy equipment off the heap and prevent compaction during construction. Allowing hauling trucks on top of the first lift while constructing the second lift may cause packing that results in puddle formation on the top of the heap and solution channeling.

The ideal stacking method involves an automated gantry system traveling above a reusable, impervious pad. It can lay down a rectangular, non-compacted pile of ore with a flat top to any height up to 15 feet. In some cases, new lifts are placed on top of previously leached heaps (Figure 6–9). The leach solution applied to the new lift percolates through, completing the extraction of any remaining precious metals in older heaps. It is a good practice to restrict truck traffic to a single middle lane on top of each new lift. The ore can be spread with a bulldozer over the previous lift. It has been shown that a bulldozer will not compact a heap as much as a

FIGURE 6-9. Heap construction using radial-arm stacker.
Reprinted by permission. D. J. A. van Zyl, I. P. G. Hutchinson, and J. E. Kiel, eds., *Introduction to Evaluation, Design and Operation of Precious Metal Heap Leaching Projects*. Society for Mining, Metallurgy, and Exploration (Littleton, CO), 1988.

rubber-tired vehicle. The entire top of the new lift, and especially the middle truck lane, should be ripped by the bulldozer (Figure 6-10).

Ripper blades as long as 5 feet, with about 4-foot penetrations, have been reported. The idea of burying up to 6-foot-diameter pipes upright in heaps has been patented. Pulling out those pipes after partial leaching of the heap may create new void patterns for percolation. The practicality and cost of the proposed method are questionable. Smaller (6-in. to 12-in. dia.) perforated pipes may be planted vertically in a heap in order to introduce fresh solution to its lower layers.

Minimizing equipment time on top of a lift minimizes the density of packing and thus increases permeability. Conveyor stacking equipment is more and more established as the optimum heap-construction method. It allows stacking a non-compressed, more or less homogeneous mass of agglomerated material. An automated, traveling gantry-conveyor stacking system can achieve a lower heap-construction cost per ton of ore, improved percolation characteristics, and increased gold recovery.

There is a natural size segregation in any pile being built by truck end-dumping or by conveyor stacking. The coarse material rolls to the bottom of the pile. If the gold values are carried in seams and veinlets of the rock, rather than in the solid boulder matrix, a concentration of gold in fines accumulates at the top of the pile. In some operations a coarse fraction (e.g., + 1.5 in.) is scalped by a screen during crushing to be used as

FIGURE 6-10. Crushed ore heap construction using end tipping and dozing.
Reprinted by permission. D. J. A. van Zyl, I. P. G. Hutchinson, and J. E. Kiel, eds., *Introduction to Evaluation, Design and Operation of Precious Metal Heap Leaching Projects*. Society for Mining, Metallurgy, and Exploration (Littleton, CO), 1988.

the lower part of each heap and improve the percolation of leach solution. The finer ore—often after agglomeration—is dumped on top of the coarser layer up to the design height of the heap.

There is no standard set of dimensions for heap construction, as a 1986 survey of the North American operations showed.

- 50% operate heaps of 50,000 tons or less
- 17% operate heaps of 50,000 to 100,000 tons
- 16% operate heaps of 100,000 to 300,000 tons
- 17% operate heaps of over 300,000 tons

Most operations leach one lift at a time, to minimize equipment time on top of a lift. The majority of the North American operations have lift heights between 7 and 12 feet. In very large heaps, heights up to 40 feet can be found. The increased height of a heap leads to a lower cost per ton of ore. The operator must make certain, nevertheless, that this "height economy" does not reduce gold recovery.

The application of solution to heaps may be achieved through flooding or ponding, sprinkling with wobblers, spraying with pressure emitters, or spraying with wigglers. Flooding is not widely practiced and can only be used if the permeability of the heap is low enough to limit percolation. In practice, the leaching solution is most commonly sprinkled over the top of the heap with wobblers or pressure emitters.

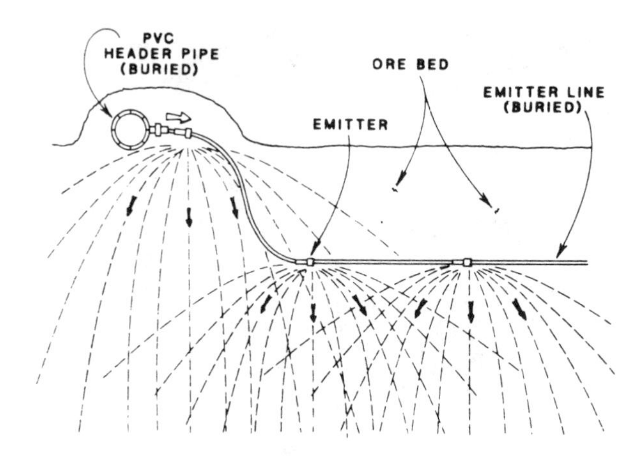

FIGURE 6-11. Leaching-solution emitters covered with ore.

Reprinted by permission. O. A. Muhtadi, in D. J. A. van Zyl, I. P. G. Hutchinson, and J. E. Kiel, eds., *Introduction to Evaluation, Design and Operation of Precious Metal Heap Leaching Projects*. Society for Mining, Metallurgy, and Exploration (Littleton, CO), 1988.

The wobbler is an off-center, rotary-action sprinkler usually mounted on a steel riser not higher than 4 feet above the surface of the heap. Pressure emitters are turbulent-flow devices that provide even solution distribution. They can be installed under the surface covered with ore for winter operation (Figure 6-11). A small number of operations do not use sprinklers, but they spray from drilled holes on the pipe. Although this method of spraying is less expensive than using sprinklers, it is doubtful that it achieves the maximum rate of extraction. Five percent of the North American operations (in a 1986 survey) have built a retaining edge around the top of the heaps and pond the leaching solution (taking the risk of fast and inefficient solution circulation, if there are channels in the heap).

As the leaching solution is sprinkled in the air, it is enriched in oxygen, which is indispensable for dissolving gold with cyanide. In some pilot heap tests, addition of hydrogen peroxide—as a booster to oxidation—was tried with no apparent effects on the extraction rate or the solution grade. Cyanide solution dissolves iron, steel, copper, or brass. Hence, the abundant availability and low cost of plastics has been a blessing for heap leaching.

Scale in the leaching solution comes from limestone bearing ores and/or the lime added to the ore for pH control. As the cyanide solution is continuously recycled from the gold-recovery system to the heaps, soluble calcium reaches saturation and precipitates. Scale inhibitors should be added to the barren cyanide solution to reduce scale formation. In order to reduce the scaling problem, the majority of the North American operations use sodium hydroxide to control the pH of the barren cyanide solution.

Precautions must be taken to ensure continuous heap leaching operations in cold winter climates. A number of procedures have been adopted to prevent freezing of solutions, including burying the solution lines, adding antifreeze to the solution, covering the heaps, operating under ice covering, or combinations of the above. Covered heaps and ponds should also be considered in locations with heavy rainfall, in order to maintain the necessary water balance of a heap-leaching system.

Gold Recovery from Pregnant Solutions

The pregnant solutions produced from heap leaching are, as a rule, very dilute, containing 0.01 to 0.10 oz./ton (0.34 to 3.40 mg/l) of gold. Gold is recovered from these dilute solutions by either cementation with zinc (in a Merrill-Crowe system) or by adsorption on activated carbon, followed by desorption and electrowinning of gold.

The preferred method of recovering gold from dilute pregnant solutions in heap-leaching operations is by adsorption on activated carbon followed by desorption and electrowinning. The pregnant solution percolates through a series of columns packed with activated carbon, which can adsorb up to 250 oz./ton of gold. This huge concentration ratio explains the very extensive use of activated carbon for the recovery of gold from dilute solutions all over the world.

Gold is desorbed (stripped) from the activated carbon, usually under high temperature and pressure, yielding a pregnant strip solution with 10 to 15 oz./ton (0.34 to 0.50 g/l) of gold. The pregnant strip solution is the electrolyte flowing into electrowinning cells where gold (and silver) are deposited on cathodes. Both the "Merrill-Crowe Zinc Cementation" and the "Activated Carbon-Electrowinning" systems of gold recovery are described and discussed in detail in Chapter 10.

Recycling the Barren Solution

The barren solution is pumped to the barren solution pond, where cyanide is added to adjust its concentration and sodium hydroxide is added, if needed, to adjust its pH. This is now the leaching solution to be sprayed on the heaps. In the event that nonferrous metals or other impurities have accumulated in the barren solution, a bleed is drawn and replaced with fresh cyanide solution. The bleed has to be treated for either cyanide recovery or destruction (see Chapter 12).

Vat Leaching

Vat leaching of low-grade gold ores has not been as widely practiced as heap leaching. Vat leaching has been successfully employed with oxidized copper ores to produce directly a pregnant leach solution of sufficient copper concentration for electrowinning.

Vat leaching is, by definition, flooding (submerging) finely crushed ore (−1 in.) in cyanide solution in large vats able to accommodate thousands of tons of ore. The degree of size reduction and agglomeration must be justified by improved gold recovery. The objective of vat leaching gold ores is to reduce the leaching time and increase gold extraction and the gold content of the pregnant solution, in comparison with relevant heap leaching situations.

Patent applications were submitted in 1985 for the concept of "tank leaching" (large tanks, 12 in. dia. by 40 ft., that could be disassembled and transported). The concept has apparently not been successful, since it did not proliferate. Cyanide is corrosive to metals and inexpensive plastic lining is inadequate for protection of large movable tanks (Anonymous, 1985).

Vat leaching has advantages over heap leaching under extreme climatic conditions (such as desert environments, high rainfall, high altitude, very low temperatures), since it is better suited for water-balance control. In principle, vat leaching has also the following advantages over heap leaching:

- Well-defined, self-contained solution circulation system
- Smaller plant site per ton of ore
- Closer metallurgical control (solution-solids contact, controlled consumption of reagents, improved washing and draining of the ore)
- Lower capital and operating costs

In vat leaching, the solution should be advanced counter currently; the barren solution is recycled to the most-leached vat and then advances toward the fresh ore. The leach solution must be circulated through the ore by either upward or downward percolation, with adequate pumping systems. Vats can be equipped with perforated bottoms to improve the uniformity of solution distribution and assist in draining the ore at the completion of the cycle.

In practice, vat leaching does not satisfy expectations for a significantly higher gold recovery than in heap leaching. In spite of the controlled circulation of solution, dead (dry) bulks of ore can occur within the vat. Heap leaching may have significant economic advantages over vat leaching, especially for high-tonnage operations not exposed to extreme climatic conditions. Agglomeration, or desliming, is required before charging the ore in a vat; this is more important than in heap leaching, since the ore for vat leaching is crushed finer than for heap leaching.

A vat-leaching system using large in-ground, plastic-lined pools has been attempted in Western Australia (at Horseshoe Lights). Finely crushed ore is stacked in the in-ground pool and flooded with cyanide solution. The solution, drawn from a central well, can be recirculated through the ore as in regular vats. This concept is an extension of the design of impermeable pads for heap leaching. The control of solution percolation throughout the stacked ore is less effective than through ore in regular vats. The costs of stacking and removing the ore and of distributing the solution are higher than in regular vat leaching, but the capital cost of the installation is significantly lower.

This concept cannot be applied in regions with high rainfall or high evaporation rates, where vat leaching is mostly needed. The system has not been widely adopted by the gold-mining industry.

A 1,000 ton/day gold ore vat-leaching operation has been thriving in Colorado. The friable lean ore (0.06 oz./ton) is crushed to −1/2 in. and agglomerated with a concentrated solution of sodium cyanide/sodium hydroxide (1% solution added). The agglomerated ore is charged into one of four leaching vats (80 ft. long by 50 ft. wide by 9 ft. high) with a 1,000-ton capacity, and allowed to cure for eight hours. It takes four hours to load a vat. The vat is flooded with cyanide solution after the curing period. Experimental evidence indicates that 90% of the gold in the ore dissolves during the curing period. The vats are run on four-day cycles. They are housed within a plant and are kept warm throughout the year.

References

Note: Sources with an asterisk (*) are recommended for further reading.

Anonymous. 1985. Tank leaching farms. *Northern Miner*, July 18, August 1, August 29.

Argall, Jr. J. O. 1985. Heap leaching at Smokey Valley Gold. *Eng. Min. J.*, December: 18–23.

Brewer, R. E. 1986. The Barrick-Mercur Leach Project. *TMS Paper* A36–86.

Dayton, S. H. 1986. Galactic pumps new life into Summitville. *Eng. Min. J.*, August: 34–39.

Dorey, R., D. van Zyl, and J. Kiel. 1988. Overview of heap leaching technology. In *Introduction to Evaluation, Design and Operation of Precious Metal Heap Leaching Projects*, edited by D. van Zyl, et al.: 3–22.

Giroud, J. P., and R. K. Frobel. 1983. Geomembrane products. *Geotechnical Fabrics Report*, Fall: 38–42.

Harper, T. G., J. A. Leach, and R. T. Tape. 1987. Slope stability in heap leach design. In *Geotechnical Aspects of Heap Leach Design*, edited by D. van Zyl. 33–40. Littleton, CO: Soc. Mining Engineers, 33–40.

Herkenhoff, E. C. and J. G. Dean. 1987. Heap leaching: Agglomerate or deslime? *Eng. Min. J.*, June: 32–39.

Hickson, R. J. 1981. Heap leaching practices at Ortiz Gold Mine. *SME Reprint Paper 81:* 347.

Hutchinson, I. 1988. Surface water balance. In *Introduction, Design and Operation of Precious Metal Heap Leaching Projects*, edited by D. van Zyl et al., 203–53.

Lastra, M. R., and C. K. Chase. 1984b. Permeability, solution delivery and solution recovery. *Min. Eng.* November: 1537–39.

*McLelland, G. E. and S. D. Hill. 1981. Silver and gold recovery from low grade resources. *Min. Congress J.*, May: 17–23.

McLelland, G. E., D. L. Pool, and J. A. Eisele. 1983. Agglomeration-heap leaching operations in the precious metals industry. *U.S.B.M. IC 8945.*

Muhtadi, O. A. 1988. Heap construction and solution application. In *Introduction, Design and Operation of Precious Metal Heap Leaching Projects*, edited by D. van Zyl et al.: 92–106.

Potter, G. M. 1981. Design factors for heap leaching operations. *Min. Eng.*, March: 227–81.

Strachan, C., and D. van Zyl. 1988. Leach pads and liners. In *Introduction, to Evaluation, Design and Operation of Precious Metal Heap Leaching Projects*, edited by D. van Zyl et al.: 176–202.

*van Zyl, D. J. A., I. P. G. Hutchinson, and J. E. Kiel. 1988. *Introduction to Evaluation, Design and Operation of Precious Metal Heap Leaching Projects.* Littleton, CO: Society for Mining, Metallurgy, and Exploration, Inc.

CHAPTER 7

Recovery of Secondary Gold

The cleanest kinds of precious metal scrap are old jewelry, metal parts of old dentures, and the dust and bits produced in the manufacture of jewelry or dentures. Old jewelry may be made of sterling[1] silver, carat[2] gold (alloyed gold), or of base metals plated with gold. Dentists often use high-quality gold, platinum, or alloys of gold with platinum, silver, and/or copper.

In some instances, gold wastes might be very dirty, like the dust collected from gold-shop exhaust systems, scrubbers, and filter bags. In a few cases, gold waste might be in liquid form, such as solutions used in gilding or plating baths.

Base Metals Present in Gold Scrap

Copper is a component of most gold alloys; it hardens the alloy and reduces its cost. Nickel is a component of many "white gold" alloys. Zinc is found in some gold alloys and in some solders used for gold alloys. Mercury is found in dental gold scrap. Lead and tin are components of a soft solder used in fastening parts of jewelry. Iron and steel are used as fine wires in holding parts of jewelry together while they are soldered. Finally, brass wire may also be used for holding jewelry parts together during soldering.

---

[1]The standard of fineness of legal British coinage is 0.500 for silver and 0.91666 for gold.

[2] $1/24$ of pure gold; pure gold is 24 carats; e.g., 20-carat gold is $20/24$ gold.

Special Cases of Gold Scrap

Green gold is an alloy of gold and silver with a high silver content, as high as 25% Ag. It does not always dissolve in aqua regia, since the silver chloride that is formed coats the particles of the alloy and prevents further dissolution. Green gold must have a high gold content to be dissolved in nitric acid, and must have a very high silver content to be dissolved in aqua regia. A base metal (copper or zinc) should be added to the molten green gold to dilute its gold content to six carats or lower, at which concentration silver can be dissolved in nitric acid, and the remaining gold in aqua regia. Alternatively, fine gold can be added to the green-gold scrap to bring the silver content to less than 10%. This high-gold alloy can then be digested in *aqua regia* after granulation.

Green-gold and platinum combinations consist of a green-gold base and a platinum top welded together. A base metal should be added to this scrap, after melting, to lower its gold content. After granulation, the low-gold-content alloy should be treated with nitric acid. The remaining residue should be treated with *aqua regia*.

Gold can be parted from platinum in a simple electroplating cell with the gold-platinum alloy as anodes in a cyanide bath. Gold is plated on copper cathodes, and a platinum residue settles on the floor of the cell. The cell should be operated at high amperage with eight to ten volts across the electrodes and the bath temperature around 65° C.

Low-grade gold scrap assays 5% or less gold and consists of metals rolled or plated with gold, or scrap containing plenty of soft (tin/lead) solder. A large proportion of scrap from the electronics industry is low-grade gold scrap. The recovery of gold from such low-grade scrap is difficult and costly, unless a large-scale operation can be justified. Usually, low-grade scrap is sold to copper refineries, where it is added to molten copper in the anode furnace. Ultimately, the precious metals are recovered from the slimes of electrorefining.

Gold can be stripped electrolytically from cheap gold-plated jewelry, which is used as an anode in a cell where the cathode is lead plate. The gold does not adhere to the lead cathode; it flakes off and settles to the bottom of the cell. This process of stripping gold is speedy and efficient.

Some gold scrap contains lead solder. Aqua regia will convert any tin from soft (tin/lead) solder that is left on the scrap into pasty metastannic acid—which may obstruct and contaminate the dissolution of gold. Therefore, it is advisable to remove any pieces of solder from gold scrap, either mechanically or, preferably, by immersion in nitric acid. Lead dissolves in the nitric acid, whereas tin cracks into little pieces. The nitric-acid solution should be rejected after dilution with water. The digestion of the scrap in nitric acid should be repeated if necessary. Solutions and washings should be rejected by careful decantation.

White gold with palladium is an expensive alloy used occasionally in dental work. *White gold with nickel* is a relatively inexpensive alloy used in dental work. Some white-gold scrap from dental work and, rarely, from jewelry may contain platinum. Goldsmiths have a simple test to distinguish nickel from palladium or platinum in white gold. They make a mark on their "streak-test stone" with the unknown white-gold piece. The mark is dissolved with one to two drops of nitric acid or *aqua regia*, and a few drops of ammonia are added for neutralization. A drop of dimethyl-glyoxime (DMG) indicator will cause a bright rose-red precipitate if there is nickel in solution.

If palladium is present, the precipitate will have a bright canary-yellow color. If nickel is suspected along with palladium, ammonia is added drop by drop, the yellow precipitate dissolves, and a rose-red precipitate appears (in the presence of nickel). To test for platinum, a few drops of stannous chloride are added to the wet mark; platinum generates a yellow to brown color.

Gold filings in complex gangue originate from dental laboratories and jeweler's shops. The gangue often contains sand, crushed glass, porcelain, or enamel. Smelting with fluxes may sometimes achieve separation of gold, but there is always the possibility of gold being attached to nonsmelting pieces of porcelain.

The best method is to heat the complex gangue with hydrofluoric acid in a lead dish. Hydrofluoric acid is very corrosive and has to be handled with extreme care. It can cause injuries when in contact with human skin or when inhaled. It should be stored in plastic containers, since it attacks glass. Hydrofluoric acid cannot dissolve gold, which constitutes, in effect, the residue of this treatment.

Some gold filings are mixed with carborundum. Carborundum is not soluble by any acid or base. When filings of gold (and/or other precious metals) are mixed with carborundum dust, they have to be submitted to gravity separation in a hand-shaked pan.

Loewen (1980) proposed a *small-scale refining of gold* produced from scrap or wastes. The scrap or other waste is first incinerated to get rid of organic matter; then it is digested with *aqua regia* to dissolve the precious and other metals present. The excess of acid is removed by evaporation. Gold is precipitated with the addition of sodium bisulfite.

References

Note: Sources with an asterisk (*) are not cited but are recommended for further reading.

*Embleton, F. T. 1981. A new gold refining facility. *Gold Bull*. 14(2): 65–68.

Loewen, R. 1980. Small scale gold refining. *Technical report No 44/1*. London: The Worshipful Company of Goldsmiths.

CHAPTER 8

Cyanidation of Gold Ores

Alchemists—driven by greed and curiosity—did not manage to produce gold from cheap materials, but they certainly initiated and advanced chemical knowledge. The discovery of *aqua regia* by the Arab alchemist Jabir Ibn Hayyan in the sixth century introduced a chemical technology for the extraction of gold from its ores.

$$6HCl + 2HNO_3 + 2Au \rightarrow 2AuCl_3 + 2NO + 4H_2O$$

Chlorine was discovered in 1774 by C. W. Scheele, who noted that it dissolved all metals, including gold. K. F. Plattner first applied chlorination for the recovery of gold from its ores in 1851.

$$2Au + 3Cl_2 \rightarrow 2AuCl_3$$

The apparent "transmutation" of iron into copper, by which copper precipitated from its solution in the presence of iron, was known by the alchemists.

$$Cu^{++} + Fe \rightarrow Cu + Fe^{++}$$

This concept of cementation was used by the chemists of the 19th century to recover gold from solution.

$$Au^{+++} + Fe \rightarrow Au + Fe^{+++}$$

An intense blue pigment—Berlin blue or Prussian blue—was accidentally discovered in 1704 by heating dried blood with potash ($K_2\,CO_3$) and treating the aqueous extract with iron vitriol ($FeSO_4$). This was the first artificially manufactured pigment, and it initiated the new field of the chemistry of (blue-generated[1]) cyanogen compounds.

Potassium ferrocyanide ($K_4[Fe(CN)_6]$) was discovered in France by P. J. Macquer (1718–84) as a product of reaction of the Berlin blue with an alkali (KOH). In 1782, Sheele heated the Berlin blue with dilute sulfuric acid and produced an inflammable gas that, dissolved in water, reacted as a weak acid; he called it "blue acid" (*Blaussare*). J. L. Gay-Lussac (1778–1850) determined the composition of the blue acid as HCN in 1811. Production of potassium cyanide started around 1834 by fusing potassium ferrocyanide with potash.

$$K_4[Fe(CN)_6] + K_2CO_3 \rightarrow 6KCN + FeCO_3$$

The product was contaminated with carbonates and cyanates, however. The formation of HCN in a blast furnace and the existence of cyanogen compounds in the coal gas were detected in 1835 (Habashi, 1987).

The solubility of gold in an aqueous solution of potassium cyanide was known by alchemists of the 18th century. J. W. Mellor (1923) specifically mentions that Scheele, in 1783, and Bagraton, in 1843, noted that aqueous solutions of alkali cyanide could dissolve gold. Potassium cyanide was mainly used to prepare the electrolyte necessary for electroplating gold and silver by the Elkington's process, patented in 1840.

However, L. Elsner was the first to realize the importance of oxygen in the dissolution of gold and silver by aqueous solution of potassium cyanide in 1846. The following is often referred to as Elsner's equation (although it was not published by Elsner).

$$4Au + 8KCN + O_2 + 2H_2O \rightleftharpoons 4KAu(CN)_2 + 4KOH$$

Metallurgists and scientists ratified Elsner's work and experimented with alkali cyanides as gold solvents for the next 40 years, obtaining results of little or no practical importance. It was J. S. MacArthur and his coworkers R. W. Forrest and W. Forrest who first grasped the practical importance of cyanide leaching of auriferous and argentiferous ores for the production of gold and/or silver, in 1887. MacArthur and the Forrest brothers patented the dissolution of gold from ground ores by a weak cyanide solution. They also patented its subsequent precipitation from the pregnant solution by zinc shavings, and thus radically changed the gold-extraction process.

Amalgamation could only be applied to high-grade ores containing coarse gold with rather mediocre (75–80%) recoveries. Chlorination has been a costly, cumbersome process. The industrial application of the cyanidation reaction and zinc precipitation have been major milestones in the extraction of gold, and have immensely broadened the definition of its ores.

---

[1]*Kyanos* (*cyanos*) is the Greek word for blue and the origin of the chemical terms *cyanogen, cyanide,* and *cyanidation*.

The MacArthur-Forrest process made a mark in extractive hydrometallurgy due to the high degree of efficiency that is attained in gold mills. The global production rate of gold, and worldwide gold deposits, increased tremendously after the application of the MacArthur-Forrest patents.[2]

Theories on Gold Cyanidation

Various theories have been proposed to explain the mechanism of gold and silver dissolution in aqueous cyanide solution. This section summarizes these theories (Habashi, 1987; Cornejo and Spottiswood, 1984).

1. *Oxygen Theory* It was recognized by Elsner in 1846 that oxygen was vital for the dissolution of gold in cyanide solution. The following is also referred to as Elsner's equation (see also page 142).

$$4Au + 8NaCN + O_2 + 2H_2O \rightleftharpoons 4NaAu(CN)_2 + 4NaOH$$

2. *Hydrogen Theory* L. Janin (1888, 1892) presented the following equation, which shows that hydrogen gas evolves during the process of gold cyanidation.

$$2Au + 4NaCN + 2H_2O \rightleftharpoons 2NaAu(CN)_2 + 2NaOH + H_2$$

3. *Hydrogen Peroxide Theory* G. Bodlander (1896) suggested that dissolution of gold by cyanide proceeds through two steps, according to these equations:

$$\begin{array}{r} 2Au + 4NaCN + O_2 + 2H_2O \rightleftharpoons 2NaAu(CN)_2 + 2NaOH + H_2O_2 \\ H_2O_2 + 2Au + 4NaCN \rightleftharpoons 2NaAu(CN)_2 + 2NaOH \\ \hline 4Au + 8NaCN + O_2 + 2H_2O \rightleftharpoons 4NaAu(CN)_2 + 4NaOH \end{array}$$

Hydrogen peroxide is formed as an intermediate product. Bodlander found experimentally that H_2O_2 was formed, and he was able to account for approximately 70% of the theoretical amount of H_2O_2 that should be formed according to his equation.

Experiments showed that the dissolution of gold and silver in NaCN and H_2O_2 in the absence of oxygen is a slow process. Therefore, Bodlander's second reaction,

$$2Au + 4NaCN + H_2O_2 \rightleftharpoons 2NaAu(CN)_2 + 2NaOH,$$

---

[2]MacArthur's financial rewards were short lived; his patents were contested by the S. A. Chamber of Mines in 1896 and annulled in South Africa on the grounds that gold had been extracted from ore samples by a solution of potassium cyanide, and it had been precipitated on zinc plates, in 1885.

which is a reduction step,

$$H_2O_2 + 2e \rightleftharpoons 2OH^-,$$

takes place only to a minor extent. In fact, the dissolution of gold is inhibited if large amounts of H_2O_2 are present due to the oxidation of cyanide ion to cyanate ion. Cyanate ion has no dissolving action on the metal:

$$CN^- + H_2O_2 \rightleftharpoons CNO^- + H_2O$$

4. *Cyanogen Formation* S. B. Christy suggested, also in 1896, that oxygen is necessary for the formation of cyanogen gas, which he believed to be the active reagent for the dissolution of gold according to these reactions:

$$O_2 + 4NaCN + 2H_2O \rightleftharpoons 2(CN)_2 + 4NaOH$$
$$4Au + 4NaCN + 2(CN)_2 \rightleftharpoons 4NaAu(CN)_2$$

Christy's two-step process mechanism adds up to Elsner's equation (see theory number 1).

5. *Corrosion Theory* B. Boonstra showed in 1943 that the dissolution of gold in cyanide solution is similar to a metal-corrosion process in which oxygen dissolved in the solution is reduced to hydrogen peroxide and hydroxyl ion. It was pointed out that Bodlander's equation should be further divided into the following steps:

$$O_2 + 2H_2O + 2e \rightleftharpoons H_2O_2 + 2OH^-$$
$$Au \rightleftharpoons Au^+ + e$$
$$Au^+ + CN^- \rightleftharpoons AuCN$$
$$AuCN + CN^- \rightleftharpoons Au(CN)_2^-$$
$$\overline{Au + O_2 + 2CN^- + 2H_2O + e \rightleftharpoons Au(CN)_2^- + 2OH^- + H_2O_2}$$

Janin's ill-conceived Hydrogen Theory (1888) is the only one among the proposed theories that does not recognize the oxygen requirements during gold and silver cyanidation. (Such requirements had been proven experimentally since the middle of the 19th century, by Faraday among others.) The equilibrium constant calculated for Janin's equation is so low that the formation of hydrogen should be considered impossible under ordinary cyanidation conditions. On the contrary, thermodynamic evaluation of the Oxygen Theory (Elsner's and Bodlander's equations) shows a very high equilibrium constant; the reaction will proceed until all the cyanide has been consumed or all the gold has been dissolved.

Habashi (1987), in discussing *The Theory of Cyanidation*, supports Bodlander's first equation,

$$2Au + 4KCN + O_2 + 2H_2O \rightarrow 2KAu(CN)_2 + 2KOH + H_2O_2,$$

as the main description of the dissolution of gold by cyanide solution. Experiments on the kinetics of Bodlander's second equation have shown it to be a slow process.

The Mechanism of Cyanidation

The process of gold dissolution in cyanide (and consequently the extraction of gold from its ores by cyanide) involves heterogeneous reactions at the solid-liquid interfaces (Figure 8-1). Hence, the following sequential steps may be assumed as leading to the dissolution of gold (from its ores) by cyanide:

- Absorption of oxygen in solution
- Transport of dissolved cyanide and oxygen to the solid-liquid interface
- Adsorption of the reactants (CN^- and O_2) on the solid surface
- Electrochemical reaction
- Desorption of the soluble gold-cyanide complexes and other reaction products from the solid surface
- Transport of the desorbed products into the bulk of solution

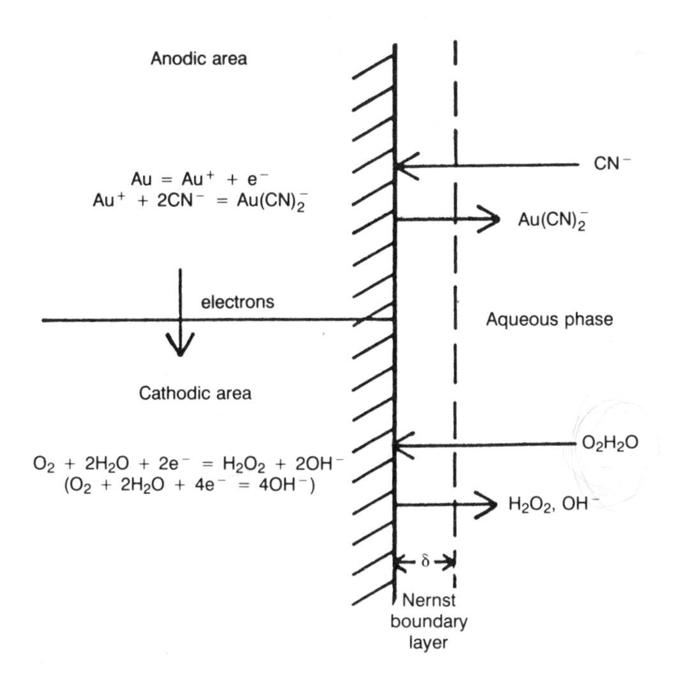

FIGURE 8-1. Sketch of the solid-liquid reaction of gold dissolution in cyanide solution.

Reprinted by permission. F. Habashi, Kinetics of gold and silver dissolution in cyanide solution, *Bull. 59*, Montana Bureau of Mines and Geology (Butte, MO), 1967.

The cyanidation process is affected by a number of influential parameters that are discussed below. These include the availability of oxygen at the solid-liquid interfaces, the pH and Eh of the solids-solvent suspension, the effect of alkalies, and foreign (other than CN^-) ions in solution.

The Effect of Oxygen on Gold Cyanidation

The importance of oxygen in the dissolution of gold cannot be overemphasized. Although oxidizing agents such as sodium peroxide, potassium permanganate, bromine, and chlorine have been used, adequate aeration will give results as good as those of chemical oxidizers, at a lower cost. The amount of oxygen dissolved in dilute cyanide solution depends on four parameters.

- The altitude (barometric pressure)
- The temperature of the solution
- The type and intensity of agitation
- The ionic strength of the solution

At low cyanide concentration, the oxygen pressure has no effect on the rate of gold dissolution. However, at high cyanide concentration, where the

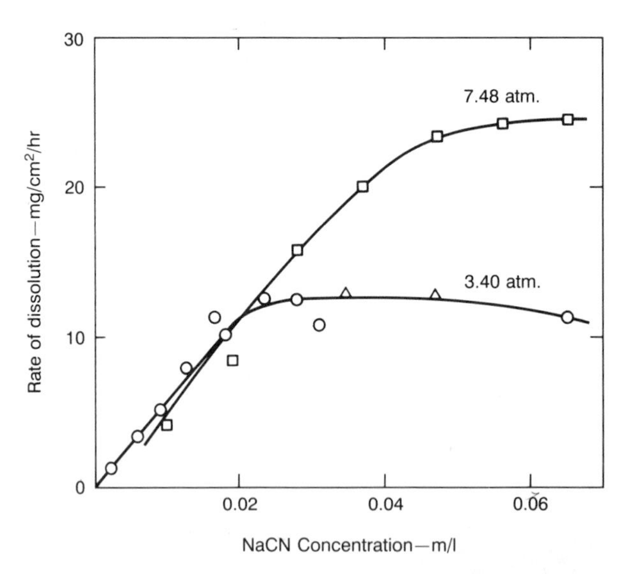

FIGURE 8-2. Rates of dissolution of silver (mg/cm/hr) at different oxygen pressures as a function of NaCN concentration at 24° C.

Reprinted by permission. G. A. Deitz and J. Halpern, Reaction of silver with aqueous solutions of cyanide and oxygen. *J. Met.* (9): 1109–16. The Minerals, Metals & Materials Society (Warrendale, PA), 1953.

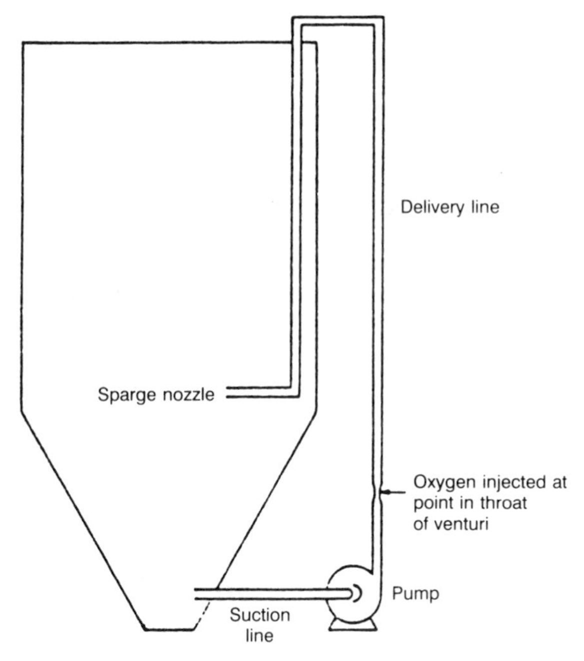

FIGURE 8-3. Sketch of Vitox installation for oxygen injection in an air-agitated tank.

Reprinted by permission. Schematic of Vitox installation on an air-agitated tank, Randol Gold Forum 88, Randol International, Ltd. (Golden, CO), 1988, p. 192.

rate of dissolution is independent of the solvent concentration, the reaction rate is dependent on oxygen pressure (Figure 8-2).

The major influence on the oxygen mass-transfer rate (to the solution) has been found to be the superficial air velocity (defined as the air flow rate per unit of the cross-sectional area of the tank). The oxygen mass-transfer rate decreases with increasing slurry density and decreasing particle size.

The use of pure oxygen in cyanidation has been studied since 1950, but its first commercial application was for pressure cyanidation in 1983. A significant number of South African mills are currently using pure oxygen during cyanidation. The Vitox oxygenation system (Figure 8-3) consists of an agitation tank with a bleed recirculation leg into which oxygen is injected through a venturi; the system achieves high oxygen-dissolution efficiencies. Efficient oxygenation is expected to increase the rate of cyanidation (saving capital costs) and gold recovery, and it should decrease the consumption of cyanide (by destroying cyanicides).

Pressure cyanidation (combining high pressure, high temperature, and optimum high concentrations of cyanide) has been employed for the treatment of certain refractory ores (see Chapter 5). Any side reactions consuming oxygen during the treatment of gold ores are detrimental to the cyanidation. For example:

$$2Fe(OH)_2 + 0.5O_2 + H_2O \rightarrow 2Fe(OH)_3$$
$$2S^{=} + 2O_2 + H_2O \rightarrow S_2O_3^{=} + 2OH^-$$

The Effect of Cyanide Concentration

At atmospheric pressure, the cyanidation is not dependent on cyanide concentration. At room temperature and atmospheric pressure, 8.2 mg/l of oxygen are dissolved in water, equivalent to 0.26×10^{-3} mole/l. Hence, the required sodium cyanide (molecular weight 49) concentration should be at least equal to $4 \times 0.26 \times 10^{-3} \times 49 = 0.051$ g/l. Higher than 0.05 g/l sodium cyanide concentrations cannot affect the rate of gold dissolution, since at atmospheric pressure it is controlled by the constant oxygen concentration in solution.

The Effects of pH and Eh on Gold Cyanidation

The redox chemistry of cyanide leaching of gold is presented in Eh *vs.* pH diagrams[3] for three systems:

- Gold-water (Figure 8-4)
- Cyanide-water (Figure 8-5)
- Gold-cyanide-water (Figure 8-6)

Auric ions (Au^{3+}) can be in equilibrium with gold metal (Au°) in the gold-water system (Figure 8-5), or with aurous-cyanide ions ($Au(CN)_2^-$) in the gold-cyanide-water system (Figure 8-7), up to about pH 1. The predominant gold ions are auric (Au^{3+}), and not aurous (Au^+), because the reaction $Au^\circ \rightarrow Au^{3+}$ lies well below the stability limit of the reaction $Au^\circ \rightarrow Au^+$. The stability field of metallic gold, at relatively low reduction potentials, covers the whole pH range, as does the stability of water (see Figures 8-4, 8-6).

Gold can form insoluble oxide species (hydrated auric oxide —$Au_2O_3.3H_2O \rightleftharpoons 2Au(OH)_3$—or gold peroxide, AuO_2) at relatively high potentials. Both oxides are thermodynamically unstable and hence powerful oxidants. Figures 8-4 and 8-6 indicate that the oxidizing power (potential) of these oxides depends on the acidity of the system and declines with increasing pH. Hydrogen cyanide (HCN) and cyanide ions (CN^-) are the stable species at very low potentials (see Figure 8-5), with the latter

[3]Potential-pH diagrams were first presented by N. P. Finkelstein (1972). Thermodynamic data and the specific equilibrium reactions are to be found in Appendices A and B.

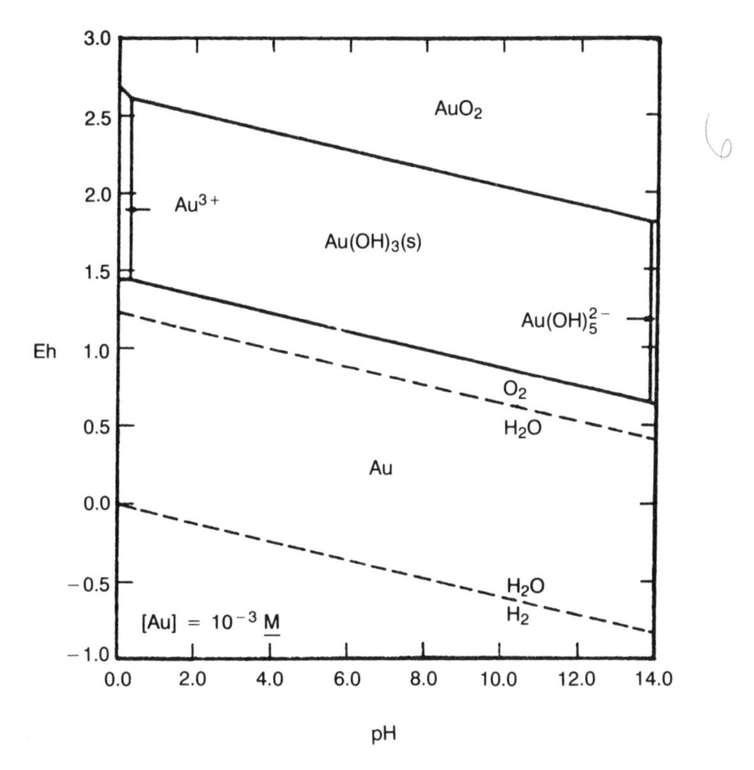

FIGURE 8-4. Redox potential vs. pH diagram for the Au-H_2O system at 25° C with 10^{-3} $Kmol/m^3$ gold concentration.

Reprinted by permission. T. Xue and K. Osseo-Assare, *Metallurgical Transcriptions B*, Vol. 16B, p. 457. The Minerals, Metals & Materials Society (Warrendale, PA), September, 1985.

being predominant at pH higher than 9.24. At high potentials, cyanate (CNO^-) is the only stable species.

Although solid aurous cyanide (AuCN) and the auric cyanide complex $[Au(CN)_4]^-$ have been reported in the literature, the aurous cyanide complex $[Au(CN)_2]^-$ is the only stable cyanide-complex ion during cyanidation (see Figure 8-5). A comparison of Figures 8-4 and 8-6 indicates that the introduction of cyanide in the aqueous system drastically reduces the stability fields of metallic gold and its oxides. The aurous cyanide complex $[Au(CN)_2]^-$ has a substantial stability field (see Figure 8-6) extending into a large area of the water/gold stability fields (as estimated in the absence of cyanide in Figure 8-4). The presence of this extensive aurous cyanide stability field—especially at a pH above 9, where formation of HCN can be totally avoided—makes the leaching of gold-bearing ores with cyanide solutions feasible.

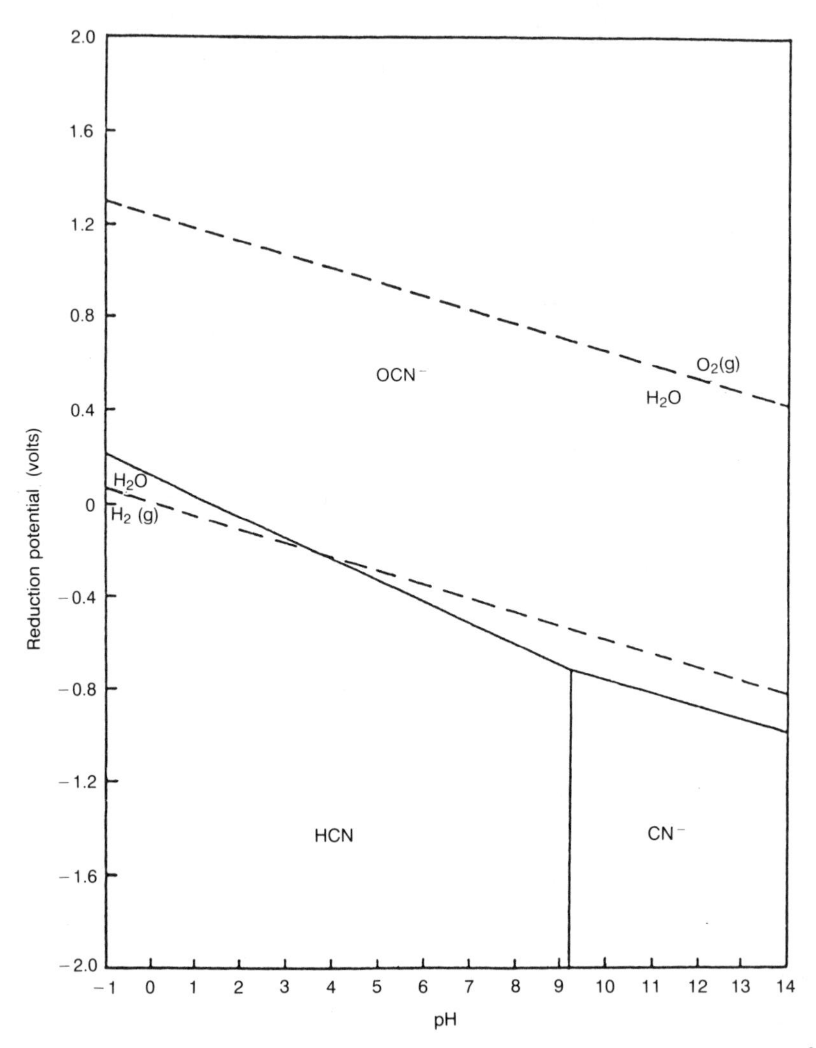

FIGURE 8-5. Redox potential vs. pH for CN-H_2O system at 25° C with 10^{-3} Kmol/m^3 cyanide concentration.

Reprinted by permission, T. Xue and K. Osseo-Assare, *Metallurgical Transcriptions B*, Vol. 16B, p. 457. The Minerals, Metals & Materials Society (Warrendale, PA), September 1985.

The Effect of Alkali Additions

The purposes of adding bases (CaO, NaOH, or Na_2CO_3) to the cyanidation process include the following:

- To prevent the loss of cyanide by hydrolysis
- To prevent the loss of cyanide by the action of carbon dioxide in the air

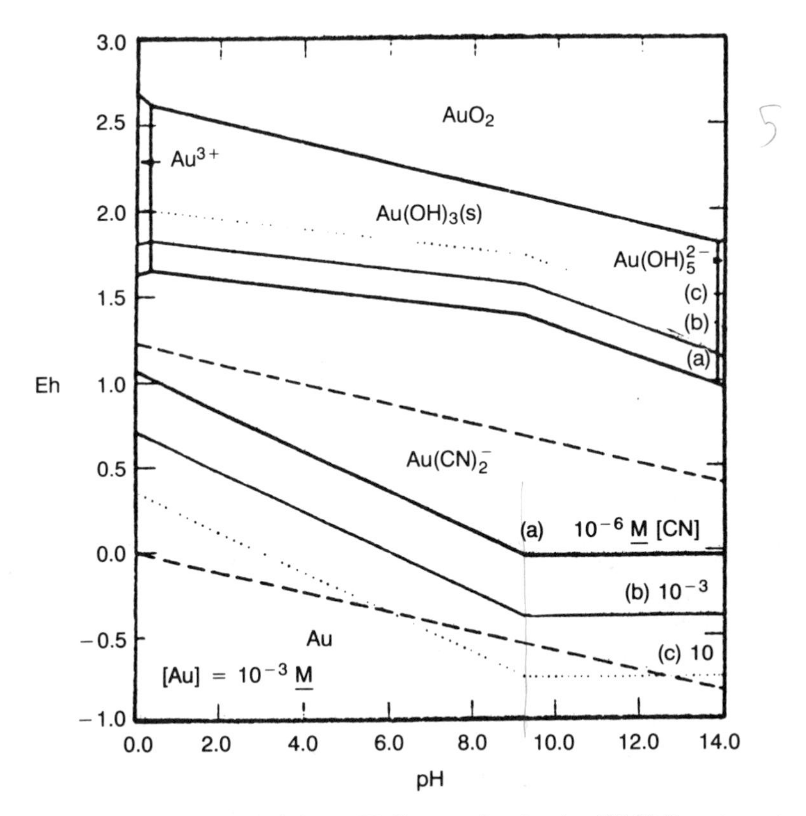

FIGURE 8-6. Redox potential vs. pH diagram for the Au-CN-H_2O system at 25° C with concentrations in Kmol/m^3: [Au] = 103. (a) [CN] = 10^{-3}; (b) [CN] = 10^{-6}; (c) [CN] = 10^{-3}

Reprinted by permission. T. Xue and K. Osseo-Assare, *Metallurgical Transcriptions B*, Vol. 16B, p. 457. The Minerals, Metals & Materials Society (Warrendale, PA), September, 1985.

- To decompose bicarbonates in the mill water before using it in cyanidation
- To neutralize acidic compounds such as ferrous salts, ferric salts, and magnesium sulfate in the mill water before adding it to the cyanide circuit
- To neutralize acidic constituents—pyrite, etc.—in the ore

In addition, the use of lime promotes the settling of fine ore particles so that clear pregnant solution can be easily separated from cyanided ore pulps.

Although the use of an alkali is essential in cyanidation, many researchers have stated that alkali such as sodium hydroxide, and particularly calcium hydroxide, retard the dissolution of gold in cyanide solutions. Barsky et al. (1934) investigated the effects of calcium hydroxide and

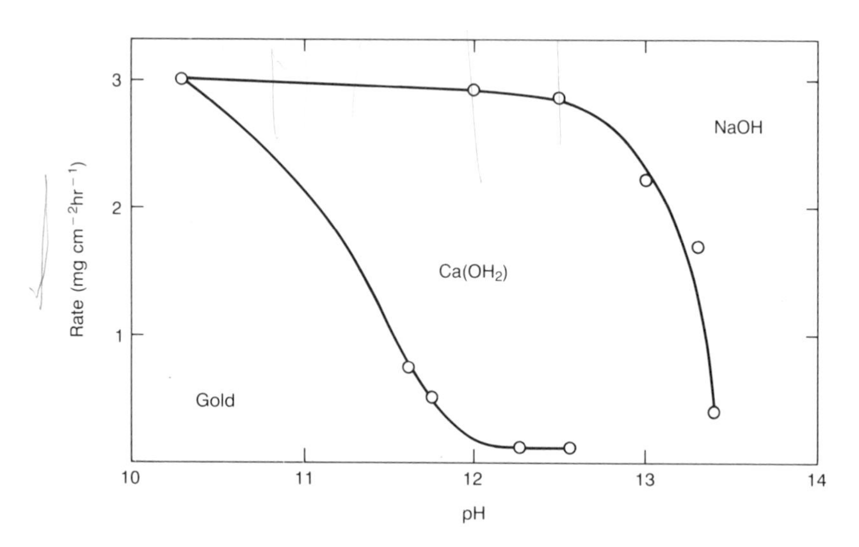

FIGURE 8-7. Retarding effect on the dissolution of gold by calcium ions at high alkalinity.

Reprinted by permission. G. Barsky, S. J. Swanson, and N. Hedley, Dissolution of gold and silver in cyanide solution, *Trans.*, A.I.M.E. (112): 660–77, 1934.

sodium hydroxide on the rate of gold dissolution in cyanide solutions containing 0.1% NaCN (Figure 8-7). They found that when calcium hydroxide was used, the rate of dissolution decreased rapidly as the pH of the solution neared 11, and dissolution was practically nil at pH 12.2. The effect of sodium hydroxide was much less pronounced; the rate of dissolution started to slow down above pH 12.5. However, dissolution was more rapid at pH 13.4 with sodium hydroxide than in a solution of the same cyanide strength containing calcium hydroxide at pH 12.2. The effect of the calcium ion on the dissolution of gold was then investigated by adding $CaCl_2$ and $CaSO_4$ to a cyanide solution. Neither of these salts affected the rate of gold dissolution to any appreciable extent. The solubility of oxygen in cyanide solutions containing various amounts of $Ca(OH)_2$ was then determined, but no appreciable difference was found between a solution containing no calcium hydroxide and one containing up to 5 percent. Therefore, it was concluded that the reduction in the rate of dissolution of gold in NaCN solutions caused by the addition of $Ca(OH)_2$ is *not* due to either lower solubility of oxygen or to the presence of calcium ions. Habashi (1967) attributed the retarding effect of $Ca(OH)_2$ to the formation of calcium peroxide on the metal surface, which prevents the reaction with cyanide. Calcium peroxide was thought to be formed by the reaction of lime with H_2O_2 accumulating in solution, according to the following reaction.

$$Ca(OH)_2 + H_2O_2 \rightleftharpoons CaO_2 + 2H_2O$$

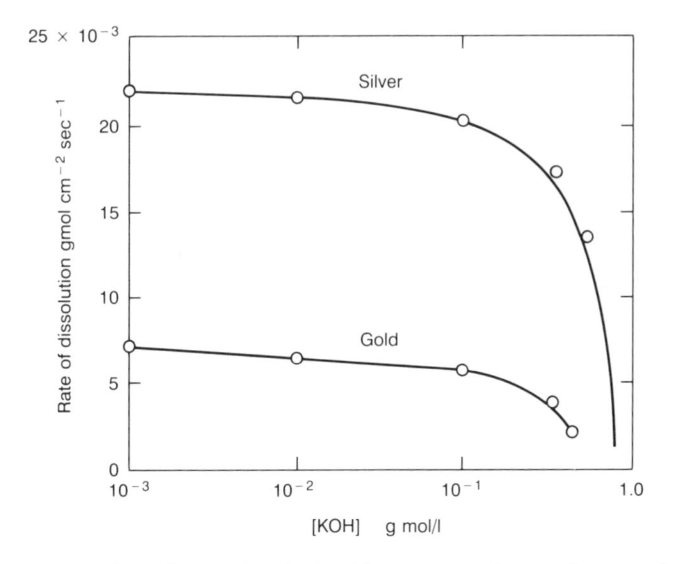

FIGURE 8-8. Effect of potassium hydroxide concentration on the rate of dissolution of gold and silver by KCN.

Reprinted by permission. I. A. Kakovski and A. N. Levedev, The influence of surface-active materials on the rate of dissolution of gold in cyanide solutions, *Doklady Physical Chemistry* (164): 686–89, 1960.

Since lime is the reagent commonly used in cyanide leaching of gold to adjust the pH of the pulp and to promote settling, the effects of its use must be very carefully monitored.

Water saturated with HCN gas and oxygen attacks gold, with the formation of insoluble AuCN and hydrogen peroxide.

$$2Au + 2HCN + O_2 \rightleftharpoons 2AuCN + H_2O_2$$

Therefore, to avoid the formation of AuCN, the cyanide solution should be alkaline during the leaching of gold to prevent two things:

1. Hydrolysis of the cyanide ion

$$CN^- + H_2O \rightleftharpoons HCN + OH^-$$

2. Cyanide decomposition by atmospheric CO_2

$$CN^- + H_2CO_3 \rightleftharpoons HCN + HCO_3^-$$

Hydrogen cyanide (HCN), which forms insoluble AuCN and not soluble $Au(CN)_2^-$, is liberated in both cases above. On the other hand, high pH has an adverse effect on the rate of gold dissolution (see Figures 8-7 and 8-8).

Therefore, the pH of the cyanide leaching solution should be carefully optimized to prevent HCN formation and achieve a high gold-leaching rate. As a rule, the optimum pH range in practice is 11 to 12.

The Effect of Foreign Ions on Cyanidation

Many investigators agree that the dissolution of gold by cyanide is diffusion controlled, but in the industrial cyanidation of ores, cyanide- and/or oxygen-consuming substances may decidedly affect the rate of gold extraction. Pyrrhotite (and, to a lesser extent, pyrite), copper, zinc (and all base metals), arsenic, and antimony minerals consume cyanide. Some of the known cyanicide reactions are below.

$$Fe^{++} + 6CN^- \rightarrow Fe(CN)_6^{4-}$$
$$2Cu^{++} + 7CN^- + 2OH^- \rightarrow 2Cu(CN)_3^{2-} + CNO^- + H_2O$$
$$ZnO + 4NaCN + H_2O \rightarrow Na_2Zn(CN)_4 + 2NaOH$$
$$Ca_3(AsS_3)_2 + 6NaCN + 3O_2 \rightarrow 6KCNS + Ca_3(AsO_3)_2$$

Base metal ions (Cu^{2+}, Fe^{2+}, Fe^{3+}, Mn^{2+}, Ni^{2+}, and Zn^{2+}) form stable complexes with cyanide, thus consuming it, reducing its activity, and retarding gold cyanidation.

In its monovalent state, copper forms a series of extremely stable, soluble complexes in cyanide solutions:

$$Cu(CN)_2^- \rightarrow Cu(CN)_3^{2-} \rightarrow Cu(CN)_4^{3-}$$

The rate of gold cyanidation is not affected by the presence of those cuprocyanide ions, as long as a significant excess of cyanide is maintained in solution [(total cyanide in solution)/(total copper in solution) above 4]. Hence, if there are high levels of cyanide-soluble copper in the ore that cannot be removed before cyanidation, a significant excess of cyanide has to be added. When the leaching solution contains more than 0.03% copper, $Cu_2(CN)_2.2NaCN$ can be precipitated by controlled acidification with H_2SO_4. Cyanide should be regenerated from this precipitate. If the copper content of the leaching solution is allowed to surpass 0.03% in industrial practice, the dissolution of gold decelerates considerably. Gold recovery from solutions containing copper has to be performed by the carbon-in-pulp process (CIP), since zinc precipitation of gold is inefficient in the presence of high levels of copper (see Chapter 10).

Sulfide minerals react with cyanide and oxygen to form thiocyanate ions.

$$S^{2-} + CN^- + 0.5O_2 + H_2O \rightarrow CNS^- + 2OH^-$$

Pyrite is the most stable (and hence the least troublesome) of iron sulfides. Marcasite decomposes more readily than pyrite and can be a cyanicide, although it is less obnoxious than pyrrhotite. The latter is a particularly harmful cyanicide in the alkaline environment of cyanidation; its rate of decomposition is faster than that of other pyritic minerals. Pyrrhotite is particularly prone to oxidation, forming ferrous and ferric sulfates. Hence, it is not only a wasteful cyanicide but also a consumer of the oxygen dissolved in solution.

$$Fe_7S_8 + NaCN \rightarrow NaCNS + 7FeS$$
$$FeS + 2O_2 \rightarrow FeSO_4$$
$$FeSO_4 + 6NaCN \rightarrow Na_4\,Fe(CN)_6 + Na_2SO_4$$

Alkali sulfide may be one of the initial products of the reaction between pyrrhotite and the alkaline cyanide of the solution. The sulfide ion is a very potent poison for the cyanidation of gold. It has been reported that even less than 0.05 ppm sulfide ion will significantly decelerate dissolution (Fink and Putnam, 1950). Alkali sulfide should be oxidized in the presence of oxygen to thiosulfate, sulfate, and/or thiocyanate.

$$2Na_2S + 2O_2 + H_2O \rightarrow Na_2S_2O_3 + 2NaOH$$
$$Na_2S_2O_3 + 2NaOH + 2O_2 \rightarrow 2Na_2SO_4 + H_2O$$
$$2Na_2S + 2NaCN + 2H_2O + O_2 \rightarrow 2NaCNS + 4NaOH$$

Thus, decomposition of sulfur-bearing minerals may result in thiosulfate and polythionate ions that can waste both cyanide and dissolved oxygen:

$$2S_2O_3^{2-} + 2CN^- + O_2 \rightarrow 2CNS^- + 2SO_4^{2-}$$

Thiocyanate is formed (as in previously listed reactions) by the reaction of cyanide with sulfite, oxy-sulfur, arseno-sulfur ions, and elemental sulfur.

Arsenic-containing minerals, arsenopyrite (FeS.FeAs) realgar (As_2S_2), and orpiment (As_2S_3) are often associated with gold ores and can react with cyanide solutions, consuming oxygen and cyanide. AsS_3^{2-}, AsO_3^{2-}, AsO_4^{2-}, $S_2O_3^{2-}$, SO_4^{2-}, and CNS^- have been identified as reaction products of orpiment and aerated alkaline cyanide solutions. Some of these ions will further consume oxygen to reach a stable state, and others, like S^{2-} and $S_2O_3^{2-}$, may be adsorbed on gold surfaces and inhibit further cyanidation.

Arsenopyrite, if contained in the ore, may cause the waste of cyanide and dissolved oxygen through its decomposition. Stibnite can be even worse than arsenopyrite in these respects; in addition to the oxygen adsorbed by the alkali sulfides formed during its decomposition, the generated thio-antimonite (SbS_3^{2-}) and sulfide (S^{2-}) ions may be adsorbed on gold surfaces and hinder their reaction with cyanide.

When pyrrhotite and ferrous sulfate—resulting from the oxidation of FeS—are present in the ore, litharge (PbO) is sometimes added in the grinding mills.

$$Fe_7S_8 + PbO \rightarrow PbS + 7FeS + 0.5O_2$$
$$FeSO_4 + PbO \rightarrow PbSO_4 + FeO$$
$$2FeO + 3H_2O + 0.5O_2 \rightarrow 2Fe(OH)_3$$

Litharge leads to the oxidation of pyrrhotite without formation of wasteful ferrocyanides. High lime addition (high pH) is detrimental to gold dissolution because, in addition to deceleration of the cyanidation (see Figure 8-7), it accelerates the decomposition of pyrrhotite and reduces the effects of lead salts.

The inhibiting actions of arsenic and antimony minerals can be overcome with lead ions, which, in addition to precipitating sulfide (S^{2-}) ions, accelerate the oxidation of thio-arsenite (AsS_3^{2-}) and thio-antimonite (SbS_3^{2-}) ions. The effects of stibnite can be controlled, at least partially, by allowing the ground ore to weather before cyanidation. An insoluble basic antimony sulfate is formed on the weathered stibnite surface on contact with water. (Such weathering of stibnite should be avoided if it contains microcrystalline gold.)

A vigorous preaeration of alkaline slurries with high content of pyrrhotite, arsenic, and/or antimony minerals is prerequisite before cyanidation.[4] The *Salsigne Process* consists of grinding ore high in pyrrhotite in water, adding a predetermined amount of lime so that after 24-hour agitation in Pachuca tanks the alkalinity of the slurry will drop to not less than 0.01% CaO, filtering, and washing the solids. Conventional cyanidation follows this pretreatment step. (After the Salsigne pretreatment, cyanide consumption is reduced significantly, with up to 50% savings in cyanide consumption reported.) In effect, the Salsigne process attempts to remove soluble sulfides as basic sulfates, oxidize Fe^{++} and precipitate $Fe(OH)_3$, and passivate any insoluble sulfides (Dorr and Bosqui, 1950).

The two lead minerals, galena (PbS) and anglesite ($PbSO_4$), behave similarly in cyanide solutions, since galena oxidizes easily to lead sulphate. Low alkali or lime concentrations should be used in the presence of lead minerals to avoid the formation of plumbite, which can be a cyanide consumer. With an excess of alkali, the following reactions occur:

$$4NaOH + PbSO_4 \rightarrow Na_2PbO_2 + Na_2SO_4 + 2H_2O$$
$$3Na_2PbO_2 + 2NaCN + 4H_2O \rightarrow Pb(CN)_2.2PbO + 8NaOH$$

With controlled low concentrations of alkali, the waste of cyanide may be avoided.

---

[4]Gold ores containing pyrite, pyrrhotite, arsenopyrite, and/or stibnite are refractory, as a rule; arsenic and antimony minerals, in addition to their high consumption of cyanide, may contain microcrystalline gold locked up in their lattices. (The treatment of refractory gold ores is disscussed in detail in Chapter 5.)

$$2Na_2PbO_2 + 2NaCN + 3H_2O \rightarrow Pb(CN)_2.PbO + 6NaOH$$
$$Pb(CN)_2.PbO + 4H_2O \rightarrow 2PbO.3H_2O + 2HCN$$

It has been reported that, during the roasting of pyritic gold ores containing even a low concentration of lead minerals, lead compounds may coat the gold particles and make them quasi-insoluble to cyanide. The recovery of gold from such calcines can be enhanced by preliminary treatment with dilute hydrochloric acid to remove the lead, followed by lime addition and cyanidation.

The lead chromite mineral crocoite ($PbCrO_4$) can waste cyanide, in the presence of lime, by oxidizing it to cyanate. In effect, the calcium chromate formed by the crocoite oxidizes the cyanide to useless cyanate, and causes reagent waste.

$$2PbCrO_4 + Ca(OH)_2 \rightarrow PbCrO_4.PbO + CaCrO_4 + H_2O$$
$$CaCrO_4 + 2NaCN + 2H_2O \rightarrow 2NaCNO + Ca(OH)_2 + Cr(OH)_2$$

The Effect of Flotation Reagents on Cyanidation

Gold-bearing ores are often concentrated by flotation before cyanidation. Thiol-type collectors (xanthates, dithiophosphates, etc.) are usually employed during flotation. If the flotation products are not washed well, the presence of small amounts of thiol reagents has a negative effect on cyanidation. Thiols are potent poisons of the cyanidation reaction as sulfide ions. Their poisonous effect increases with their concentration and chain length. The adverse effect of thiols can be reduced with increased cyanide concentration (Ashurst and Finkelstein, 1970).

Two mechanisms have been proposed to interpret the behavior of the thiol reagents. First, the adsorption of thiols on the mineral surfaces makes them hydrophobic and thus resistant to the diffusion of the aqueous solvent. Hydrophobic particles flocculate and accumulate at the liquid-air surface. Alternatively, the adsorbed collector passivates the gold surfaces.

Cationic reagents that are used for the flotation of pyrite from gold-bearing ores have been shown to affect cyanidation adversely. Significant retardation occurs for reagent concentration as low as 0.01 ppm. However, the effect of these reagents may be largely reduced in the presence of certain finely ground silicate minerals.

Other surface-active reagents such as ketones, ethers,and alcohols are claimed to accelerate the rate of dissolution of gold in cyanide solutions when present in concentrations of the order of 10^{-2} kmol/m^3. At higher concentrations, the rate of gold dissolution is found to decelerate. It has been reported that these reagents compete with oxygen for sites at the gold surface. At low concentrations, this competition has the effect of depassivating the surface, but as the concentration increases, the rate of dissolution is retarded, owing to a shortage of oxygen at the surface.

Other Sources of Cyanide Loss

Silver Leaching. The amount of cyanide that combines with gold in an average ore is negligible. However, the leaching of silver sulfide consumes a significantly higher proportion of cyanide:

$$Ag_2S + 5NaCN + 0.5O_2 + H_2O \rightarrow 2NaAg(CN)_2 + NaCNS + 2NaOH$$

Decomposition Loss. When insufficient free alkali is present, carbon dioxide in the air can cause a loss of cyanide:

$$2NaCN + CO_2 + H_2O \rightarrow 2HCN + Na_2CO_3$$

In very dilute solutions, there is always a tendency for hydrolysis:

$$NaCN + H_2O \rightarrow HCN + NaOH$$

In tailing dams, where the supernatant solution is exposed to sunlight and to the atmosphere, the cyanide decomposes quickly and its concentration falls to approximately 0.001% within 24 to 36 hours.

Mechanical Losses. In addition to losses of cyanide solution by leakages and runovers, there is always loss of cyanide adsorbed in the residue, in spite of washing the tailings through counter-current decantation (CCD) thickeners.

The Kinetics of Gold Cyanidation

The longer the time required for cyanidation to achieve a desired recovery from a gold ore, the larger the capacity required for the leaching tanks, and hence the larger the capital cost of the plant. In practice, a wide range of residence time (10 to 72 hours) is encountered in gold mills.

The dissolution of gold in alkaline cyanide solution is a heterogeneous reaction occurring at the solid-liquid interface. The rates of mass transfer of reactants from the liquid to the interface, and of the reaction products from the interface to the liquid bulk, may have an important (if not controlling) effect on the overall kinetics of gold cyanidation. Thus, the rate of dissolution not only depends on the rate of the chemical reaction at the solid-liquid interface but also on mass-transfer rates within the phases. The rate will also depend on the area of the reaction interface, which is continuously diminishing during gold dissolution.

Experimental studies have shown that the dissolution of gold is controlled by the diffusion (mass transfer) of both the dissolved oxygen and the cyanide ions through the boundary layer of the solid-liquid interface (see Figure 8-1). The dissolution rate increases with oxygen concentration and the intensity of agitation. However, when oxygen concentration and agitation are increased beyond certain levels, the gold may become passive, and its dissolution rate may decrease to a lower constant level.

The major influence on the oxygen mass-transfer rate in Pachuca tanks has been found to be the superficial gas velocity, defined as the air flow rate per unit of a cross-sectional area of the tank. The oxygen mass-transfer rate was found to decrease with increasing slurry density and decreasing average particle size.

At a low cyanide concentration, oxygen pressure has no effect on the rate of gold dissolution, while at a high cyanide concentration—where the reaction is not controlled by the cyanide—the reaction rate depends on the oxygen pressure (see Figure 8-3). Habashi's formulation of the rate of cyanidation is

$$\text{Rate} = \frac{2A.D_{CN}.D_{O_2}.[CN^-].[O_2]}{\delta\{D_{CN}.[CN^-] + 4D_{O_2}[O_2]\}} \qquad \text{in g. equiv/sec}^{-1}$$

where A = surface area of the gold particle in cm^2

$[CN^-]$, $[O_2]$ = concentrations (in moles/ml) of cyanide and dissolved oxygen, respectively

D_{CN^-} = 1.83×10^{-5}: the diffusion coefficient of the cyanide ion ($cm^2\ sec^{-1}$)

D_{O_2} = 2.76×10^{-5}: the diffusion coefficient of dissolved oxygen ($cm^2\ sec^{-1}$)

δ = the thickness of the boundary layer, between 2 and 9×10^{-3} cm depending on the intensity and method of agitation

It can be derived from the rate equation that, when

$$D_{CN^-}.[CN^-]_b = 4D_{O_2}.[O_2]_b \quad \text{or} \quad \frac{[CN^-]_b}{[O_2]_b} = 4\frac{D_{O_2}}{D_{CN}}$$

the cyanidation rate is

$$\text{Rate} = \frac{\sqrt{D_{O_2}.D_{CN^-}}}{2}.A.[O]_b^{1/2}.[CN]_b^{1/2}$$

Hence, at this ratio of concentrations of cyanide and oxygen, the rate of dissolution is equally dependent on both.

As shown in Table 8-1, the average ratio D_{O_2}/D_{CN^-} is 1.5. Hence,

$$\frac{[CN^-]_b}{[O_2]_b} = 4\frac{D_{O_2}}{D_{CN^-}} = 6$$

Therefore, the limiting rate is reached when the ratio of the concentration of cyanide to that of oxygen in bulk solution is equal to six.

In practice, the ratio of the two concentrations is of great importance. If solutions with excessive cyanide concentrations—in respect to the oxygen dissolved—are used, the excess cyanide is, in effect, wasted. On the other

TABLE 8-1. Estimated values of diffusion coefficients.

Temp. °*C*	*KCN* %	D_{KCN} *sq. cm/sec.*	D_{O_2} *sq. cm/sec.*	$\frac{D_{O_2}}{D_{KCN}}$	*Investigator*
18	—	1.72×10^{-5}	2.54×10^{-5}	1.48	White, 1934
25	0.03	2.01	3.54	1.76	Kameda, 1949b
27	0.0175	1.75	2.20	1.26	Kudryk and Kellogg, 1954.
			Average:	1.5	

Reprinted courtesy of Montana Bureau of Mines and Geology. F. Habashi, Kinetics and mechanisms of gold and silver dissolution cyanide solution, *Bull. 59*, 1967. Butte, MT: Montana Bureau of Mines and Geology.

hand, if oxygen saturation is achieved in a solution deficient in free cyanide, the rate of cyanidation will suffer. For the maximum cyanidation rate, and hence the most productive plant, it is imperative to control the concentrations of both cyanide and dissolved oxygen to their optimum molar ratio (equal, or very close, to six). On-line cyanide analyzers are very useful in controlling the optimum concentration of cyanide in series of leaching tanks (see Figure 8-9) where intensive aeration supplies the maximum dissolved oxygen under the local conditions.

Use of Pure Oxygen in Cyanidation

The use of pure oxygen in gold leaching has been studied since the mid-1950s, but the first commercial application was in 1983, in a pipe reactor (at Consolidated Murchison in South Africa) for the treatment of a refractory ore. In order to achieve saturation of the leaching solution with oxygen, a large number of minute oxygen bubbles should be created and dispersed in the slurry, deep enough and long enough to allow for adequate dissolution.

The Vitox system for dissolving oxygen in aqueous solutions and/or slurries, patented by the British Oxygen Company in the early 1970s, is shown in Figure 8-3. Although the system was developed for the sewerage-treatment industry, where it has been very successful, it is currently in use in a few cyanidation plants. The system can achieve up to 90% oxygen dissolution efficiencies, with up to 2 kg oxygen dissolved per kwh.

The Effect of Temperature on Cyanidation Rate

When cyanide solution containing gold is heated, the rate of further gold dissolution is affected in two ways. First, the increase in temperature

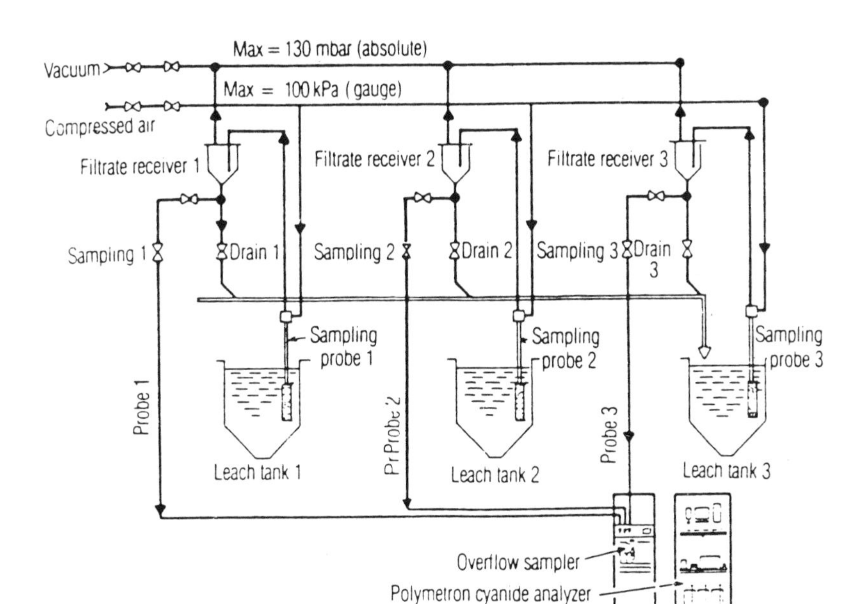

FIGURE 8-9. Flow diagram of automatic sampling and analysis for cyanide content of process solutions.

increases the activity of the solution, and thus increases the rate of dissolution. Second, the amount of dissolved oxygen decreases, since the solubility of gases decreases with increasing temperature.

The solubility of gold in 0.25% KCN solution as a function of temperature is given in Figure 8-10. The rate of dissolution is maximum at about 85° C, although the oxygen content of the solution at this temperature is less than half the oxygen content at 25° C. The rate of dissolution of gold at 100° C is only slightly less than the maximum, although the solution contains no dissolved oxygen at this temperature. This is attributed to the lower capacity of an electrode to absorb or retain hydrogen at its surface in a heated solution, as compared to in the cold. Hence, the maximum opposing electromotive force (EMF) due to polarization becomes less and less as the solution becomes heated, until the EMF of the dissolving gold overbalances polarization, and dissolution of the gold can proceed without oxygen. Thus, polarization can be prevented by either oxygen, which oxidizes the hydrogen at the surface of the gold and permits dissolution of gold at low temperatures, or by heat, which dislodges the hydrogen from the gold surface and permits the gold to dissolve without using oxygen.

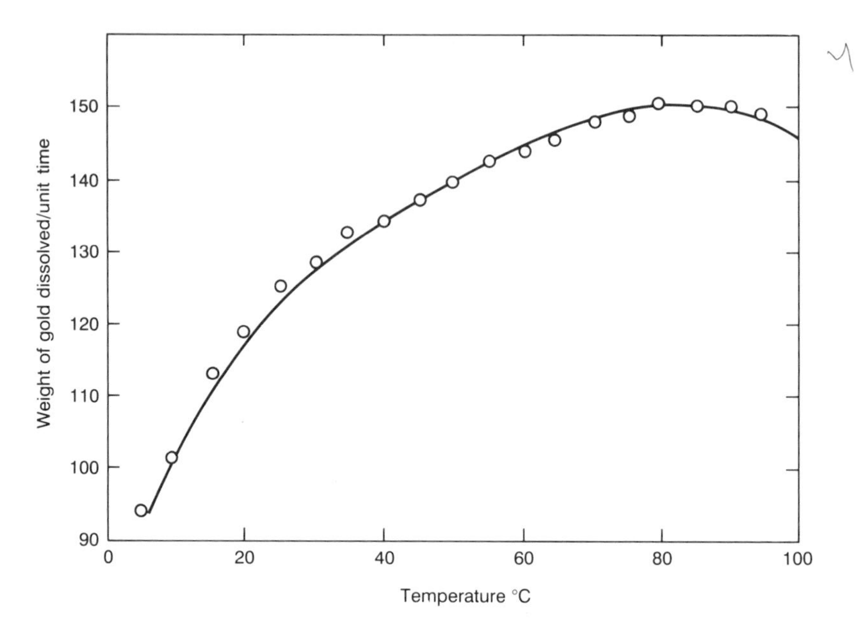

FIGURE 8-10. Effect of temperature on the rate of dissolution of gold in 0.25% KCN.

H. F. Julian and E. Smart, *Cyanidation of Gold and Silver Ores*, 1921. Reproduced by permission of the publishers, Richard Griffin (High Wycombe) Ltd., 30 Disraeli Crescent, High Wycombe, Bucks HP13 5EJ, England.

The Effect of Particle Size

A gold ore has to be ground fine to liberate the precious metal particles and make them amenable to cyanide leaching. Under the ideal conditions with respect to aeration and agitation, it has been determined that the maximum rate of dissolution of gold is 3.25 mg per cm^2 per hour. This equals a penetration of 1.68 microns on each side of a flat 1-cm^2 gold particle, or a total reduction in thickness of 3.36 microns per hour. Therefore, a piece of gold 37 microns thick would take about 11 hours to dissolve. Coarse free gold particles are generally removed by gravity concentration prior to cyanidation, since these coarse particles might not be completely dissolved in economically acceptable time for cyanidation.

Chemical Enhancement of Cyanidation

The presence of small amounts of lead, mercury, bismuth, or thallium salts accelerates the dissolution of gold (Figure 8-11). Gold can actually displace the ions of only these four metals, as can be concluded by calculation of their respective electrode potentials in cyanide solutions. The rapid

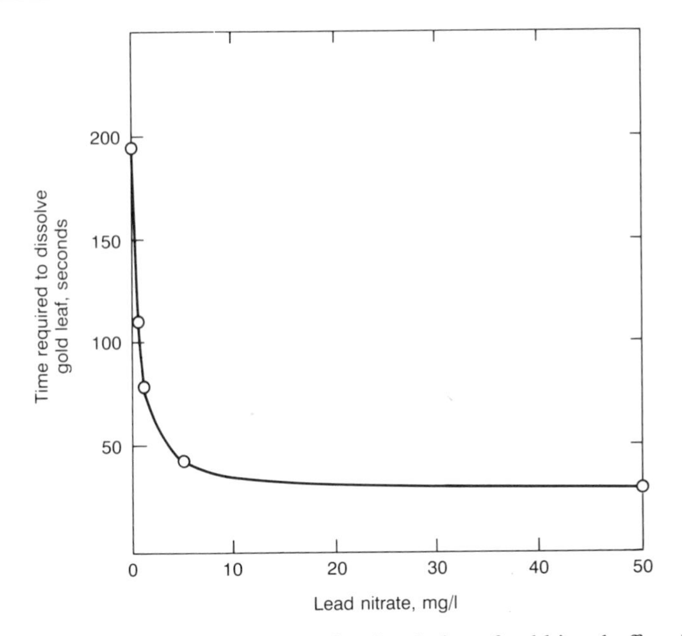

FIGURE 8-11. Effect of lead ion on the dissolution of gold in a buffered 0.1% NaCN solution.

Reprinted by permission. C. G. Fink and G. L. Putnam, The action of sulphide ion and of metal salts on the dissolution of gold in cyanide solutions, *Transcripts of American Institute of Mining, Metallurgical, and Materials Engineers* (187): 925–55, 1950.

dissolution of gold in the presence of these ions might be due to an alteration of its surface characteristics caused by alloying with the displaced metals. Since dissolution is a diffusion-controlled process, it is possible that changes in the surface character of the metal may lead to a decrease in the thickness of the boundary layer through which the reactants diffuse until they reach the metal surface.

Hydrogen peroxide is an intermediate product in the accepted mechanism for gold dissolution in alkaline cyanide solutions. The rate of dissolution of gold is enhanced by addition of some peroxygen compounds. The concentration of peroxygen compounds in cyanide solution must be carefully controlled to minimize its consumption of cyanide.

Physical Enhancement of Cyanidation

Ultrasonic Treatment of Pulp. Application of ultrasonics to an aqueous medium containing a ground ore can produce further breakage, cause a widening of secondary fissures in the ore, and allow exposure of new surfaces that include microscopic gold particles. Contact of these new surfaces with the leaching reagents, along with the effect of ultrasonics, could enhance the rate and degree of dissolution of gold from the ore.

Mixing Efficiency. The second physical aspect of leach enhancement involves the improvement of agitator and Pachuca mixing efficiency and control, as well as the availability of a technique to measure the efficiency of continuous leaching devices. Several references have been made by operators—particularly those using flat-bottomed, air-agitated Pachucas or Pachucas with very shallow cone bottoms—to inefficient mixing and problems with Pachucas sanding out. A partially sanded-up Pachuca will offer a greatly reduced residence time to a leaching pulp, because its effective capacity has been reduced (Hallett, Mohemius, and Robertson, 1981).

A row of leaching vessels arranged for continuous leaching of a pulp is subject to short-circuiting, whereby an ore particle finds a way through the system with a shorter residence time than designed. Pulp residence time in leach agitators arranged for continuous feed can be assessed by dosing the pulp with a radioactive tracer and measuring its radioactivity as it leaves the Pachuca. In order to avoid the reduction in effective capacity of leach agitators through sanding out and short-circuiting, the use of large leaching vessels in a batch mode (rather than continuous feed) is advocated.

Pressure Cyanidation. Pressure in combination with a high temperature for cyanidation is being applied by the Calmet Process,[5] which claims high gold extractions from complex ores containing sulfides, tellurium, selenium, arsenic, antimony, and bismuth. The consumption of oxygen is low, indicating limited oxidation of sulfides and other contaminants. One has to conclude that the cyanidation is enhanced by increased solubility of oxygen in the solution. Oxidation of 15% of the contained sulfur is normally sufficient. Typical process times are about three to four hours total for a 2.0–2.5-ton batch, which includes 1.5 hours of autoclave time. It is obviously a high-cost process addressed to high-grade refractory ores.

Pietsch et al. (1983a) patented the pressure leaching of gold and silver ores with cyanide. The patent proposes the compression of the ore-cyanide-solution slurry in a pipe reactor under high oxygen pressure (25 to 130 atmospheres). Retention times of 10 to 25 minutes in the pipe reactor were reported to result in 94% to 97% gold extractions (Figure 8-12). Cyanide decomposition is not significant even under 50 atmospheres of oxygen pressure (Pietsch et al., 1983).

Intensive Cyanidation. Intensive cyanidation is mostly applied in the treatment of gold mills' gravity concentrates. Gravity concentrates have

[5]A proprietary process, with scarce operating conditions disclosed, owned by the Calmet Corporation.

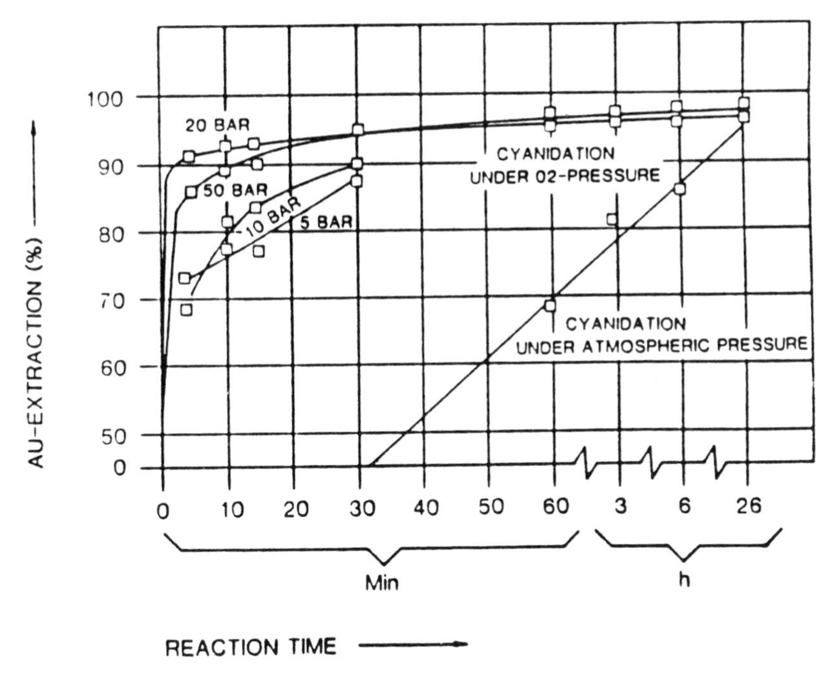

FIGURE 8-12. Gold extraction by cyanidation under oxygen pressure.
Reprinted by permission of VCH Publishers, Inc., 220 East 23rd Street, New York, NY 10010 from: *Erzmetall* 39, Nr. 2, 1986, by R. G. Schulze.

been upgraded, as a rule, by amalgamation in the past. Concern about the environmental aspects of amalgamation and the significant security risk of handling gold-mercury amalgam were the main incentives for the development of intensive cyanidation (Dewhirst et al., 1984).

The partial pressure of oxygen in solution is the most critical factor determining the cyanidation rate. Hence, the use of oxygen, rather than air, significantly increases the dissolution rate of gold. An increased dissolved-oxygen content allows the use of a higher NaCN solution than in normal leaching. Vigorous agitation of the pulp accelerates mass transfer between liquid and solids. Temperatures on the range 30° to 35° C were found to give the best results. Figure 8-13 shows a batch intensive-cyanidation reactor, and Figure 8-14 shows an intensive-cyanidation flow sheet.

Cyanide Regeneration

The formation of cyanide or complex cyanogen compounds of base metals are the main causes of cyanide consumption. Cyanide solutions with a high content of such compounds are inefficient as solvents of gold. Precipitation

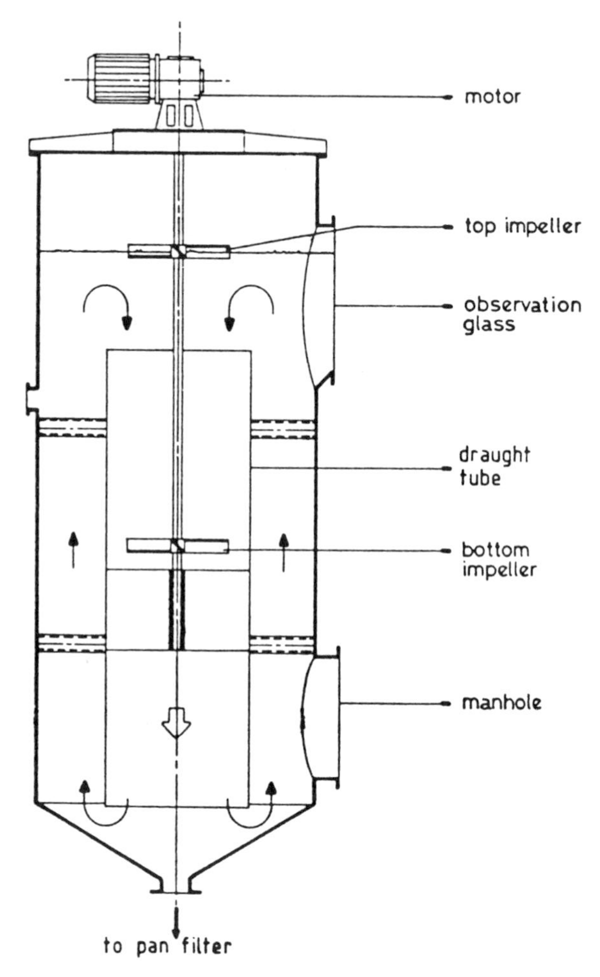

FIGURE 8-13. Selection of intensive cyanidation reactor.

Reprinted by permission. R. F. Dewhirst, S. P. Maulf, and J. A. Coetzee, *J. S. Afr. Inst. Mining and Metallurgy* (84)6: 159–63, 1984.

of gold from cyanide solution by zinc results in the formation of soluble, complex cyanide ions.

The general method for cyanide regeneration is by acidification of its solutions with sulfuric acid, SO_2, or CO_2. The acidified solution is transferred to a closed tank, in which it is stripped by aeration. The hydrogen cyanide is entrained by air, which is mixed with alkaline solution, preferably in a series of absorption columns. The acidification step produces colloidal precipitates of base metal-cyanide complexes which are very

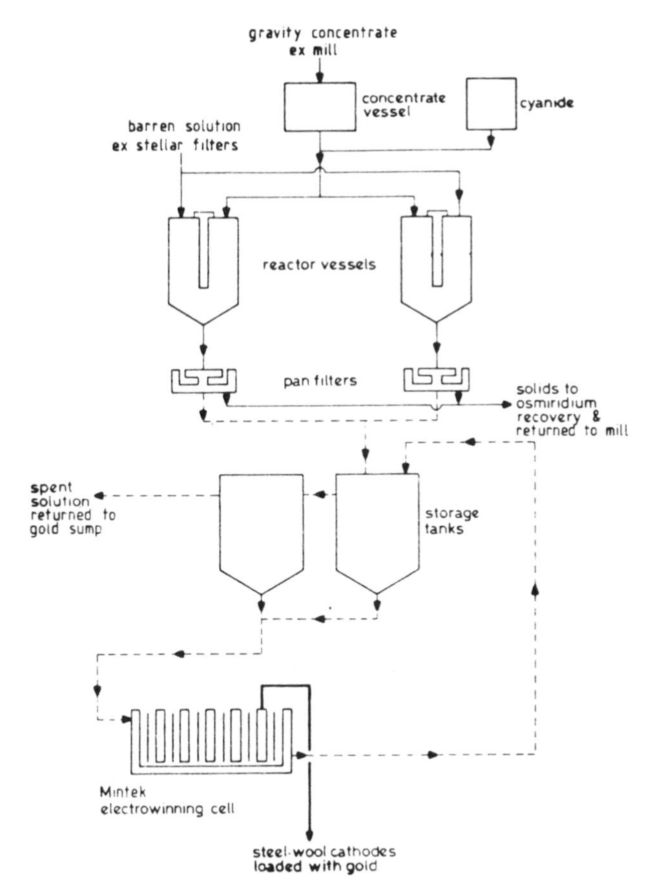

FIGURE 8-14. An intensive-cyanidation circuit.

Reprinted by permission. R. F. Dewhirst, S. P. Maulf, and J. A. Coetzee, *J. S. Afr. Inst. Mining and Metallurgy* (84) 6: 159–63, 1984.

difficult to remove and may interfere with the stripping and caustic-absorption operations.

Golconda Engineering & Mining Services (GEMS, 1988) has patented procedures for the filtration of the colloidal precipitates resulting from their process of cyanide regeneration (Figure 8-15). Other systems of cyanide regeneration propose ion-exchange columns to recover cyanide from gold-mill effluent waste solutions. Cyanide and base-metal complex cyanide ions are adsorbed on the ion-exchange resin. During elution, an upflowing acidic solution causes the desorption of the cyanide complexes. The acidic eluate is contacted counter-currently with air in a packed

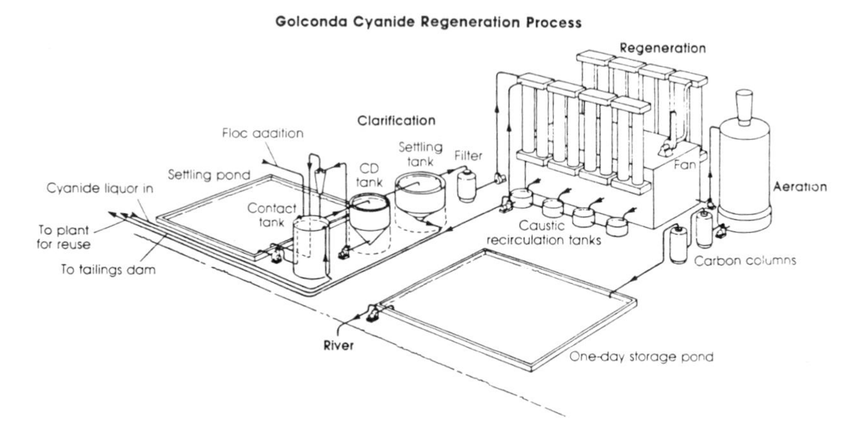

FIGURE 8-15. Flow diagram of the Golconda Cyanide Regeneration Process. Reprinted by permission. Golconda Engineering and Mining Services, New process regenerates cyanide from gold and silver leach liquors, *Eng. Min. J.*, June, 1988, 55.

stripper column. The cyanide (as HCN) is scrubbed from the acidic solution and carried by the air stream into an absorption column, inundated with down-flowing alkaline solution.

Destruction of Cyanide

Cyanide is bio-degradable, decomposing to carbon dioxide and ammonia. Cyanide, cyanate, and thiocyanate can be destroyed by ozone.

$$2CN^- + 2O_3 \rightarrow 2CNO^- + 2O_2$$
$$2CNO^- + O_3 + H_2O \rightarrow N_2 + 2HCO_3^-$$
$$2SCN^- + 2O_3 + 2H_2O \rightarrow 2CN^- + 2H_2SO_4$$

The biodegradation of cyanide is accelerated by sunlight and is hindered by low winter temperatures. The oxidation of cyanide complexed with some metal ion (Ni, Fe^{2+}, Fe^{3+}, Au, or Ag) has to be assisted by ultraviolet light, since their respective dissociation constants are very low (on the order of 10^{-31} to 10^{-42}).

Oxidation with chlorine can destroy free cyanides according to the following reactions:

$$10NaOH + 5Cl_2 \rightarrow 5\,NaOCl + 5NaCl + 5H_2O$$
$$2NaCN + NaOCl \rightarrow 2\,NaCNO + 2NaCl$$
$$2NaCNO + 3NaOCl + H_2O \rightarrow 2\,CO_2 + N_2 + 10\,NaCl + 2NaOH$$

$$2NaCN + 5Cl_2 + 8\,NaOH \rightarrow 2CO_2 + N_2 + 10\,NaCl + 4H_2O$$

References

Note: Sources with an asterisk (*) are suggested for further reading.

Ashurst, K. G., and N. P. Finkelstein. 1970. The influence of sulphydryl and cationic flotation reagents on the flotation of native gold. *J. S. Afr. I.M.M.* (70): 243-58.

Barsky, G., S. J. Swanson, and N. Hedley. 1934. Dissolution of gold and silver in cyanide solution. *Trans. A.I.M.E.* (112): 660-77.

Bodlander, G. 1896. Die chemie des cyanidverfahrens. *Z. Angew. Chem.* (9): 583-87.

Boonstra, B. 1943. Uber die losungsgeschwindigkeit von gold in kalium cyanid-losungen. *Korros. Metallschutz* (19): 146-51.

Christy, S. B. 1896. The solution and precipitation of the cyanide of gold. *Trans. A.I.M.E.* (26): 735-72.

*Cornejo, L. M., and D. J. Spottiswood. 1984. Fundamental aspects of the gold cyanidation process: A review. In *Mineral & Energy Resources*, Colorado School of Mines.

Deitz, G. A., and J. Halpern. 1953. Reaction of silver with aqueous solutions of cyanide and oxygen. *J. Met.* (5): 1109-16.

Dewhirst, R. F., S. P. Maulf, and J. A. Coetzee. 1984. Intensive cyanidation for the recovery of coarse gold. *J. S. Afr. I.M.M.* (84)6: 159-63.

Dorey, R., D. van Zyl, and J. Kiel. 1988. Overview of heap leaching technology. In *Introduction to Evaluation, Design and Operation of Precious Metal Heap Leaching Projects*, edited by D. van Zyl, et al.: 6.

Dorr, J. V. N., and F. L. Bosqui. 1950. *Cyanidation and Concentration of Gold and Silver Ores.* New York: McGraw-Hill, 238-78.

Elsner, L. 1846. Uber das verhalten verschredener metalle in einer brigen losung von cyankalium. *J. Prakt. Chem.* (37): 441-46.

Fink, C. G., and G. L. Putnam. 1950. The action of sulphide ion and of metal salts on the dissolution of gold in cyanide solutions. *Trans. S.M.E.-A.I.M.E.* (187): 952-55.

Finkelstein, N. P. 1972. The chemistry of gold compounds. In *Gold Metallurgy in South Africa*, edited by R. J. Adamson. Johannesburg: Chamber of Mines of S. Afr., 285-320.

GEMS (Golgonda Engineering & Mining Services). 1988. New process regenerates cyanide from gold and silver leach liquors. *Eng. Min. J.*, June: 55.

Habashi, F. 1967. Kinetics and mechanism of gold and silver dissolution in cyanide solution. *Bulletin 59.* Butte, MT: Montana Bureau Mines and Geology.

*Habashi, F. 1987. One hundred years of cyanidation. *C.I.M. Bull.* 80(905): 108-14.

Hallett, C. J., A. J. Mohemius, and D. G. C. Robertson. 1981. Oxygen mass transfer in Pachuca tanks. In *Extraction Metallurgy 81.* London: I.M.M., 21-23.

Janin, L., Jr. 1888. Cyanide of potassium as a lixiviant agent for silver ores and minerals. *Eng. Min. J.* (46): 548–49.

Janin, L., Jr. 1892. The cyanide process. *Mineral Industry* (1): 239–72.

Julian, H. F., and E. Smart. 1921. *Cyanidation of Gold and Silver Ores*. London: Griffin.

Kakovski, I. A., and A. N. Levedev. 1960. The influence of surface-active materials on the rate of dissolution of gold in cyanide solutions. *Doklady Physical Chemistry* (164):686–89.

MacArthur, J. S., R. W. Forrest, and W. Forrest. 1887/1889. Process obtaining gold and silver from ores. Brit. Patent 14174 (1887); U.S. Patent 403,202 (1889); U.S. Patent 418,137 (1889).

Mellor, J. W. 1923. Comprehensive Treatise on Inorganic and Theoretical Chemistry: Longmans, New York, vol. 3, p. 499.

Pietsch, H. B., W. Turke, and E. Bareuther. 1983a. Leaching of gold and silver. German Patent DE3126234.

Pietsch, H. B., W. Turke, and G. H. Rathje, 1983b. Research on pressure leaching of ores containing precious metals. *Erzmetall*: 261–65.

Randol. 1988. Vitox installation in an air-agitated tank. Randol Gold Forum 88. Golden, CO: Randol International Ltd., 192.

*Xue, T., and K. Osseo-Assare. 1985. Heterogeneous equilibria in the Au-CN-H_2O and Ag-CN-H_2O systems. *Metal. Trans. B* (16B): 455–63.

CHAPTER 9

Alternative Leaching Reagents for Gold

Sodium cyanide has been the preponderant leaching reagent for gold due to its excellent extractions from a great variety of ores and its low cost. Although cyanide is poisonous, it has safe industrial and environmental records at the high pH range required in cyanide leaching. In addition, the low concentrations of cyanide used in gold milling and heap leaching are quickly degradable. Although cyanide is a powerful lixiviant for gold and silver, it is not selective and forms complex compounds with a multitude of metal ions and minerals. (See the list of cyanicides in Figure 9-1.)

Gold cyanidation rates are relatively slow, and the industry has been searching for faster gold leaching reactions that are able to achieve a very high extraction of gold. Due to the high value of the yellow metal, even small improvements in recovery are always preferable to enhancements in the leaching rate.

In spite of significant R & D efforts, the list of alternative reagents (other than cyanide) for gold leaching is limited, and most of those reagents are addressed to refractory gold ores (non-amenable to simple cyanide leaching). Leaching of gold with acidic reagents can be advantageous when treating oxidized ores and concentrates, which are, as a rule, high lime consumers for cyanide leaching.

Thiourea Leaching of Gold

Thiourea leaching of gold ores was developed as an alternative to cyanidation by Soviet scientists (Plaskin & Kozhukhova, 1941, 1960) in the 1940s. Compared to cyanide as a gold lixiviant, thiourea has certain advantages.

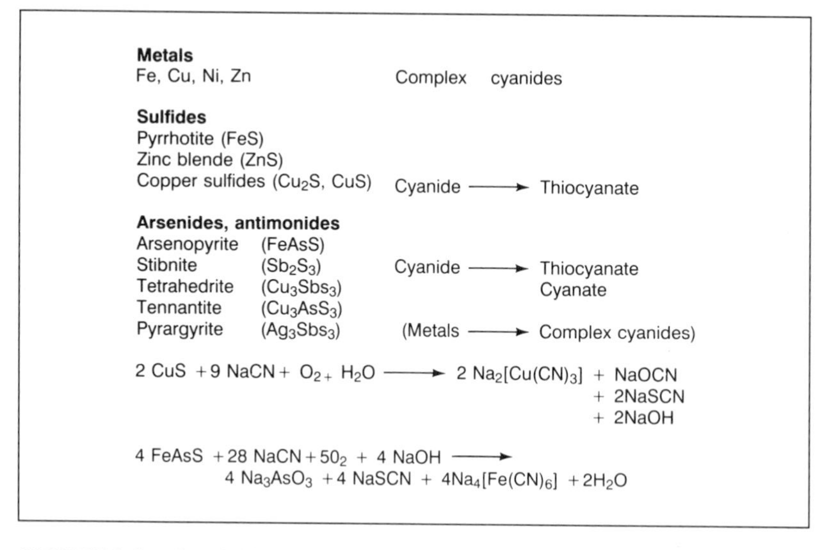

FIGURE 9-1. Cyanicides.

Reprinted by permission of VCH Publishers, Inc., 220 East 23rd Street, New York, NY 10010. From *Erzmetall* 39, Nr. 2, 1986 by R. G. Schulze.

- Low sensitivity to base metals (Pb, Cu, Zn, As, Sb)
- Low sensitivity to residual sulfur in calcines
- High gold recovery from pyrite and chalcopyrite concentrates
- Satisfactory recovery of gold from carbonaceous (refractory) ores (Chen et al., 1982)

Thiourea, $S{=}C\begin{matrix}\diagup NH_2\\ \diagdown NH_2\end{matrix}$ or $CS(NH_2)_2$, and gold form a single complex anion in an acid solution.

$$Au^\circ + 2CS(NH_2)_2 \rightleftharpoons Au[CS(NH_2)_2]_2^+ + e^- \quad E = -0.38V \quad (9.1)$$

This reaction is fast, and extractions of gold up to 99% can be achieved. The major disadvantage of thiourea gold leaching is higher reagent costs than in cyanide leaching, due to high consumptions of thiourea and acid (Groenewald, 1977). The use of thiourea has to be seriously considered in leaching gold ores containing significant proportions of cyanide consumers (see Figure 9-1).

Thiourea Leaching Chemistry

The gold leaching reactions with cyanide and with thiourea, shown in Figure 9-2, generate the following comments.

- The thiourea reaction employs condensed phase oxidants (ferric iron compounds, Fe^{+++}, or hydrogen peroxide), while cyanidation takes place in the presence of gaseous oxygen from the air. Hence, kinetically, the rate of thiourea leaching is a function of concentrations of both thiourea and the oxidant.

- Thiourea dissolves gold by forming a complex gold cation, while leaching with cyanide results in a complex gold anion.
- Thiourea disproportionates to polythionic acids and cyanimide.

At anodic potentials greater than 0.30V, thiourea is oxidized to formamidine disulfide (Groenewald, 1976).

$$2CS(NH_2)_2 \rightleftharpoons NH_2(NH)C\text{–}S\text{–}S\text{–}C(NH)NH_2 + 2H^+ + 2e^- \quad (9.2)$$

Formamidine disulfide is an active oxidant and, as such, promotes gold dissolution. The combination of reactions 9.1 and 9.2 yields the suggested overall equation for gold dissolution (Pyper & Hendrix, 1981).

$$2Au^\circ + 2CS(NH_2)_2 + NH_2(NH)CSSC(NH)NH_2 + 2H^+ \rightleftharpoons 2Au(CS(NH_2)_2)_2 \quad (9.3)$$

$$E_\circ = +0.04V, \Delta G^\circ = -1845 \text{ cal/mol}$$

For satisfactory gold and silver extractions the consumption of thiourea has to be high (Figure 9-3).

In a sulfate medium containing ferric ion, the oxidant is the Fe^{3+}/Fe^{2+} couple.

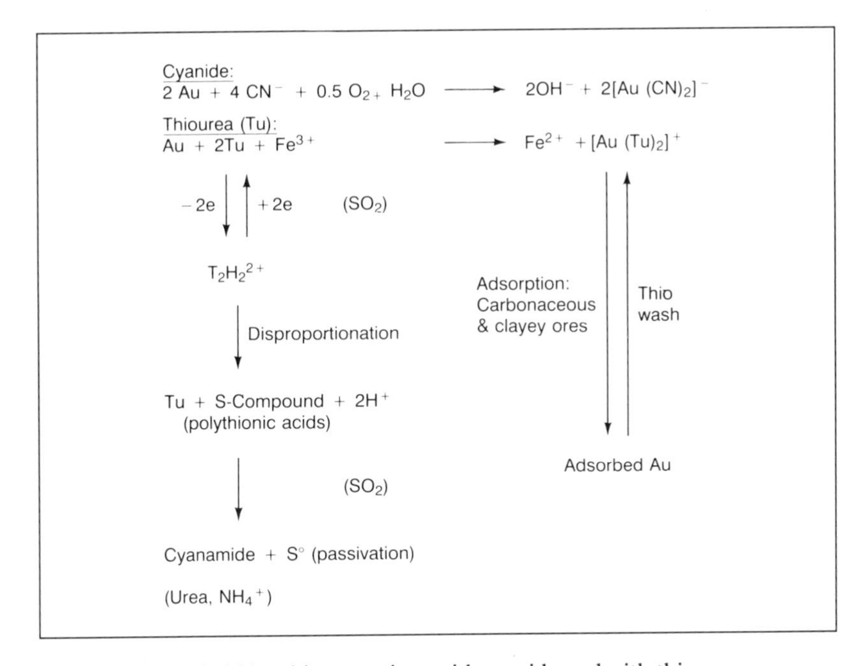

FIGURE 9-2. Gold leaching reactions with cyanide and with thiourea.

Reprinted by permission of VCH Publishers, Inc., 220 East 23rd Street, New York, NY 10010. From *Erzmetall* 39, Nr. 2, 1986 by R. G. Schulze.

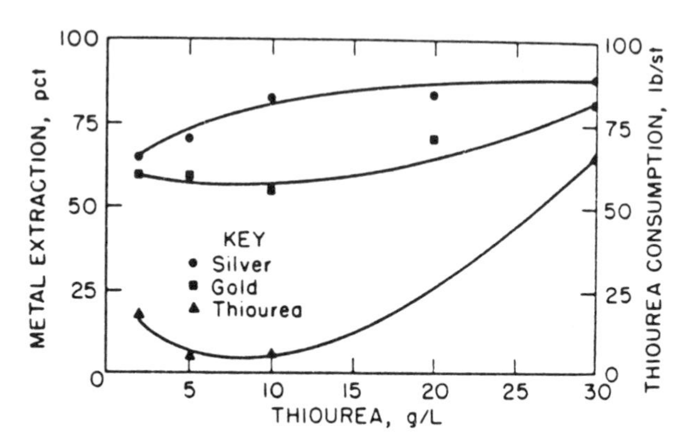

FIGURE 9-3. Gold and silver extraction and thiourea consumption, as a function of reagent concentration.

Reprinted by permission. R. G. Sandberg and J. L. Huiatt in *The Minerals, Metals and Materials Society Paper No. A86-50* (Warrendale, PA), 1986.

$$Fe^{3+} + e^{-} \rightleftharpoons Fe^{2+} \qquad E_{\circ} = +0.77V \tag{9.4}$$

The combination of equations 9.4 and 9.1 yields the reaction of gold dissolution with ferric ion as the oxidant.

$$Au^{\circ} + Fe^{3+} + 2CS(NH_2)_2 \rightleftharpoons Au[CS(NH_2)_2]_2^{+} + Fe^{2+} \tag{9.5}$$
$$E_{\circ} = +0.39V, \Delta G^{\circ} = -8994 \text{ cal/mol}$$

A review of the kinetics of the thiourea dissolution of gold by Pyper and Hendrix (1981) concluded the following. (See Fig. 9-5.)

- The use of ferric ion in sulfuric acid is the most effective system. (Ferric ion is beneficial in slowing the oxidation/degradation reaction, 9.2.)
- The leaching rate is dependent on thiourea and oxidant concentrations.
- Ferric ion, although beneficial in slowing thiourea oxidation, ties up thiourea in iron-thiourea complexes.
- The rate of gold dissolution is strongly dependent on pH (Figures 9-4 and 9-5, optimum range is pH less than 1.0).

In addition to the above, Hiskey (1984) emphasized that the use of thiourea for the treatment of many precious metals containing ores and concentrates is uneconomic due to excessive reagent consumption. Complexes with impurity metals and oxidation to useless byproducts aggravate the consumption of thiourea.

The oxidation of thiourea (9.2) has more consequences than the loss of reagent. Elemental sulfur is also generated in fine adhesive form, possibly from decomposition of formamidine disulfide. Such fine adhesive sulfur

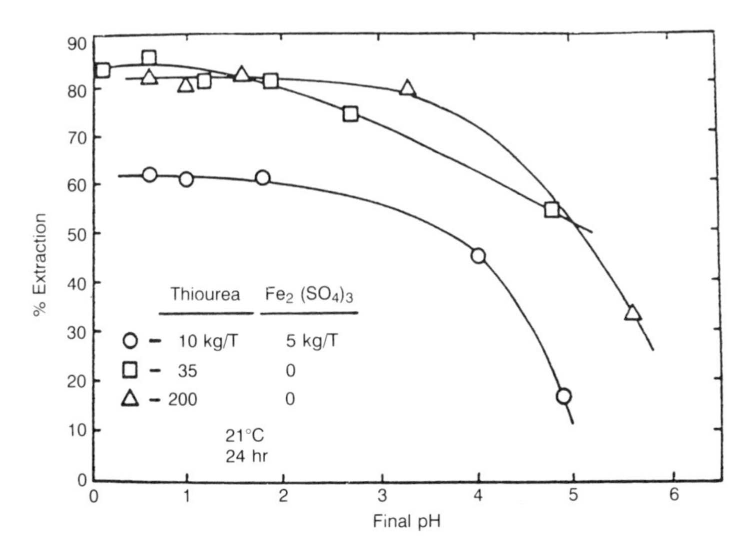

FIGURE 9-4. Effect of final pH on gold extraction by thiourea.

Reprinted by permission of The Institution of Mining and Metallurgy. R. A. Pyper and J. L. Hendrix in *Extraction Metallurgy 81*, The Institution of Mining and Metallurgy (London), 1981.

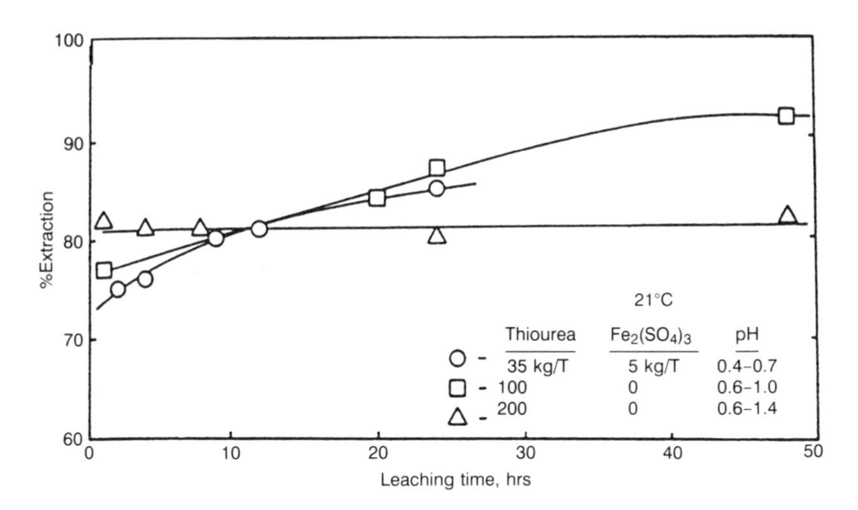

FIGURE 9-5. Effect of leaching time on gold extraction by thiourea.

Reproduced by permission of The Institution of Mining and Metallury. R. A. Pyper and J. L. Hendrix in *Extraction Metallurgy 81*, The Institution of Mining and Metallurgy (London).

may cover gold surfaces and passivate them (Schulze, 1984). Sulfur dioxide (SO_2) acts as a highly specific reductant that does not react with other oxidants as long as formamidine disulfide is present. Schulze (1984) has proposed bubbling SO_2 during thiourea leaching of precious metals to prevent decomposition of formamidine disulfide and coating of gold surfaces by fine adhesive sulfur.

The advantage of an acid prewash of the ore has been advocated to reduce thiourea consumption (Chen et al., 1982; Groenewald, 1976), but Pyper and Hendrix (1981) did not observe such an effect with Carlin ore. Acid prewash may be advisable with high acid-consuming ores.

Recovery of Gold from Thiourea Solution

Most of the methods and means available for the recovery of gold from cyanide solutions may be applicable for its recovery from thiourea solutions. However, further R & D on the subject is required to establish the optimum method.[1] Schulze (1986) proposed adsorption of gold from thiourea solutions by activated carbon (see Figure 9-6).

Cementation of Gold. Aluminum metal has been used to recover gold from thiourea pregnant solution (van Lierde et al., 1982). The results reported (2% Au in the cement) were not appealing. The U.S. Bureau of mines nevertheless reported a 98.99% recovery of gold and silver from thiourea pregnant solutions with aluminum powder, with consumption of 6.4 kg of aluminum per kilogram of precious metals (Simpson et al., 1984).

Iron powder was considered as a precipitant of gold from thiourea solutions. However, the Chinese Institute of Non-ferrous Metallurgical Industry reported better cementation results with lead-in-pulp than with iron-in-pulp (Wen, 1982). Fine lead powder was used in HCl-thiourea solutions by Tataru (1968).

$$\frac{n}{2}Au[CS(NH_2)_2]_2\,Cl + Pb \rightarrow Pb[CS(NH_2)_2]_nCl_2 + \frac{n}{2}Au + \left(\frac{n}{2} - 1\right)Cl_2$$

Activated carbon adsorption of gold and silver from thiourea solution was studied as early as 1968 by Russian scientists (Lodeishchikov et al., 1968). Recovery of gold from thiourea pregnant solution by activated carbon (Figure 9-7) has been practiced at New England Antimony Mines in New South Wales. High carbon loadings were reported, and gold was sold as a concentrate on carbon, 6-8kg/ton (Hisshion & Waller, 1984).

High recovery of gold from thiourea pregnant solution can be obtained with activated carbon. A satisfactory maximum recovery, say 90% from a

[1]Most of the work done on the gold-thiourea system is on leaching; very little is on recovery from thiourea solutions.

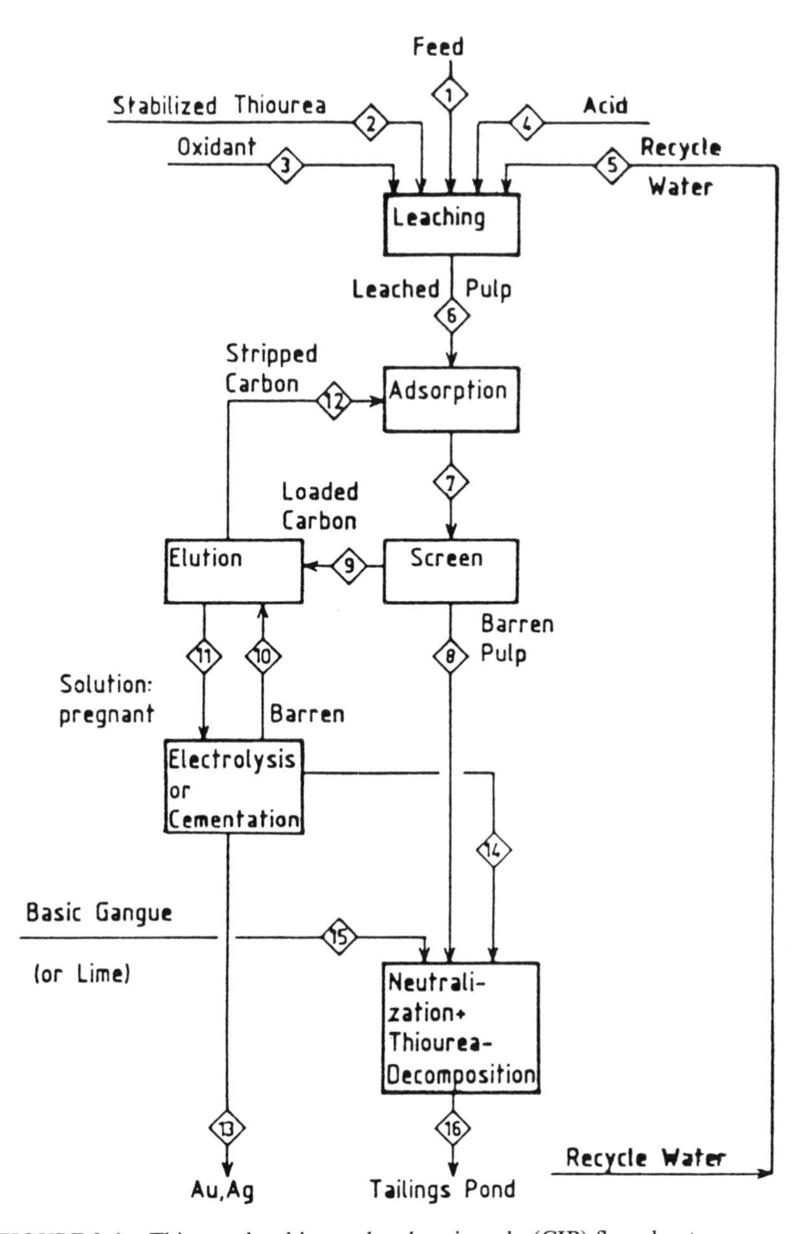

FIGURE 9-6. Thiourea leaching and carbon-in-pulp (CIP) flow sheet.
Reprinted by permission of VCH Publishers, Inc., 220 East 23rd street, New York, NY 10010. From *Erzmetall* 39, Nr. 2, 1986 by R. G. Schulze.

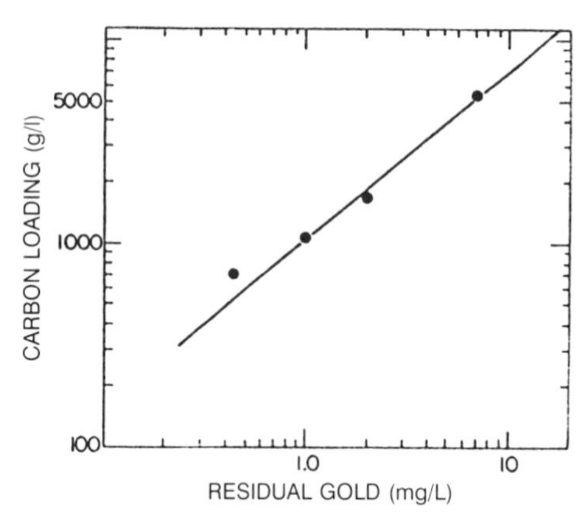

FIGURE 9-7. Carbon loading from thiourea pregnant solution (feed: 32 mg Au/1, 60 min. contact, room temp.).

Reprinted by permission. G. Deschenes, In *Gold Metallurgy* (CIM 26th Conference of Metallurgists, 1987). The Canadian Institute of Mining and Metallurgy (Montreal), 1987.

dilute pregnant solution (27 ppm Au) with 20 g/l of carbon can be penalized nevertheless by a high loss of thiourea (up to 30%) adsorbed on the carbon (Schulze, 1986). Thiourea adsorbed on carbon can be partially recovered by warm-water wash, as a rather dilute solution (Figure 9-8).

Ion-Exchange Resins. Groenewald (1977) has reported that strong cation-exchange resins are as effective as charcoal in extracting the gold(I)-thiourea complex. Higher gold recovery from thiourea solutions can be obtained with cationic resins than with activated carbon. An alkaline sodium thiosulfate solution and a solution of bromine in hydrochloric acid have been reported as the "most promising" eluting solutions for the resins. A flow sheet employing thiourea and cyanide solutions for the elution of silver and gold is shown in Figure 9-9.

Cation-exchange resins have given better adsorption of the gold(I)-thiourea complex than anion-exchange resins, as should be expected. Gold concentration and pH affect the rate of exchange and the recovery (loading). Inorganic ion exchangers (zirconium phosphate, tin phosphate, molybdenum, and titanium ferrocyanide) have been tested in the U.S.S.R. since the early 1970s, but they have not been proposed for practical applications so far.

Electrowinning. The gold(I)-thiourea complex can be reduced electrolytically, according to the following reaction (Groenewald, 1977).

$$Au[CS(NH_2)_2]_2^+ + e^- \rightarrow Au + 2CS(NH_2)_2$$

This is a diffusion-controlled reaction with a cathodic potential range of -0.15 to -0.35V. Groenewald reported that thiourea, as such, does not

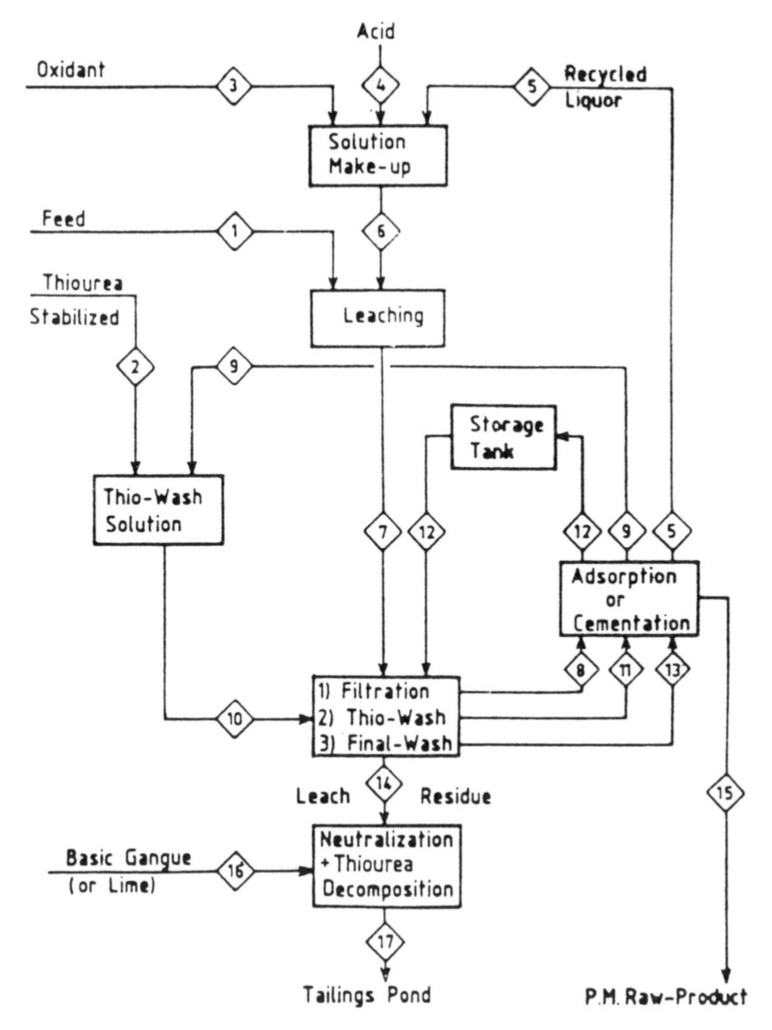

FIGURE 9-8. Flow sheet of leaching gold ore with thiourea (gold and silver are to be recovered from the loading adsorbents).

Reprinted by permission of VCH Publishers, Inc., 220 East 23rd Street., New York, NY 10010. From *Erzmetall* 39, Nr. 2, 1986 by R. G. Schulze.

contribute to the cathodic reaction of gold, but its oxidation product, formamidine disulfide, can be reduced on the cathode surface.

Separate anodic and cathodic compartments—through the use of ionic diaphragms—are needed in order to prevent two conditions:

- The anodic decomposition of thiourea and the resulting contamination of the gold deposit with sulfur
- The dissolution of (deposited) gold by anodic products

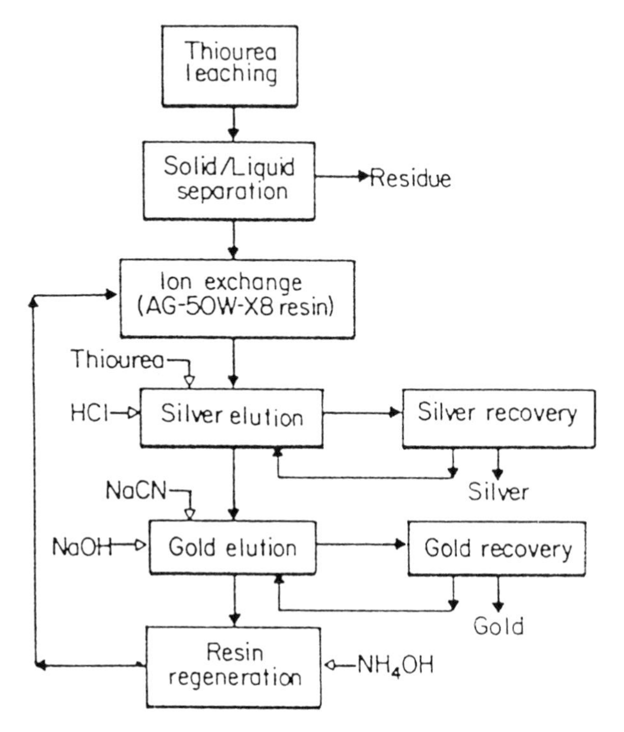

FIGURE 9-9. Recovery of gold and silver from acidic thiourea leach solution by ion-exchange resin.

Reprinted by permission. W. W. Simpson, L. Peterson, and R. G. Sandberg, Gold and silver recovery from thiourea leach solutions. U.S. Bureau of Mines, 1986.

Low current density gives better current efficiency. The use of a lead anode minimizes the thiourea decomposition (oxidation of thiourea is minimized mainly by isolating it from the anode compartment). Maximum deposition of gold is obtained with high catholyte circulation rates (Bek and Zamyatin, 1984).

Zamyatin et al. (1983) have developed a two-stage electroextraction process from thiourea solutions of silver and gold. The first cell (stage) uses titanium cathodes and recovers most of the precious metal content. The second cell uses porous graphite (extended-surface) cathodes to recover the low content of gold and deliver a tail solution with 0.03 mg/l.

Thiosulfate Leaching of Gold

Gold dissolves slowly in alkaline thiosulfate, and the process is enhanced by the presence of copper ions. The extraction of gold from auriferous

TABLE 9-1. Comparison of cyanidation and thiourea leaching.

	Cyanidation	*Thiourea leaching*
Reagents	NaCN, CaO	$CS(NH_2)_2$, H_2SO_4, SO_2, $NaHSO_3$
Main reagent min. consumption[a] g/g Au	0.50	0.77
Oxidants	O_2 from the air	Fe^{3+}, H_2O_2
pH	10–11	1-2
Feed	Most ores and concentrates	Special ores[b] and concentrates
Product	Anionic gold complex	Cationic gold complex
Gold recovery from solution	Zn cementation, carbon, IX resin	Al cementation, carbon, IX resin
Gold extraction	90–95%	95–98%
Rate of extraction	Multihour leaching	Significantly faster than cyanidation
Applicability to heap leaching	Applicable	Not applicable

[a]Assuming no side reactions.
[b]Thiourea is used with recalcitrant—almost refractory—gold ores and concentrates.

silver ores, when these were treated by chlorination, depended on the formation of double thiosulfate salts (Rose & Newman, 1937).

$$2AuCl_3 + 8Na_2S_2O_3 \rightarrow Au_2S_2O_3.3Na_2S_2O_3 + 2Na_2S_4O_6 + 6NaCl$$

The double thiosulfate may be separated with the addition of alcohol. The solubility of gold in sodium thiosulfate is greatly accelerated by ferric chloride and other oxidizing agents (Rose & Newman, 1937).

Kerley (1981) proposed the stabilization of thiosulfate during leaching of gold with the addition of SO_2 or bisulfite (HSO_3^-):

$$Au + 2Na_2S_2O_3 + H_2O \rightarrow Na_3Au(S_2O_3)_2 + NaOH + 0.5H_2$$

Kim and Sohn (1989) studied the effects of pH, temperature, and pulp density on the rate of leaching a gold ore with ammoniacal thiosulfate solution. Von Michaelis (1987) reports that there are no commercial-scale gold leaching operations using thiosulfate, although it was tested for "at least one *in situ* leach project."

Leaching with Halogens and Halides

With the exception of fluorine, halogens[2] have been tested and/or used for the extraction of gold. Almost any chloride, bromide, or iodide will dissolve gold in the presence of an oxidizing agent. The classical example is the dissolution of gold in *aqua regia*.

$$Au + HNO_3 + 4HCl \rightarrow HAuCl_4 + NO + 2H_2O$$

Chlorine can be generated in slurries and solutions by electrolysis of NaCl solution or by addition of MnO_2 to hydrochloric acid.

$$MnO_2 + 4HCl \rightarrow MnCl_2 + 2H_2O + Cl_2$$

Gold is leached rapidly by chlorine at low pH.

$$2Au + 3Cl_2 \rightleftharpoons 2AuCl_3$$

Filmer et al. (1984), after comparing the economics of leaching with chlorine or thiourea, recommended testing the acidified chlorine leaching for ores containing less than 0.5% sulfur (with higher sulfide content the consumption of chlorine cannot be afforded).

Bromine was first described as a solvent for gold in 1846 (von Michaelis, 1987). The rate of leaching gold by bromine is greatly enhanced when a protonic cation (e.g., NH_4^+) and an oxidizing agent are added (Kalocsai, 1984). Bromine can be added to the slurry (as bromide) along with chlorine or hypochlorite (as oxidants), which convert the bromide to bromine.

$$2Br^- + Cl_2 \rightleftharpoons 2Cl^- + Br_2$$
$$2Br^- + ClO^- + 2H^+ \rightleftharpoons Br_2 + Cl^- + H_2O$$

The free bromine, after reaction with gold and with gangue species such as sulfides or carbon, is converted to bromide ions.

Iodine forms the most stable gold complexes of all the halogens. Although the redox potential at which gold dissolves in an iodine solution is about half that of dissolution in acidified chlorine, it is considerably higher than in cyanide solution.

Iodine is an expensive element, and its future as an alternative solvent for gold is not bright. Iodine leaches gold over a wide pH range. Iodine reacts with sulfides below pH 7, and it is converted to iodide ions (I^-) without combining with sulfur. McGrew and Murphy (1979) demonstrated that iodide ions can be regenerated electrolytically to iodine.

[2]See the numerical properties of halogens and their tabulated characteristics in Appendix H.

Miscellaneous Solvents of Gold. *Sulfocyanides, ferrocyanides*, and some other double cyanides dissolve gold either at ordinary or elevated temperatures.

$$3Au + K_4Fe(CN)_6 + 2H_2O + O_2 \rightarrow 3KAu(CN)_2 + Fe(OH)_3 + KOH$$

References

Note: Sources with an asterisk (*) are recommended for further reading.

Bek, R. Y., and A. P. Zamyatin. 1984. Experimental study of the factors governing the efficiency of porous cathodes in metal extraction from dilute solutions. *Electrokimiya* 20(6): 854–57.

Chen, C. K., T. N. Lung, and C. C. Wan. 1982. A study of the leaching of gold and silver by acidothioureation. *Hydromet*. 5: 207–12.

Deschenes, G. 1987. Investigation of the potential techniques to recover gold from thiourea solution. In *Gold Metallurgy*, edited by R. S. Salter, D. M. Wyslouzi, and D. W. McDonald. New York: Pergamon Press, 359–77.

Filmer, A. O., P. R. Lawrence, and W. Hoffman. 1984. A Comparison of Cyanide, Thiourea and Chlorine as Lixiviants for Gold. Australian Inst. of Mining and Metallurgy, Regional Conference: Proceedings on Gold Mining, Metallurgy and Geology, October 1984, pp. 1-8.

Groenewald, T. 1976. The dissolution of gold in acidic solutions of thiourea. *Hydrometallurgy* #1, 277–90.

Groenewald, T. 1977. Potential applications of thiourea in the processing of gold. J. S. Afr. I.M.M. 77 (11): 216–23.

Hiskey, J. B. 1984. Thiourea leaching of gold and silver—Technology update and additional applications. *Miner. & Metal. Proc*. November: 173–79.

Hission, R. J., and C. G. Waller. 1984. Recovering gold with thiourea. *Min. Mag*., September: 237–43.

Kalocsai, G. I. Z. 1984. Improvements in or relating to the dissolution of noble metals. Austral. Provisional Patent 30281/84.

Kerley, B. J. 1981. Recovery of precious metals from difficult ores. U. S. Patent 4,269,622, May 26.

Kim, S. K., and Yu S. Sohn. 1989. Extraction of gold from a gold ore by ammoniacal thiosulfate leaching. Paper presented at the 1989 TMS annual meeting, 27 Feb. to 2 March, at Las Vegas, NV.

Lodeishchikov, V. V., et al. 1968. Use of thiourea as a solvent for ore gold. *Irkutsk Gos. Hauch. Issled. Inst. Redk. Tsvet. Metal.*, No. 19, 72–84. (Through CA *71*:15, 23lr.)

McGrew, K. J., and J. W. Murphy. 1979. Iodine leach for the dissolution of gold. U.S. Patent 4,557,795, 10 December.

Plaskin. I. N., and M. Kozhukhova. 1941. The solubility of gold and silver in thiourea. *Dok. Ak. Nauk. SSR 31*: 671–74.

Plaskin, I. N., and M. Kozhukhova. 1960. Dissolution of gold and silver in solutions of thiourea. *Sb. Nauch. Tr. Ins. Tsvt. Met.* 33: 107.

*Pyper, R. A., and J. L. Hendrix. 1981. Extraction of gold from finely disseminated gold ores by use of acidic thiourea solution. In *Extractive Metallurgy 81*. London: I.M.M., 57–75.

Rose, T. K., and W. A. L. Newman. 1937. *The Metallurgy of Gold*. Philadelphia, PA: J. B. Lippincott Co., 18, 76.

Sandberg, R. G., and J. L. Huiatt. 1986. Recovery of silver, gold and lead from a complex sulfide ore, using ferric chloride, thiourea and brine leach solutions. Paper presented at the 115th TMS Annual Meeting, 2–6 March, at New Orleans, LA.

Schulze, R. G. 1984. New aspects in thiourea leaching of precious metals. *J. Met.*, June: 62–65.

Schulze, R. G. 1986a. Increasing the amenability of gold ores for thiourea leaching. Paper presented at the 115th TMS Annual Meeting, 2–6 March, at New Orleans, LA.

Schulze, R. G. 1986b. Thiourea leaching of precious metals. *Erzmetall.* 39(2): 57–59.

Simpson, W. W., L. Peterson, and R. G. Sandberg. 1984. Gold and silver recovery from thiourea leach solutions. Paper presented at the Pacific Northwest Metals and Miner. Conference, April 1984, at Portland, OR.

Tataru, S. 1968. Precipitation par cementation de l'or en solutions acid. *Rev. Roum. Chim.*: 1043–49.

van Lierde, A., Ollivier, P., and Lesoille, M. 1982. Development du nouveau procede de traitement pour le mineral Salsigne. *Ind. Min., Les Tech*. 1a: 399–410.

*von Michaelis, H. 1987. The prospects for alternative leach reagents (for gold). *Eng. Min. J.*, June: 42–47.

Wen, C. D. 1982. Studies and prospects of gold extraction from carbon bearing clayey gold ore by the thiourea process. *Proceedings XIV Internat. Min. Proc. Congress*, October 1982: 17–23.

Zamyatin, A. P., A. F. Zherebilov, and V. K. Varentov. 1983. Two-stage electroextraction of precious metals from thiourea solutions. *Sov. J. Non-ferrous Met*. 6: 4–46.

CHAPTER 10

Recovery of Gold from Solutions

Cyanidation has been used to extract gold (and silver) from ores, concentrates, and calcines since the 1890s. The precipitation of gold from cyanide solutions by zinc cementation was patented in 1884 and was applied industrially as early as the cyanidation process.

Some interest in the adsorption of gold from solutions by activated carbon can be traced as far back as 1880, when W. N. Davis patented the use of wood charcoal for the recovery of gold from chlorination solutions. Soon after the cyanidation patent of MacArthur and the Forrest brothers (see Chapter 8), the recovery of gold from cyanide solutions by wood charcoal was patented by W. D. Johnson (1894).

The most significant characteristic of gold cyanide pregnant solutions is their gold content, which depends on the grade of the leach feed and the leaching system itself (milling with agitation or coarse ore leaching by percolation). The pregnant solutions obtained by agitation may contain from 2 to 15 ppm gold, whereas the leach liquor obtained by percolation in heap leaching may contain 1 ppm gold or less. The free CN^- ion concentration is generally in the 20–200 ppm range. Gold can be recovered from pregnant solutions by one (or a combination) of the following four processes:

1. Zinc cementation
2. Activated carbon adsorption
3. Ion-exchange/solvent extraction
4. Electrowinning

Zinc Cementation

The concept of cementation was used to recover gold from solution in the nineteenth century:

$$Au^{+++} + Fe \rightarrow Au + Fe^{+++}$$

The electrochemical order of metals in cyanide solution dictates their relative solubilities in that solvent. The published determinations of the electrochemical order of metals in potassium cyanide solution indicate the following sequence, from positive to negative: Mg, Al, Zn, Cu, Au, Ag, Hg, Pb, Fe, Pt. Any metal in this sequence would tend to dissolve in cyanide solution more readily than the metal to its right, and would displace those metals from solution and precipitate them. Thus copper will precipitate gold, silver, mercury, etc. Magnesium or aluminum will precipitate gold and silver more readily than will zinc.

When cyanidation was adopted on a large scale, as in the extraction process of gold from its ores in the 1890s, MacArthur used zinc shavings to precipitate gold from cyanide solutions. Addition of soluble lead salts—in a controlled concentration[1]—to create a zinc-lead alloy on the zinc particles proved beneficial because it inhibits passivation of zinc surfaces. Gold precipitation became more efficient when the use of zinc dust (yielding an immensely extensive zinc surface) was introduced by C. W. Merrill in 1904 (Habashi, 1987).

$$2Au(CN)_2^- + Zn \rightarrow 2Au + Zn(CN)_4^{2-} \qquad (10.1)$$

$$2Au(CN)_2^- + Zn + 3OH^- \rightarrow 2Au + HZnO_2^- + 4CN^- + H_2O \qquad (10.2)$$

Zinc can also react in alkaline cyanide solutions to produce hydrogen.

$$Zn + 4CN^- + 2H_2O \rightarrow Zn(CN)_4^{2-} + 2OH^- + H_2 \qquad (10.3)$$

$$Zn + 2H_2O \rightarrow HZnO_2^- + H^+ + H_2 \qquad (10.4)$$

Hence, it is possible that the precipitation of some gold may not proceed directly (reactions 10.1 and 10.2), but through the intermediate formation of hydrogen.

$$Au(CN)_2^- + H_2 \rightarrow Au + 2H^+ + 2CN^- \qquad (10.5)$$

It is known, nevertheless, that gold is not precipitated from cyanide solutions by hydrogen at atmospheric pressure. At higher pressures and temperatures, the reduction of aurous ions by hydrogen takes place at a

[1]About 10 ppm; with high concentrations of lead, say, 100 ppm, the recovery of gold is reduced.

relatively low rate. Barin et al. (1980) proposed the following overall chemical reaction for the cementation of gold by zinc.

$$Zn + Au(CN)_2^- + H_2O + 2CN^- \rightarrow Au + Zn(CN)_4^{2-} + OH^- + 0.5H_2 \quad (10.6)$$

Cementation is a hetergeneous redox system in which aurocyanide and cyanide ions have to transfer to the zinc surface; the reactants have to be adsorbed on the zinc surface; the reduction reaction takes place at the zinc surface; the products of the reaction are desorbed; and the products of the reaction are transported to the bulk of the solution. The rate of any and all of the preceding steps is proportional to the available zinc surface, as the introduction of zinc dust instead of zinc shavings has proven in practice.

Obviously, the slowest of the above steps will control the rate of gold cementation. Barin et al. (1980) confirmed Nicol et al.'s (1979) experimental conclusion that the gold cementation rate is controlled by the transfer rate of $Au(CN)_2^-$ ions (the first step above).

Finkelstein (1972) has discussed the potential reactions of the system $Zn–H_2O–CN$ and presented the respective equilibria in the potential *vs.* pH diagrams (Figure 10-1). The reduction of the aurocyanide ions by zinc on an industrial scale was further improved when T. B. Crowe removed air and the dissolved oxygen from the pregnant solution, by vacuum, before introducing the zinc dust. The presence of oxygen in solution decelerates the reduction reaction and increases the consumption of zinc. After deaeration in a Crowe vacuum tower, typical pregnant solutions contain only 0.6–1.3 ppm of oxygen.

Effects of Solution Composition

A minimum "critical" cyanide concentration is required for gold cementation [reported as 0.002 M (0.1 g/l NaCN) by Nicol et al. (1979) and 0.035 M (1.7 g/l NaCN) by Barin et al. (1980)]. The gold concentration has a direct influence on the rate of cementation, which is essentially a first-order reaction controlled by the mass transfer of $Au(CN)_2^-$ ions. Although a change in the pH of the solution within the range of 9 to 12 has no appreciable effect on the rate of cementation, higher pH may cause the formation of $Zn(OH)_2$ as an intermediate compound (Figure 10-2). Precipitation of $Zn(OH)_2$ tends to take place at the surface of the zinc particles and can retard or even stop the zinc cementation.

Finkelstein (1972) reported that sulfide, sulfate, thiosulfate, and ferrocyanide anions in the concentration range 10^{-3}M–10^{-2}M may reduce the gold recovery by 1–2% from 10^{-3}M cyanide solutions. The sulfate ion may precipitate as gypsum on zinc particles and reduce their reactivity. Nicol et al. (1979) found that sulfide ions can passivate the zinc surface even at

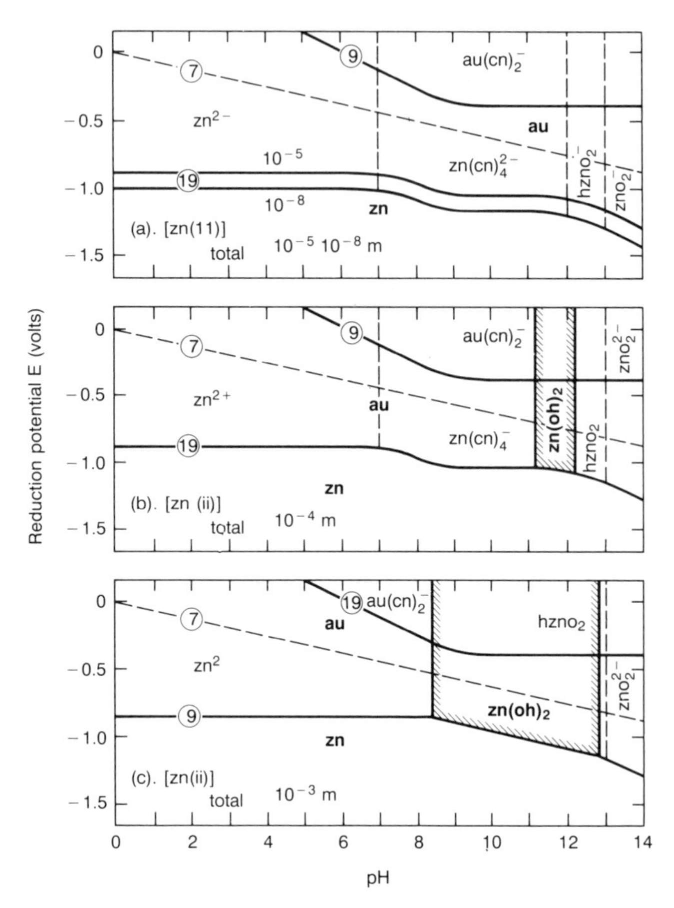

FIGURE 10-1. Potential-pH diagram for the system Zn-H_2O-CN at 25° C (including equilibria between gold, cyanide, and water).

Reprinted by permission. N. P. Finkelstein, Recovery of gold from solutions. In *Gold Metallurgy in South Africa*, edited by R. J. Adamson. Chamber of Mines of South Africa (Johannesburg), 1972.

concentrations as low as 1×10^{-4} M, whereas thiosulfate at 1 to 5×10^{-4} M slightly accelerates the cementation.

Low concentrations of copper cyanide (6×10^{-3}M), antimony (1.7×10^{-4} M), or arsenic (2.3×10^{-4} M) will stop the cementation. Even at concentrations of 10^{-6} M, those ions will significantly reduce the gold recovery.

Efficient Cementation Practice

The cardinal requirements for the efficient cementation of gold from cyanide solution with the addition of zinc powder are as follows.

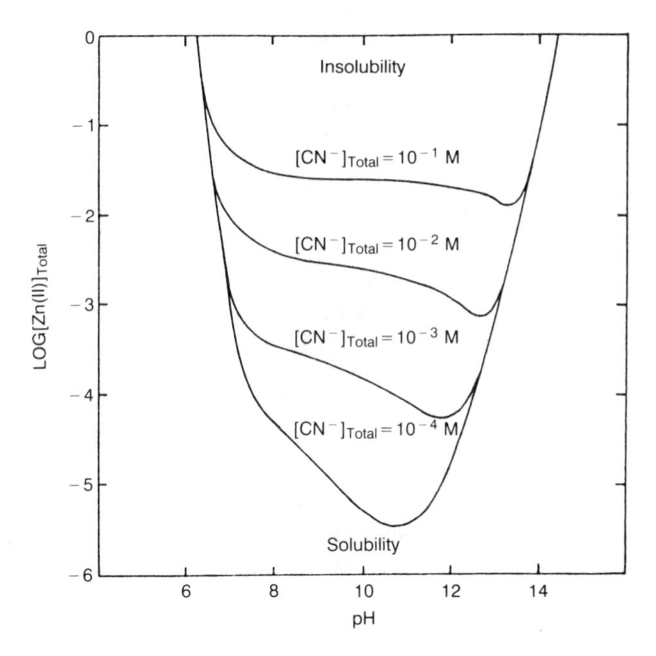

FIGURE 10-2. Domains of solubility and insolubility of Zn(II) in cyanide solutions at 25° C.

Reprinted by permission. N. P. Finkelstein, Recovery of gold from solutions. In *Gold Metallurgy in South Africa*, edited by R. J. Adamson. Chamber of Mines of South Africa (Johannesburg), 1972.

1. The pregnant cyanide solution should
   - be clarified to less than 5 ppm solids
   - be deaerated to about 1 ppm oxygen
   - have a free cyanide concentration higher than 0.035 M
   - have a pH in the range of 9 to 11 (with adequate lime addition)
   - contain an adequate proportion of lead nitrate (about 0.5 to 1 part of lead nitrate to 1 part of gold) and not a high concentration of lead.
2. Adequate addition of high-purity zinc dust (5 to 12 parts of zinc per part of gold).

The Merrill-Crowe Process

The addition of soluble lead salts, the use of zinc dust, and the deoxygenation of the pregnant solution (Crowe, 1918) were incorporated in an industrial technique of recovering gold from cyanide solutions, the Merrill-Crowe Process, developed in the United States. The process consists of four basic steps.

1. Clarification of the pregnant cyanide solution
2. Deaeration
3. Addition of zinc powder and lead salts
4. Recovery of zinc-gold precipitate

Perfect clarification of the pregnant solution is the most important single factor in obtaining efficient precipitation of gold. The cloudy pregnant solution (from a counter-current decantation, a CCD series of thickeners, or rotary filters) is pumped to a storage tank, which also serves as a settler. Precoated leaf filters or candle filters are used for the final clarification of the gold pregnant solution (Figures 10-3, 10-4, and 10-5). Precoat pressure clarification provides the best operational results. In addition to the removal of colloid solids, partial adsorption of the dissolved oxygen takes place during the flow through the coat of diatomaceous earth of precoated filters. The solution has to be crystal clear before contacting the zinc. Fine silica in suspension may coat zinc surfaces and reduce its reactivity.

Efficient and complete precipitation of gold (and silver) is enhanced by removal of the dissolved oxygen in the Crowe vacuum tower. Splash plates

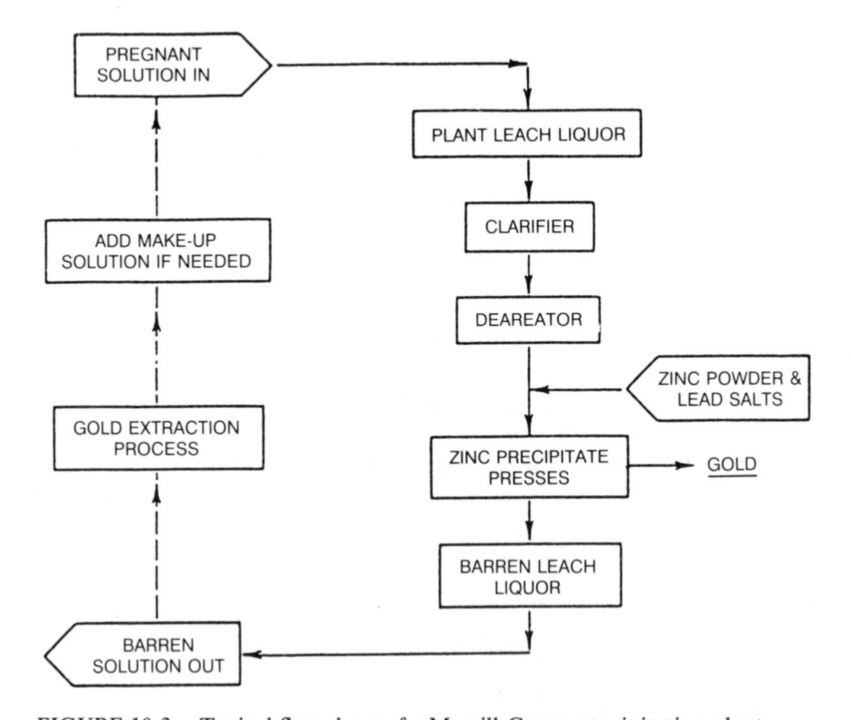

FIGURE 10-3. Typical flow sheet of a Merrill-Crowe precipitation plant.

Reprinted by permission. O. A. Muhtadi, in *Introduction to Evaluation, Design and Operation of Precious Metal Heap Leaching Projects*, edited by D. J. A. van Zyl, I. P. G. Hutchinson, and J. E. Keil. Society for Mining Engineers (Littleton, CO), 1988.

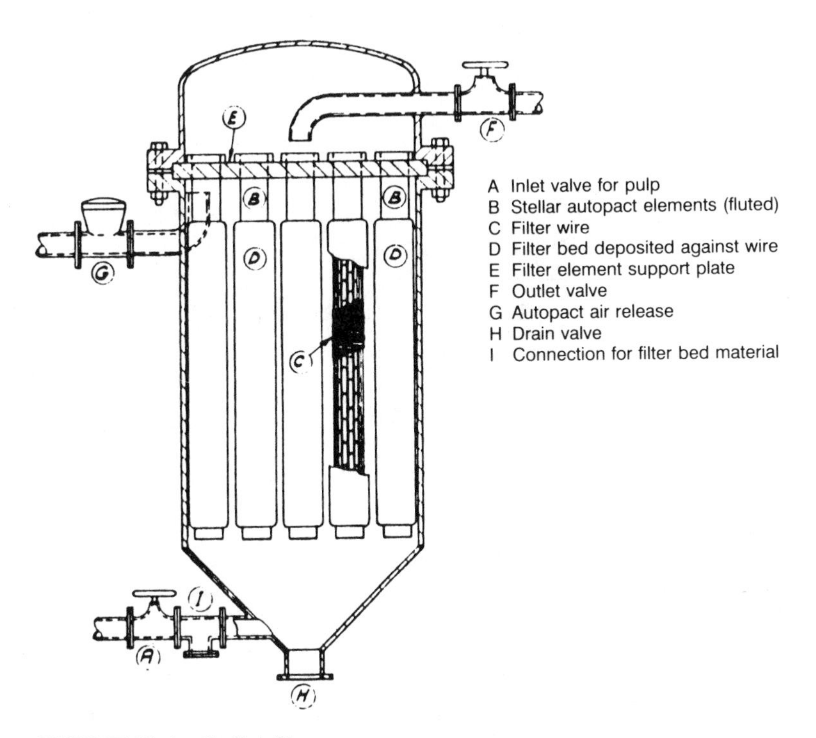

FIGURE 10-4. Stellar filter.

Reprinted by permission. N. P. Finkelstein, Recovery of gold from solutions. In *Gold Metallurgy in South Africa*, edited by R. J. Adamson. Chamber of Mines of South Africa (Johannesburg), 1972.

and cascade trays within the tower increase the surface of the solution, thus contributing to complete deaeration by the applied vacuum. Even minute traces of oxygen have a deleterious effect on gold precipitation; they passivate zinc surfaces and contribute to the dissolution of precipitated gold. The hydrogen evolved during the dissolution of zinc (reactions 10.3, 10.4, and 10.6) nullifies the effects of any traces of oxygen remaining in solution.

Clarification and deaeration of the pregnant solution are followed by continuous addition of zinc dust and the precipitation of gold (without exposing the solution to atmospheric air). The deaerated solution cannot be stored, since with standing, colloid hydrates of alumina, magnesia, and/or iron may precipitate. The clarification, deaeration, and gold precipitation in the Merrill-Crowe process have to be continuous, with no (or minimal) interruptions. A liquid-sealed centrifugal pump withdraws the solution from the Crowe tower. Inflow of solution to the tower—and hence its solution level—are controlled by an automatic float valve. Suitable grids

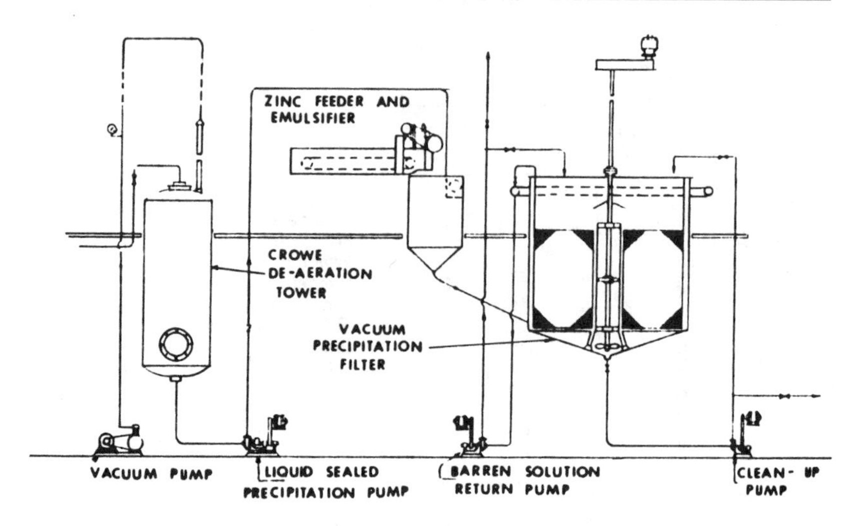

FIGURE 10-5. Merrill-Crowe precipitation plant.

Reprinted by permission. N. P. Finkelstein, Recovery of gold from solutions. In *Gold Metallurgy in South Africa*, edited by R. J. Adamson. Chamber of Mines of South Africa (Johannesburg), 1972.

within the tower break the solution flow into minute streams and thin films, thus enhancing the removal of the dissolved oxygen. A dry vacuum pump is connected with the top of the Crowe tower and maintains the required high vacuum. A zinc feeder introduces zinc dust as the solution flows to the precipitation filters. Pressure candle filters are used for the filtration of zinc-gold slime. A precoat of diatomaceous earth is applied, as in the clarification filters, followed by a secondary coating of zinc dust. The pregnant solution has to percolate through the layer of fine zinc particles, which create a very extensive surface for the solid-liquid precipitation reaction.

The totally enclosed Stellar filters are very secure; the operators do not come in contact with the precious precipitate at any time. Sintered metal candle filters can also be used for the filtration of zinc precipitates. No precoating is required, and automatic backwash prevents clogging of the pores.

In a few instances, gold solutions eluted from activated carbon are treated with zinc cementation. It was found that such solutions do not have to be deaerated, since their oxygen level is very low due to prior hot elution and depressurizing.

Other Precipitants of Gold

Other precipitants of gold from cyanide solution, such as aluminum and charcoal, have been tried. *Aluminum* can be used only with caustic soda (NaOH) solutions, since the latter is essential to the precipitation reaction.

$$
\begin{array}{rcl}
Al & \rightleftharpoons & Al^{3+} + 3e \\
Al^{3+} + 3OH^- & \rightleftharpoons & Al(OH)_3 \\
Al(OH)_3 + Na^+ + OH^- & \rightleftharpoons & AlO_2^- + Na^+ + 2H_2O \\
3Au^+ + 3e & \rightleftharpoons & 3Au^\circ \\
\hline
Al + 4OH^- + Na^+ + 3Au^+ & \rightleftharpoons & 3Au^\circ + Na^+ + AlO_2^- + 2H_2O
\end{array}
\tag{10.7}
$$

Attempts to use aluminum dust as an alternative to zinc have been encumbered by fouling of filters with calcium aluminates (when calcium minerals occur in the ore) and difficulties of smelting the precipitate. In a sodium "regime"—in the absence of Ca^{++}—aluminum precipitation works well. Since the soluble aluminate forms in preference to the insoluble aluminum hydroxide, the reaction proceeds without being hindered by the formation of surface films.

Vieille Montagne, in France, is known to use aluminum to precipitate silver from an acidic thiourea pregnant solution (von Michaelis, 1987). It has been proven that the recovery of silver by aluminum is more effective than the recovery of gold by aluminum.

Soluble reducing reagents such as H_2S, SO_2, $NaSO_3$, and $FeSO_4$ have been used on a commercial scale to precipitate gold from chloride solutions. Hydrogen sulfide precipitates gold as the auric sulfide, but the other reagents reduce it to the metal. It must be emphasized that these reducing reagents will not precipitate gold quantitatively from cyanide solutions.

Activated Carbon Adsorption

Activated carbon is a highly porous material with very large intraparticulate area per unit of mass. The term is generic for a family of substances with distinct adsorptive properties. These substances have neither a definite structural formula nor an identical chemical analysis.

Precharred carbonaceous raw materials (charcoal, charred coconut shells, or fruit stones) have to be submitted to an "activation" at high temperatures (800° to 1100° C) under mildly oxidizing gaseous agents such as steam, carbon dioxide, or mixtures of steam and/or carbon dioxide with air. The controlled combustion burns the more reactive constituents of the carbon structure and thus develops pores and significantly increases the surface area of the "activated" product. Carbon produced from coconut shells have proved to be best suited for liquid-phase adsorption. Surface area is the most important physical property of activated carbon. The bulk density of an activated carbon, along with its specific adsorptive capacity (for a given substance), can be used to design an adsorption system.

Gold complexes with either chloride or cyanide are strongly adsorbed by activated carbon. The adsorption of gold cyanide complexes into porous charcoal particles involves diffusion into pores and attraction on active sites. It constitutes a very important unit operation in modern gold mills.

Activated carbon was introduced as an adsorbent for the recovery of gold and silver from cyanide solutions around 1880. Davis (1880) patented a process in which wood charcoal was to be used for the recovery of gold from chlorination leach liquors. Johnson (1894) patented the use of charcoal for the recovery of gold from cyanide solutions, soon after the discovery (in 1890) that cyanide was an excellent solvent for gold. Nevertheless, gold adsorption on carbon could not compete with filtration and zinc cementation until the early 1950s, when Zadra (1950) developed a process for stripping gold and silver from loaded carbon.

Activated carbon is not a homogeneous material. The softer a carbon is, the higher its activity. Hence, the softer carbon lost due to attrition is the most active portion. When selecting carbons, the resistance to attrition should be evaluated and the carbon with the highest activity after attrition should be selected.

Activated-carbon recovery systems have gained a very wide acceptance in the gold industry, especially over the last 15 years. The major advantage of these systems is *not* having to treat the leach liquor (pregnant solution) prior to recovery. Even in the carbon-in-column system, which is most commonly used with heap leaching, the pregnant solution is run through a bed of carbon without any pretreatment.

The mechanism of the adsorption of gold from cyanide solutions has not been fully explained as yet. The main theories advanced include the following:

- The $Au(CN)_2^-$ ion is adsorbed as such and held by electrostatic or Van der Waal's forces
- The gold compound is altered to some other form during the adsorption process
- Metallic gold is precipitated on the carbon

There is much controversy in the literature as to the mechanism by which activated carbon is loaded with $Au(CN)_2^-$. The fact that activated carbon is not directly amenable to investigations by infrared spectroscopy or X-ray diffraction explains why so very little is known about the adsorbent itself. McDougall et al. (1980) attempted to elucidate the mechanism of the adsorption of gold cyanide by X-ray photoelectron spectroscopy, but they were not able to positively identify the composition(s) of the adsorbate.

The recovery of gold (and silver) from pregnant solutions by activated carbon consists of three distinct operations (Figure 10-6).

1. Loading: Adsorption of gold (and/or silver) from solution onto the carbon

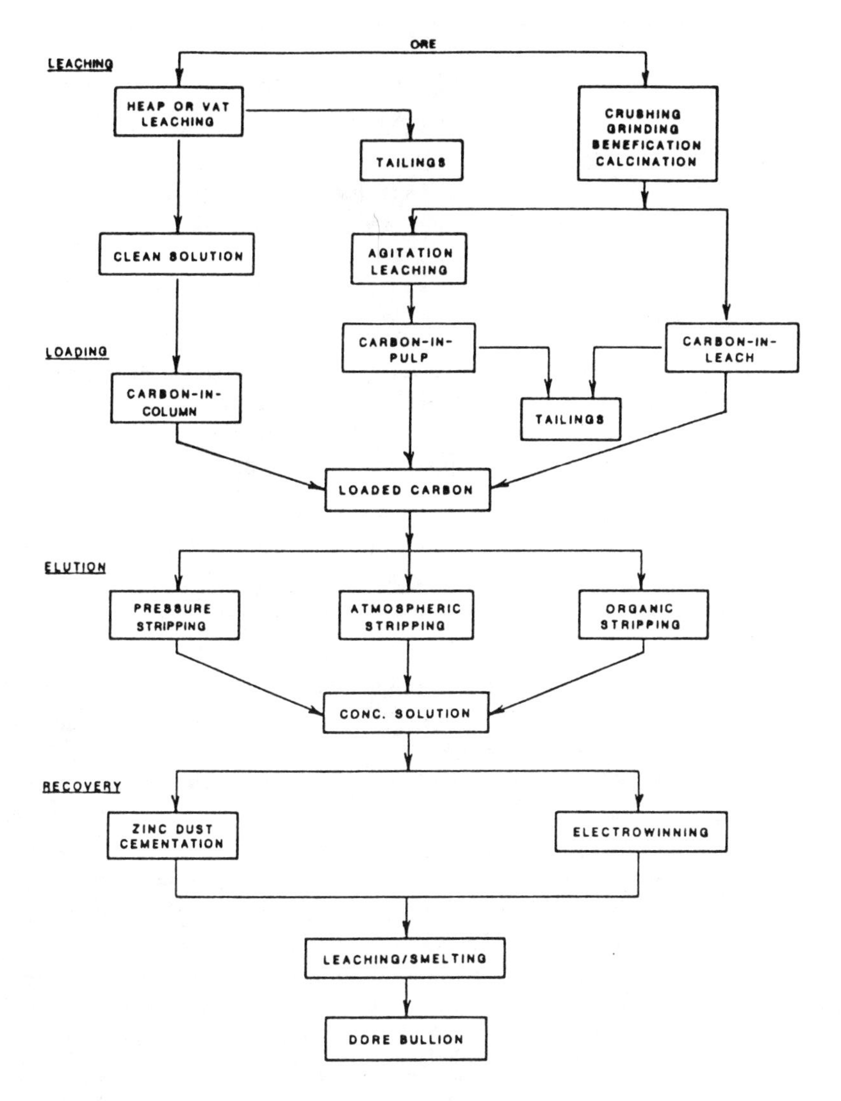

FIGURE 10-6. General flow sheet for recovery of gold by carbon adsorption systems.

Reprinted by permission. O. A. Muhtadi, in *Introduction to Evaluation, Design and Operation of Precious Metal Heap Leaching Projects*, edited by D. J. A. van Zyl et al. Society for Mining, Metallurgy, and Exploration (Littleton, CO), 1988.

2. Elution or stripping: Desorption of gold (and/or silver) from carbon into a more concentrated [than the pregnant] solution[2]
   2a. Carbon regeneration
3. Metallic gold production: Electrowinning or cementation of gold from the concentrated (eluate) solution.

The barren solution is either rejected or, after a certain bleed—depending on its content of impurities—is returned to the elution step. After stripping, the activated carbon is submitted to regeneration.

Loading of Carbon

Although the feasibility of adsorbing soluble gold and silver compounds by "activated" charcoal was reported by Lazowski in 1848, a carbon-cyanidation leach process was first proposed by Chapman and his coworkers in 1949 (Chapman, 1939). The proposed process included the following.

- The dissolution of gold by cyanide from a fine ore pulp containing activated carbon, and the simultaneous adsorption of gold by the carbon
- The counter-current movement of activated carbon, in stages, versus the flow of the cyanided pulp

The major breakthrough in the use of activated carbon for gold adsorption occurred in the early 1950s, when Zadra, at the U.S. Bureau of Mines, demonstrated the feasibility of eluting gold as aurocyanide from loaded carbon with hot caustic-cyanide solutions, and of electrowinning gold from such solutions.

The industry was slow in adopting the elegant Zadra process, relying on the time-proven zinc cementation process. The Homestake Mining Co., in association with the U.S. Bureau of Mines, developed and installed the first carbon-in-pulp plant in 1973. Since then, the use of activated carbon proliferated in the gold extraction industry, all over the world. Carbon is the essential material of the following gold recovery techniques (see Figure 10-6).

- *Carbon-in-pulp (CIP):* The activated carbon is directly mixed with the leached gold-ore pulp
- *Carbon-in-leach (CIL):* The activated carbon is added to the ore pulp in the mixing tanks as cyanidation takes place
- *Carbon in columns:* Clarified or semi-clarified gold solutions percolate through columns packed with activated granular carbon.

---

[2]In effect "Activated Carbon Adsorption" recovers the gold cyanide compounds from dilute production solutions and it delivers them as concentrated gold solution (and not as metal).

Carbons with a very high abrasion resistance are preferred, especially with the CIP and CIL processes, to avoid gold loss in carbon fines. The carbon has to be coarse to allow separation from the fine ore pulp (in the CIP and CIL processes) and free flow of solution in columns. As a rule 6 × 12, 6 × 16, and 8 × 16 U.S. mesh carbons are used in the pulp processes, with only the two coarser sizes used for column packing.

Hydraulic transportation of activated carbon, in process solutions or in water, is in universal use, since it provides simplicity of operation, low maintenance costs, and relatively low capital costs. The carbon is put in contact with the leached ore pulp, counter-currently, in a series of mildly agitated adsorption tanks. The carbon added to the last tank is separated from the fine pulp by screening in each tank as the two streams move counter-currently. After screening, the loaded carbon is removed from the head tank, acid washed, and submitted to the elution process. Gold is usually removed from the eluate by electrowinning. The spent carbon is stored in a tank and is continuously fed to a carbon-regeneration furnace.

As a rule, gold ores contain contaminations of metallic elements (copper, mercury, lead, zinc, arsenic, antimony, cobalt, nickel) as oxides, sulfides, arsenides, or antimonides. Most of these contaminations are at least partially soluble in cyanide and are adsorbed on activated carbon along with the gold. Gold will load in preference to copper. Mercury will follow the gold and must be retorted from the final product. Lead, zinc, cobalt, nickel, arsenic, and antimony will also be adsorbed, but usually they are at very low concentrations and do not cause much of a problem.

Factors Affecting Gold Adsorption

The effect of the ionic strength of solution on carbon loading is significant. Carbon loading from gold solution in de-ionized water is low. The ionic strength of the gold solution plays an important role in the adsorption mechanism.

Davidson (1974) proved experimentally that the degree to which various cyanide complexes of gold are adsorbed on activated carbon depends on their cation. Calcium aurocyanide appears to be the most strongly adsorbed complex in the following series:

$$Ca^{++} > Mg^{++} > H^{+} > Li^{+} > Na^{+} > K^{+}$$

The Effect of Cyanide Concentration. High free cyanide concentrations have a detrimental effect on gold adsorption. Nevertheless, it is well known that free cyanide prevents soluble copper from loading on the carbon. Hence, a compromise on free cyanide concentration is recommended for copper-containing gold ores.

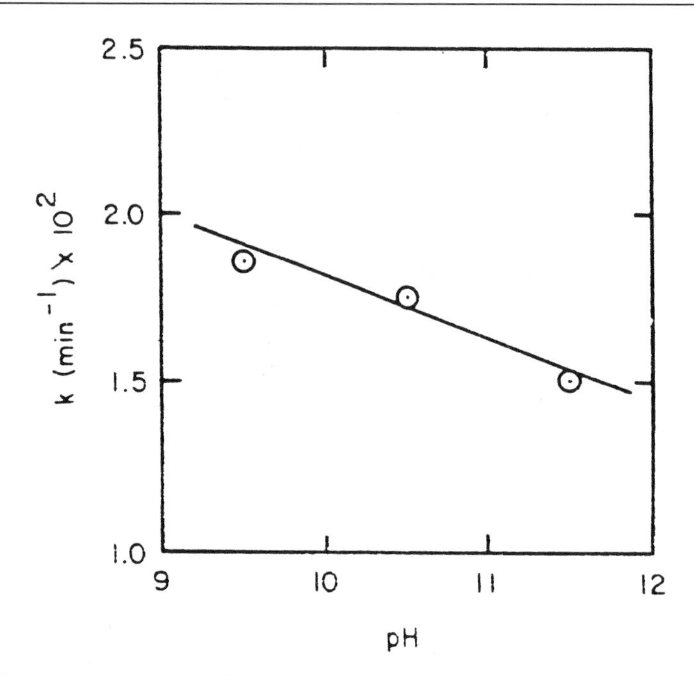

FIGURE 10-7. Effect of pH on the rate of gold adsorption. $[Au_o] = 3$ ppm; d = 2.18 mm.

Reprinted by permission. M. C. Fuersteneau, G. O. Nebo, J. R. Kelso, and R. M. Zaragosa, in *Mineral and Metallurgical Processing*. Society for Mining Engineers (Littleton, CO), 1987.

The Effect of pH. Since both hydrogen and hydroxide ions tend to be adsorbed by carbon, pH affects the adsorption capacity of the carbon. The equilibrium gold loading on the carbon increases with lower pH (Figure 10-7). The effects of calcium cation and hydroxyl anion are antithetic in practice, where pH is adjusted with lime; high calcium concentrations will enhance gold adsorption, while high pH will decrease it.

The Effect of Particle Size. The adsorption rate is affected significantly by the size of the carbon particles. Smaller particle size, for the same mass of carbon, enhances the gold adsorption (as expected by the increase in surface area; Figure 10-8).

Elution or Stripping

The activated carbon, being an excellent adsorbent for soluble gold, does not relinquish the adsorbed gold easily. As already mentioned, Zadra's discovery of a method for successfully eluting gold and silver from a loaded carbon has promoted the use of activated carbon in the extraction of gold

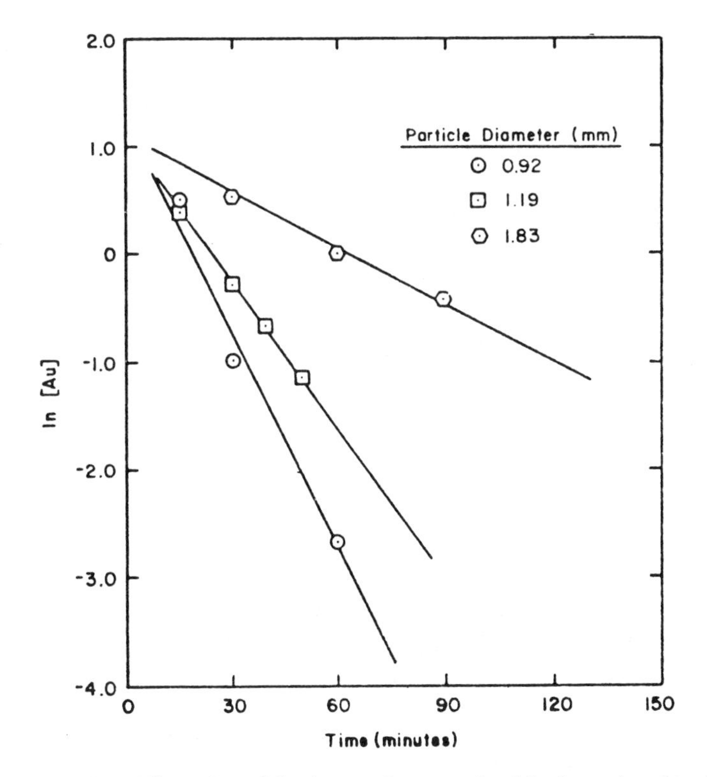

FIGURE 10-8. Effect of particle size on the rate of gold adsorption. $[Au_o]$ = 3 ppm; d = 2.18 mm.

Reprinted by permission. M. C. Fuersteneau, G. O. Nebo, J. R. Kelso, and R. M. Zaragosa, in *Mineral and Metallurgical Processing*. Society for Mining Engineers (Littleton, CO), 1987.

from its ores. A number of other techniques requiring shorter stripping times than Zadra's were subsequently developed.

All elution techniques are based on mass transfer of a soluble gold compound, which is generated by a favorable concentration gradient and high temperatures. The following elution methods are commonly used. The selection of the elution process is dictated mainly by the scale of the operation and local economic conditions.

Chemical treatment, before or after the elution, may be required to remove slimes and deposits from the pores of carbon. Calcium carbonate and/or sulfate often precipitate in the pores of carbon during loading. In South African plants, a hot dilute acid (3% HCl) leaching at approximately 90° C precedes the elution. The acid leaching helps remove calcium and silica from the carbon and enhances the elution kinetics. Acid washing may also help in removing some of the base metals, such as copper, zinc, and nickel, loaded on the carbon (Figure 10-9). Precautions should be

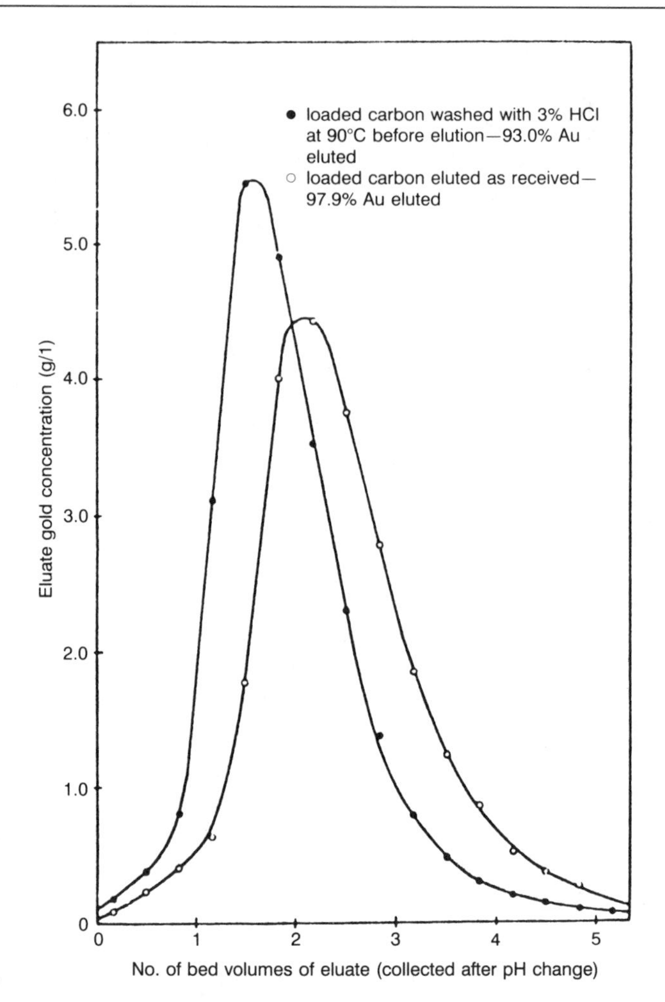

FIGURE 10-9. Effect of acid washing on carbon sample at 110° C.
Reprinted by permission. R. J. Davidson and V. Veronese, in *J. S. Afr. Inst. Mining and Metallurgy* (Johannesburg), October 1979.

taken to thoroughly water-wash the carbon before acid leaching and to remove any HCN gas formed during acid washing.

In some North American plants, acid washing of the carbon takes place, if necessary, after its elution and before its reactivation. In at least one South African plant the carbon is acid-washed twice, before stripping and after regeneration. A caustic neutralizing wash is required before the carbon is recycled to the adsorption circuit. Davidson et al. (1982) reported copper sulfate solutions as effective for chemical treatment of activated carbon.

The atmospheric Zadra process was developed by Zadra at the U.S. Bureau of Mines in the early 1950s. As already mentioned, this process was a major breakthrough for the use of activated carbon in the gold extraction industry.

The process elutes gold from loaded carbon with a solution containing 0.1–0.2% NaCN and 1% NaOH heated at 85°–95° C. Precious metals are stripped by percolating the hot solution through the loaded carbon bed. Gold and silver are electrowon from the stripping solution. The barren solution is then reheated and recycled through the carbon. The process is carried out at atmospheric pressure and requires 24 to 60 hours.

The simplicity and the low capital and operating costs of this process make it suitable for small-scale operations. Its long cycle time is a serious drawback, especially for larger-scale gold mills.

The alcohol stripping process was developed by H. J. Heinen et al. at the U.S. Bureau of Mines in 1976. This process is an extension of the Zadra process; it proposes a 20% vol. alcohol addition to the stripping solution (0.1 NaCN, 1% NaOH, and 20% ethanol at 80° C). The alcohol addition reduces the elution cycle time to five or six hours, and thus the size of the stripping section is drastically reduced. The major disadvantages of this system are the high fire and explosion risks associated with alcohol, and the higher operating cost due to alcohol loss by volatilization. Effective vapor recovery equipment, and safety features to minimize the fire risk, are incorporated in the design of this stripping process. The carbon stripped by this technique does not require frequent regeneration.

The high-pressure stripping process was developed by Ross et al. (1973) at the U.S. Bureau of Mines in the early 1970s. The process uses high pressure and high temperature to scale down the stripping time to two to six hours. It involves stripping loaded carbon with a solution containing 0.1% NaCN and 1% NaOH at 160° C and 350 kPa (50 psi). Use of high pressure reduces reagent consumption, carbon inventory, and the size of the stripping section. The high pressures and temperatures employed in this process, and the extensive cooling of solutions required before pressure reduction, impose a high capital cost on the installation.

The Anglo-American stripping process was developed by R. J. Davidson and involves preconditioning the loaded carbon with half a bed volume of 5% NaOH and 1% NaCN for 30 minutes to one hour, and then stripping the carbon with five bed volumes of hot (110° C) water at a flow rate of three bed volumes per hour. The high operating temperature imposes an operating pressure of 50-100 kPa. Figure 10-10 shows typical elution profiles and the effect of the eluant temperature. The total cycle time, including the acid prewashing, is nine hours, considerably shorter than that of the atmospheric Zadra treatment.

The Anglo-American process has advantages similar to those of high-pressure stripping. The use of elevated temperatures and pressures imposes

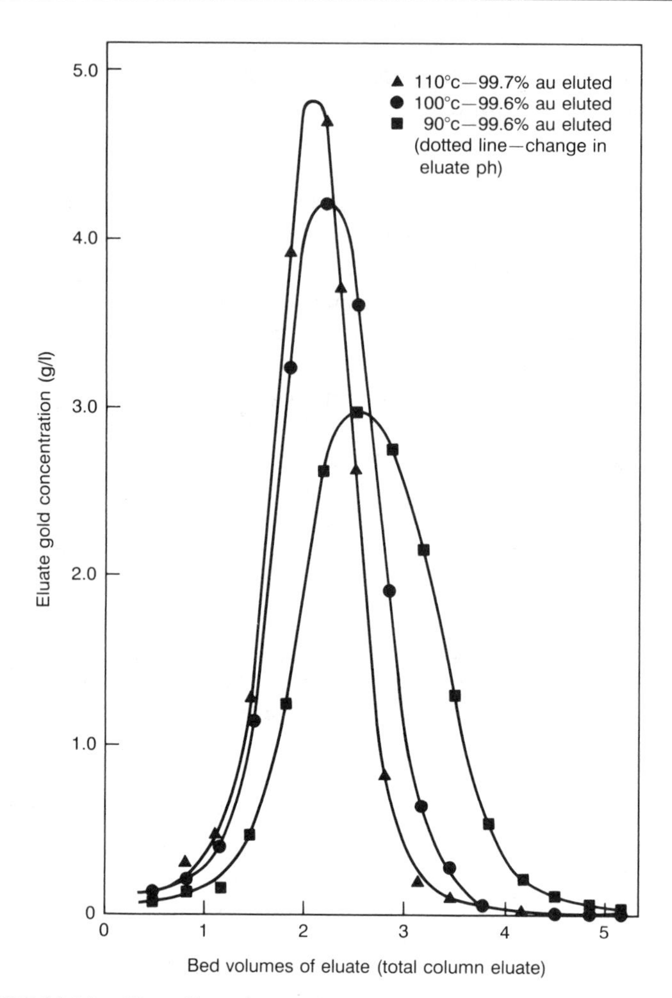

FIGURE 10-10. The effect of temperature on the elution of a carbon sample. Reprinted by permission. R. J. Davidson and V. Veronese, in *J. S. Afr. Inst. Mining and Metallurgy* (Johannesburg), October 1979.

a high capital cost on the installation, and the requirement for multiple streams increase the complexity of the circuit.

The Micron gold desorption process uses the principle of conditioning the carbon in a strong, alkaline cyanide solution followed by desorption in a fractionating column, using the carbon as the column packing and methanol as the desorbing agent (Figure 10-11). The system offers the following advantages.

- The processes of desorption, methanol recovery, and reactivation are all carried out in a single column, with the wet carbon as the column packing.

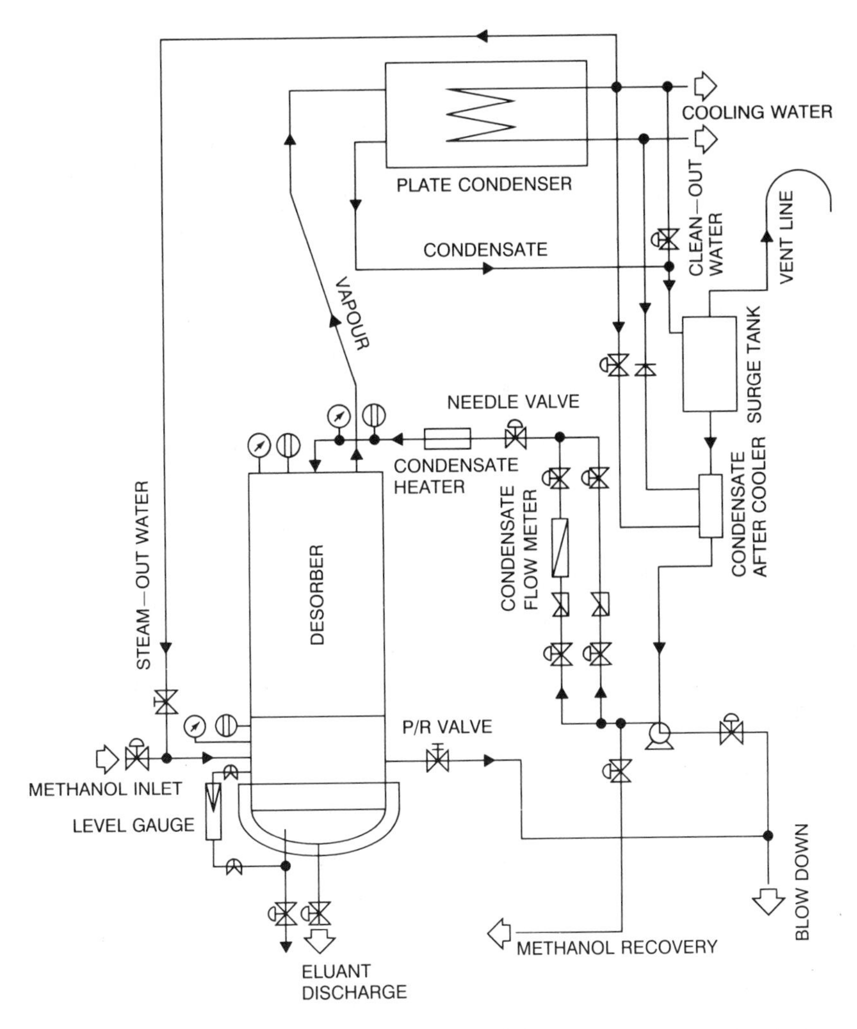

FIGURE 10-11. Flowchart of the Micron gold desorption process (sketch distributed by Micron Research, W.A.). The patented process provides a method to rapidly desorb gold from activated carbon using alcohol as solvent at low temperature and pressure. The gold is recovered from the pregnant eluant by electrolytic cell.

Reprinted by permission. Micron Research, Metallurgical Services Laboratory, Western Australia.

- The metal values are concentrated into the aqueous cyanide content of the carbon, and no other water is used except as condenser coolant. This produces very high-grade gold and silver solutions, which give high current efficiencies in the electrowinning step. These high gold concentrations allow almost complete separation of gold and silver from copper in the electrolytic cell.

- The process is completed to a tailing assay of less than 200 g Au/ton carbon in 15–20 hours, allowing turnaround of carbon batches on a daily basis.
- During electrowinning, the metals are plated onto aluminum foil that is later dissolved in sodium hydroxide solution. The gold product can be shipped to a mint, making plant furnace facilities unnecessary.
- The treated carbon is automatically reactivated to 60–80 percent of the new carbon value.

In gold mills where the pregnant solution is heavily contaminated with mercury, the loaded carbon is presoaked for three to six hours in the strip solution, for mercury removal before elution.

A continuous gold elution process was patented by A. M. Stone (1985). The continuous elution takes place by counter-flow of the stripping liquor *vs.* the loaded carbon in a pressurized, heated vessel.

Carbon Regeneration

Thermal reactivation of carbon, on a periodic basis after the elution step, is necessary in order to remove mostly organic matter and inorganic impurities that are picked up by the carbon during adsorption and are not removed by elution and acid washing. The process involves heating the wet carbon to 650°–750° C, in the absence of air, for up to 30 minutes.

Fleming (1982) and Nicol (1979) have shown that heating the wet carbon results in partial combustion.

$$C + H_2O \rightarrow CO + H_2 \tag{10.8}$$

$$C + 2H_2O \rightarrow CO_2 + 2H_2 \tag{10.9}$$

Some burn-off of the carbon is beneficial, since its capacity for adsorption has been found to increase slightly after successive reactivation treatments. The initial U.S. Bureau of Mines tests used quenching in water after reactivation, but Homestake Mining established that air cooling of the carbon (in the cooling section of the kiln or in a hopper) results in higher carbon activity. The reactivated carbon has to be screened (at 200 mesh) to remove fines and conditioned with water before recycling to the adsorption circuit.

The reactivation of carbon is conducted, as a rule, in rotary kilns that are heated externally by electric power. The steam generated from the water in the wet carbon excludes air from the kiln. The amount of water charged to the kiln with the carbon has to be minimized in order to reduce the cost of reaching temperatures of 650°–700° C, which are necessary for efficient regeneration of the carbon.

Vertical regeneration kilns were introduced in the early 1980s, and a number of vertical kiln designs are now available. A patented vertical furnace heated with horizontal electrical resistance heaters ("globars";

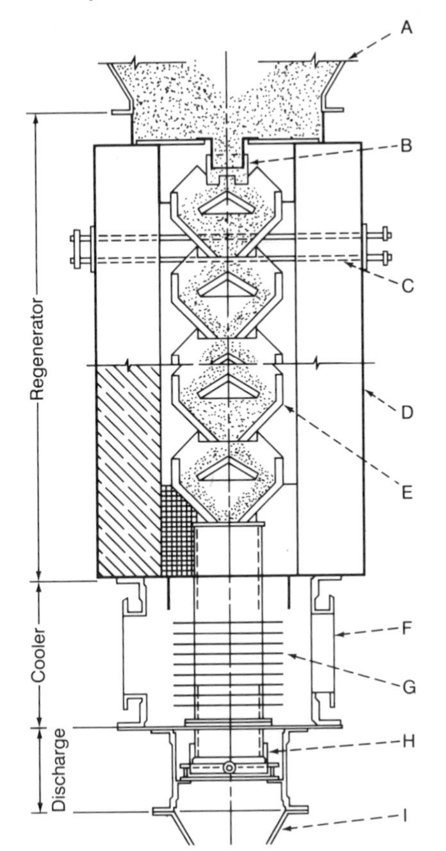

FIGURE 10-12. Vertical regeneration kiln (with electrical resistance heater).
Reprinted by special permission from *Chemical Engineering* (Marques and Nell, Inc., 1984), McGraw-Hill, Inc., New York, NY 10020.

Figure 10-12) can reach temperatures of up to 1090° C. The carbon flows "gently" along the furnace with minimum burning and attrition. The atmosphere within the furnace is regulated by a vent and damper to maintain about 1% oxygen. In the lower section of the furnace, the carbon is externally cooled before exiting. The capital cost of this furnace is significantly lower than rotary horizontal regenerators.

The South African Council for Mineral Technology (MINTEK) developed a process of carbon regeneration by direct electrical heating of the

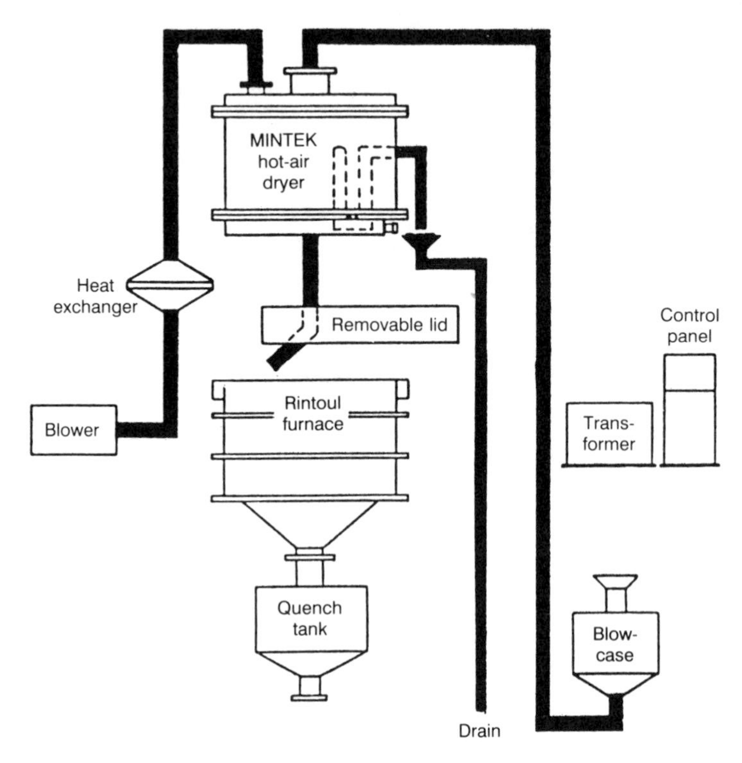

FIGURE 10-13a. Flowchart of carbon regeneration by direct electrical heating of the carbon.

Reprinted by permission. P. M. Cole, D. S. von Broemsen, and P. A. Laxen, in *Gold 100: Proceedings of the International Conference on Gold*. South African Institute of Mining and Metallurgy (Johannesburg), 1986.

carbon. After drying, the carbon is charged into a vertical furnace with a "floating" or common neutral. Voltage is applied—with two electrodes—across the dry, granular activated carbon (which has limited electrical conductivity), so resistive heating is generated (Figures 10-13a and b). It has been shown that this furnace can be used to reactivate heavily contaminated industrial carbon (Cole et al., 1986).

Electrowinning of Gold from Cyanide Solutions

Electroplating of Gold

Although there is evidence that gold plating was carried out in ancient times, the electroplating of silver and gold was developed in the 19th century (Fisher & Weimer, 1964). Practically all of the systems for plating

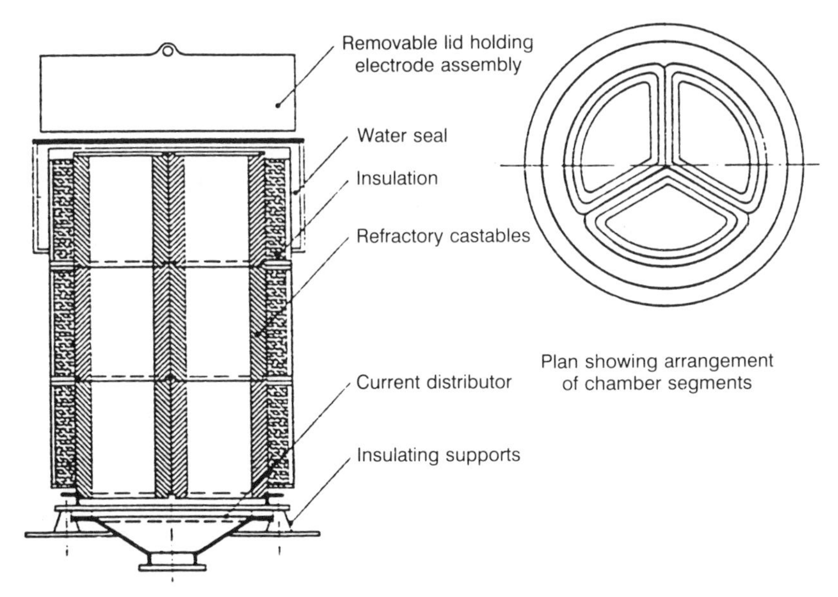

FIGURE 10-13b. Sections of the Rintoul furnace.

Reprinted by permission. P. M. Cole, D. S. von Broemsen, and P. A. Laxen, in *Gold 100: Proceedings of the International Conference on Gold*. South African Institute of Mining and Metallurgy (Johannesburg), 1986.

gold are based on gold(I) cyanide baths (Lowenheim, 1977). Gold(I) baths are, of course, more efficient in terms of electricity consumed than gold (III) baths.

Gold is commonly plated from an electrolyte in which the gold is present as $KAu(CN)_2$, with rolled gold anodes or insoluble anodes. The use of insoluble anodes, usually stainless steel (18% Cr/8% Ni with a low content of Mo or Ti), is a popular way to minimize capital investment. The concentration of the electrolyte can be maintained by adding gold salts in proportion to the gold-plating rate. The spacing between the anode and cathode is not critical.

Oxygen is evolved at the anode of a gold electrowinning cell, while hydrogen is evolved along with gold deposition at the cathode.

$$4OH^- \rightleftharpoons O_2 + 2H_2O + 4e^- \quad (10.10)$$
$$2e^- + 2H_2O \rightleftharpoons H_2 + 2OH^- \quad (10.11)$$
$$KAu(CN)_2 \rightleftharpoons K^+ + Au(CN)_2^- \quad (10.12)$$
$$Au(CN)_2^- \rightleftharpoons Au^+ + 2CN^- \quad (10.13)$$
$$e^- + Au^+ \rightleftharpoons Au \quad (10.14)$$

Wilkinson (1986) proposed a conceptual mechanism of gold deposition from aurocyanide solutions.

> The cathode attracts predominantly positive ions to a region near its surface, known as the Helmholtz double layer. Negatively charged complex ions, $Au(CN)_2^-$, which approach this layer, become polarized in the electric field of the cathode. The distribution of the ligands around the metal is distorted and the diffusion of the complex ion into the Helmholtz layer is assisted. Finally, within the Helmholtz layer, the complex breaks up. Its component ligand ions or molecules are freed and the metal released in the form of the positively charged metal cation which is deposited as the metal atom on the cathode.

The aurocyanide complex anion has a high stability constant (Wilkinson, 1986).

$$\frac{[Au(CN)_2^-]}{[Au^+][CN^-]^2} = 10^{38.3}$$

Also, the concentration of $[Au^+]$ is extremely low in the electrolyte.

$$[Au^+] = \frac{[Au(CN)_2^-]}{[CN^-]^2} \cdot 10^{-38.3}$$

Hence, it would appear that the reasonable rates of gold deposition from gold(I) cyanide solutions are only possible because of the polarization of $Au(CN)_2^-$ ions, which approach the cathode surfaces and are distorted according to Wilkinson's model. Although gold cyanide electroplating baths can be operated under alkaline or neutral conditions, in acidic (pH of 3.1 to 7.0) baths the concentration of $[Au^+]$ increases, since the equilibrium of reaction (10.11) is affected by the formation of undissociated HCN (Wilkinson, 1986):

$$H^+ + CN^- \rightleftharpoons HCN \qquad (10.15)$$

A solution of potassium aurocyanide is stable with no evolution of gaseous HCN, down to a pH of 3.1 (Fisher & Weimer, 1964). The permissible current density—and the cathode-current efficiency achieved—increase with the gold concentration of the solution in electroplating (electrowinning). Therefore, in order to obtain fast rates of deposition from a gold-plating bath, high concentrations of gold are recommended.

Following are some typical bath compositions in electroplating, according to Fisher and Weimer (1964).

	g/1	g/1	g/1
Gold (as cyanide)	2.1	8.4	10
Potassium cyanide	15	11	12
Disodium phosphate	4	—	—
Alkali carbonate	various	various	various

In the past, the gold content has been kept low (1–3 g/l) to control the "captive" capital of the electrolyte. However, the use of dilute solutions may constitute a false economy. More concentrated baths have been used for fast rates of deposition.

Gold (as potassium aurocyanide)	8–14 g/l
Free KCN	15–30 g/l
Temperature	60–70° C
Current density	2–10 amp/sq. ft.

A gold-plating bath for fast rates of deposition used in the U.S.S.R. has the following composition (Fisher & Weimer, 1964).

$KAu(CN)_2$	15–25 g/l as Au
Free KCN	8–10 g/1
K_2CO_3	up to 100 g/l
KOH	up to 1 g/l
Current density (cathodic)	20–40 amp/sq. ft.
Current density (anodic)	10 amp/sq. ft.
Temperature	55–60° C
Anodes	gold

Increasing the temperature of the bath significantly increases the current efficiency of gold deposition, according to Raub et al. (1986).

Electrowinning of Gold from Pregnant Gold-Mill Solutions

Electrolytic recovery of gold from pregnant cyanide solutions was applied on an industrial scale in the late 19th century with the Siemens-Halske electrolytic method (Adamson, 1972). Gold in cyanide solution was electrolytically deposited on lead foil cathodes that were removed periodically, melted into lead ingots, and cupelled to recover the gold. Lead was recovered from the litharge, produced by cupellation, and rolled into foil to provide new cathodes. With a high gold content in the cyanide electrolyte, the electrodeposition efficiency was high, but with low-grade solutions it fell significantly. The industrial success of zinc precipitation at this time prevented further developments of gold electrowinning.

Electrowinning of gold from pregnant cyanide solutions—as compared to gold precipitation by chemical reagents—has two major inherent advantages. First, due to the electrolytic reduction, no chemical reagents are added to the solution, which can thus be further recycled in the extraction process. Second, the electrowon gold is, in principle, of high purity. Nevertheless, the very low gold content of mills' pregnant solutions makes them poor electrolytes, with a very low current density, a low current efficiency, and, most important, an extremely slow rate of deposition per unit of cathode surface.

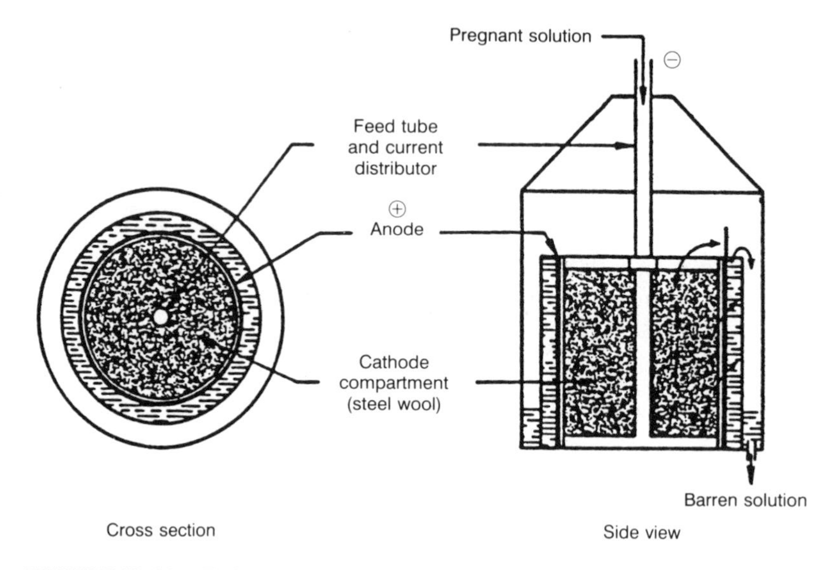

FIGURE 10-14. Zadra electrolytic cell.

Reprinted by permission. J. A. Eisele et al. Staged heap leaching and direct electrowinning. U.S.B.M. IC 9059, 1986.

$$2Au + 4NaCN + O_2 + 2H_2O \rightarrow 2NaAu(CN)_2 + 2NaOH + H_2O_2$$

The cyanidation of gold ores is controlled by the oxygen concentration of the leaching solution, which is low. Leaching solutions contain proportionally low cyanide concentrations, to preserve the reagent and avoid soluble-in-cyanide contaminations. Gold pregnant solutions have low gold contents, even after concentration by carbon adsorption. Rich strip solutions range from 0.035 to 0.900 g/l (Arnold & Pennstrom, 1987). Such solutions are significantly lower in gold content than the average electrolyte used in gold electroplating.

The permissible current density—and the current efficiency obtained—during electrowinning of gold from cyanide solutions is proportional to the gold content of the electrolyte. Hence, the recovery of gold by electrowinning of pregnant gold mill solutions has to cope with a very low current density (and a very slow rate of deposition on conventional flat cathodes).

Zadra (1950) and Zadra et al. (1952) proposed and tested a "cathode compartment" packed with steel wool. He used 1.68 pounds of stainless steel wool per cubic foot of cathode compartment, and estimated 9 to 14 square feet of surface per pound of wool (Figure 10-14). Thus, with the extended cathode surface, a significant increase in the rate of gold electrowinning was obtained, in spite of the low gold concentration of the pregnant solutions or eluates (and the resulting low current densities permissible with such solutions).

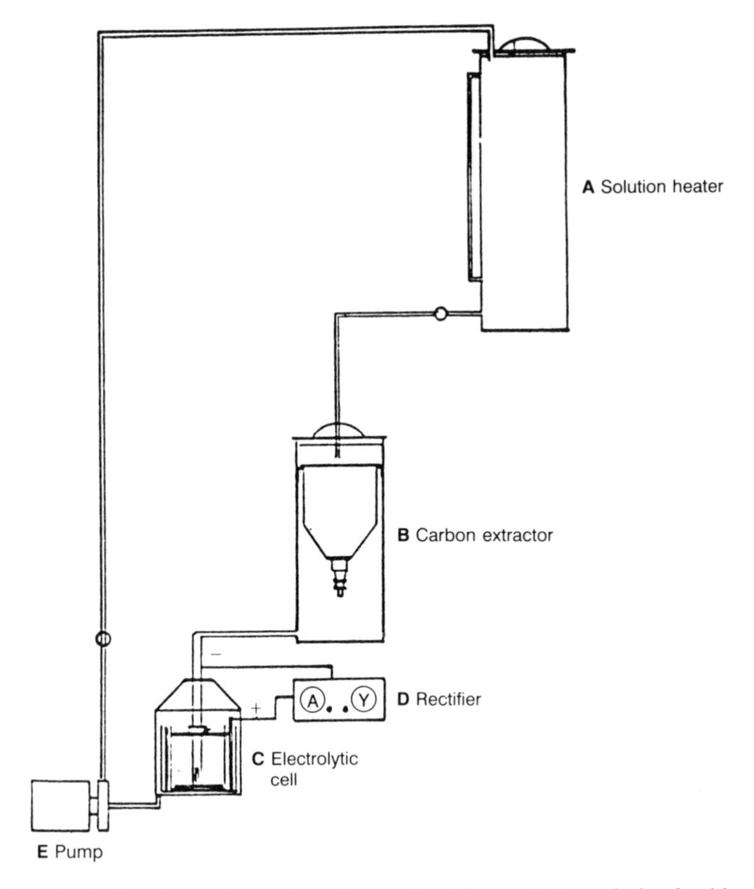

FIGURE 10-15. Flow sheet of continuous leaching and electrolysis of gold- and silver-laden activated carbon.

Reprinted by permission. J. A. Eisele et al. Staged heap leaching and direct electrowinning. U.S.B.M. IC 9059, 1986.

Zadra (1950) was the first to propose pervious cathodes (in a boxlike frame) filled with steel wool, where the rate of deposition is very high due to the extensive area of the cathode. "A thick sheet of gold can readily be deposited (on solid plate stainless steel cathode)," Zadra reported, but at a low rate of deposition due to the restricted area of this type of cathode. "The use of this (type of) cathode necessitates construction of a long electrolytic cell to accommodate enough cathodes to provide the desired deposition area" for a satisfactory plating rate. Zadra's gold electrowinning cell was proposed as part of a flow sheet for continuous leaching and electrolysis of gold- and silver-laden activated carbon, reproduced in Figure 10-15 and operating at Homestake Mining for years.

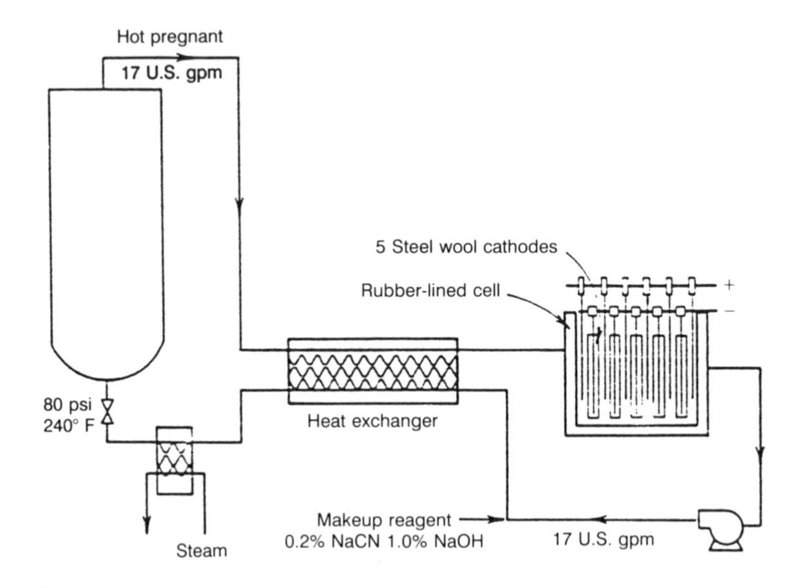

FIGURE 10-16. Flowchart of gold elution and electrowinning.

Reprinted by permission. D. A. Milligan et al., in *Introduction to Evaluation, Design and Operation of Precious Metal Heap Leaching Projects*, edited by D. J. A. van Zyl et al. Society for Mining, Metallurgy, and Exploration (Littleton, CO), 1988.

Zadra's sound approach to electrowinning from dilute gold solutions was extended to other cell designs (Figure 10-16). The Anglo-American Research Laboratories (AARL) modified the Zadra cell by using a cation-permeable membrane to divide it to anode and cathode compartments with different anolyte and catholyte circulations. The pregnant electrolyte flows through the cathode basket, whereas a strong alkaline solution circulates through the anode compartment (Fleming, 1982).

The South African Council for Mineral Technology (MINTEK) developed a "sandwich-type cell" with a permanent graphite-chip cathode separated from the anode compartment by an ion-exchange membrane (Paul et al., 1982). The gold was first deposited on the graphite, then was dissolved off as potassium aurocyanide by reversal of the polarity, and was finally redeposited on titanium sheets in a small external cell, from which the gold could be stripped and melted. However, on occasions the graphite became passivated during anodic dissolution and it was necessary to remove the graphite from the cell and smelt it to recover the gold.

MINTEK designed and tested a rectangular cell with pervious cathodes packed with steel wool and stainless steel perforated anodes—or, preferably, stainless steel mesh anodes—to ensure a uniform flow of solution through the cell (Paul et al., 1982; Briggs, 1983). The electrodes are sized

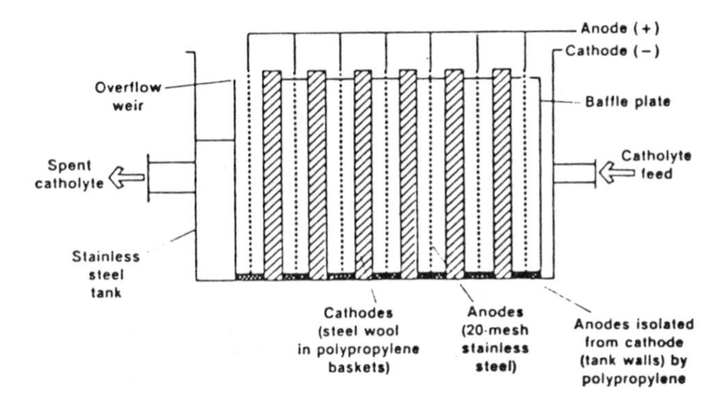

FIGURE 10-17. The MINTEK steel wool cell.

Reprinted by permission. A. P. W. Briggs, in *J. S. Afr. Inst. Mining and Metallurgy* (Johannesburg), October 1983.

to fit snugly down the sides and along the bottom of the rectangular tank, so that the electrolyte has to flow through the electrodes during its continuous circulation (Figure 10-17).

It is possible for 2 kg of gold to be deposited onto 0.5 kg of steel wool in each cathode compartment before the cell's current efficiency drops or the cathode becomes blocked by the gold deposit. The loaded cathodes are calcined at 700° C for 20 hours, according to Briggs (1983). The calcine, mixed with 40% borax, 30% Na_2CO_3 and 25% silica, is melted at 1300° C. Filmer (1982) reported gold deposition up to 20 times the weight of the steel wool packed into the cathode basket of a Zadra cell at a density of about 35 g/1.

All electrowinning cells with pervious, packed-bed cathodes can be divided into two groups, those that operate with the flow of the electrolyte at a right angle to the current flow, and those that operate with parallel solution and current flows (Figure 10-18). The MINTEK cell—with parallel solution and current flows—claims more uniform packing in the cathodes and hence a more uniform flow of pregnant solution than the cylindrical cells. The Zadra-type cells may have the disadvantages of uneven packing of the cathode, uneven flow distribution, and inefficient use of the cell volume.

An "Improved Mass Transfer" (IMT) cell has been investigated by the U.S. Bureau of Mines (Eisele et al., 1986). The high recirculation rate of the electrolyte and the quasi-parallel solution and current flows are claimed to increase the cell's efficiency. Gold-winning rates up to 200% greater than those obtained without using recirculation were reported (Eisele et al., 1986).

Electrowinning of gold from cyanide solutions onto steel wool cathodes has become a very strong competitor of zinc precipitation, and is the

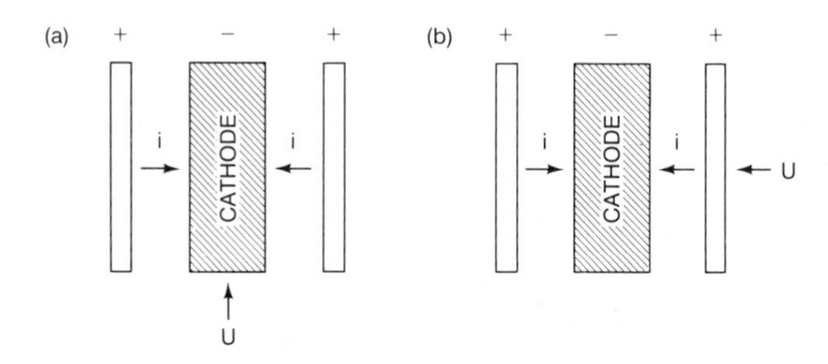

FIGURE 10-18. Two groups of gold electrowinning cells, with the flow of solution u: (a) at right angles to the flow of current i, (b) parallel to the flow of current i.

standard procedure of gold recovery for plants employing CIL or CIP. The gold-laden cathodes are washed, dried, and smelted, or treated with acid to remove the excess of iron and then smelted. Hence, iron slag with a high gold content is ground and gravity-enriched, and the gravity tails are recycled into the mill.

The gold electrowinning/replate circuits used at Gold Fields' Ortiz and Mesquite operations avoid the pitfalls of smelting the steel wool cathodes and treating gold-containing slags (Arnold & Pennstrom, 1987). Strip solution (35 to 260 mg/l) is sent to two electrowinning cells in parallel (eight perforated pervious cathodes, filled with 40G stainless steel wool per cell). Ninety-nine percent of the gold is electrowon, and the stainless steel wool cathodes are fully loaded after two to four days. The gold-laden steel wool electrodes are transferred into an electrorefining (replating) cell, where they become anodes, with stainless steel plates as cathodes. After two to four days the gold in the replating cell is collected as gold foil on the cathodes, granular product over the foil, and electrolytic mud.

Table 10-1 lists, in detail, the conditions of electrowinning and replating (electrorefining) in the Mesquite gold room. The main advantages of the two-step electrodeposition are, of course, the high-grade bullion obtained by melting the replate and the restricted proportion (3%) of gold recycled with the slag. However, the two-step process prolongs the production of gold in the electrolytic room by about three days, on the average.

Staged Heap Leaching and Direct Electrowinning

A typical heap-leaching pregnant solution contains 0.75–2.00 ppm (g/ton), compared to 50–100 ppm of gold in carbon stripping solutions from which, as a rule, gold is electrowon. Although the carbon adsorption-desorption step can efficiently concentrate heap-leaching pregnant solutions and remove undesirable metal contaminations, it significantly increases the operating and capital costs of a heap-leaching operation. Eisele et al. (1986)

TABLE 10-1. Mesquite gold room operating parameters.

Electrowinning	
Days operated/week	6–7
Hours/day	18–24
No. of cells operating	2
Cathode transfers/week (to replating)	1–2
No. of cathodes/transfer	8 per cell, 16 total
Average electrolyte flow rate	38 l/min per cell 76 l/min total (10 GPM/cell; 20 GPM total)
Electrolyte feed avg. assay	141 g/tonne
Electrolyte return avg. assay	1.6 g/tonne (0.05 oz Au/ton solution)
% recovery gold (electrowinning)	99%
NaCN concentration (electrolyte)	0.1%
NaOH concentration (electrolyte)	1.0%
Cell amperage	800 amps
Cell voltage	2–4 volts
Electrolyte temperature	77° C (170° F)
Avg. cathode loading, au oz/cathode	2.2 kg/cathode (72 oz/cathode)
Replating	
Days operated/week	7
Hours/day	24
No. of cells operating	1–2
Cathode transfers/week	1–2
No. of cathodes/transfer	16
No. of scrapes/week	1–2
Average electrolyte flow rate (recirculating)	6 l/min (20 GPM)
Electrolyte avg. assay	2.8 g au/tonne solution (0.1 oz/ton)
Avg. cathode loading in	2.2 kg (72 oz)/cathode
Avg. cathode loading out	31.1 g (1.0 oz)/cathode
% recovery gold (replating)	98.6%
NaCN concentration (electrolyte)	0.5%
NaOH concentration (electrolyte)	1.0%
Cell amperage, 1st 24 hours after transfer	150–200 amps
Cell amperage, remaining time interval	500 amps
Cell voltage, 1st 24 hours after transfer	1.8 volts
Cell voltage, remaining time interval	2–4 volts
Electrolyte temperature	77° C (170° F)

(*continued*)

TABLE 10-1 (continued). Mesquite gold room operating parameters.

Average scrape	35.7 kg (1150 oz)
Average foil scrape	30.2 kg (970 oz)
Melting	
Melts/week	1–2
Hours/melt (includes charging and pouring)	4
Avg. charge/melt (from replate)	37 kg (1190 oz)
Flux makeup ratio 6 borax: 2 miter: 2 silica: 1 soda ash: 2 slag	
Avg. flux/melt	13.6 kg (30 lbs)
Pour temperature	1038° C (1900° F)
No. of bars of bullion/melt	4–5
Oz. bullion/bar	250
Avg. melt losses to slag	3.0%
Avg. bullion assay	96% Au, 3% Ag
Avg. doré kg/melt	35.8 kg (1150 oz)

Reprinted by permission. J. R. Arnold and W. J. Pennstrom, The gold electrowinning/replate circuits at Gold Fields' Ortiz and Mesquite operations. *Min. & Metal. Proc.* Society of Mining and Metallurgical Engineers (Littleton, CO), May 1987: 65–67.

proposed the "Staged Heap Leaching/Direct Electrowinning" process (Figure 10-19a) and the "Improved Mass Transfer Electrolytic Cell" (Figure 10-19b) as an alternative to upgrading dilute pregnant solutions by carbon adsorption-desorption. Eisele's scheme increases the solution pumping cost and the impurities concentration in the pregnant solution. Most importantly, it aggravates the gold loss due to solution seepage and dissipation.

Industrial Uses of Activated Carbon

Carbon-in-Pulp (CIP)

The original development of CIP is credited to Professor T. G. Chapman of the University of Arizona in the late 1930s. Chapman initially added fine activated carbon to cyanide-leached gold ore pulp; after adequate agitation, he separated the fine carbon from the ore by flotation. Gold was recovered by burning the loaded carbon. In an alternative system, Chapman used coarse granules of activated carbon in cylindrical screen baskets that were rotated into the cyanided gold ore pulp. Stripping the coarse carbon with hot cyanide solution resulted in a concentrated gold solution. In the early 1950s, Zadra and his coworkers at the U. S. Bureau of Mines

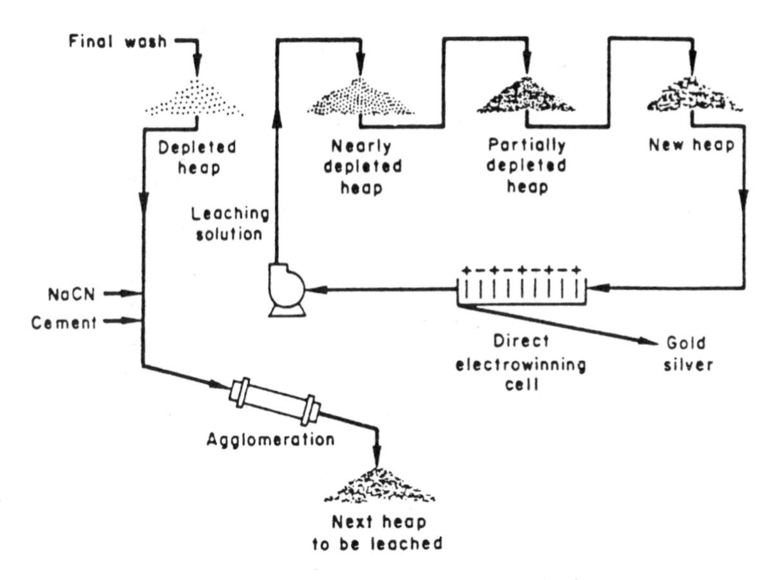

FIGURE 10-19a. Staged heap leaching/direct electrowinning.

Reprinted by permission. J. A. Eisele et al. Staged heap leaching and direct electrowinning. U.S.B.M. IC 9059, 1986.

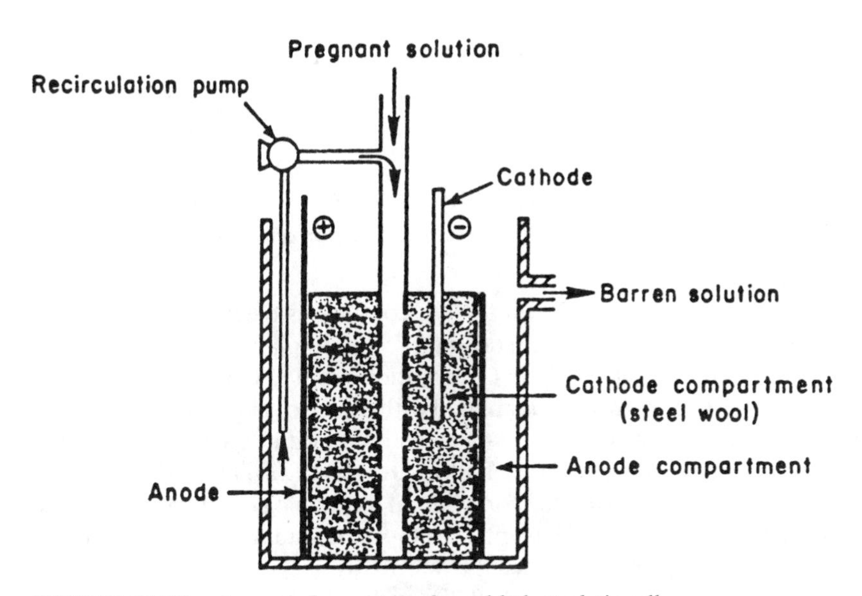

FIGURE 10-19b. Improved mass transfer gold electrolytic cell.

Reprinted by permission. J. A. Eisele et al. Staged heap leaching and direct electrowinning. U.S.B.M. IC 9059, 1986.

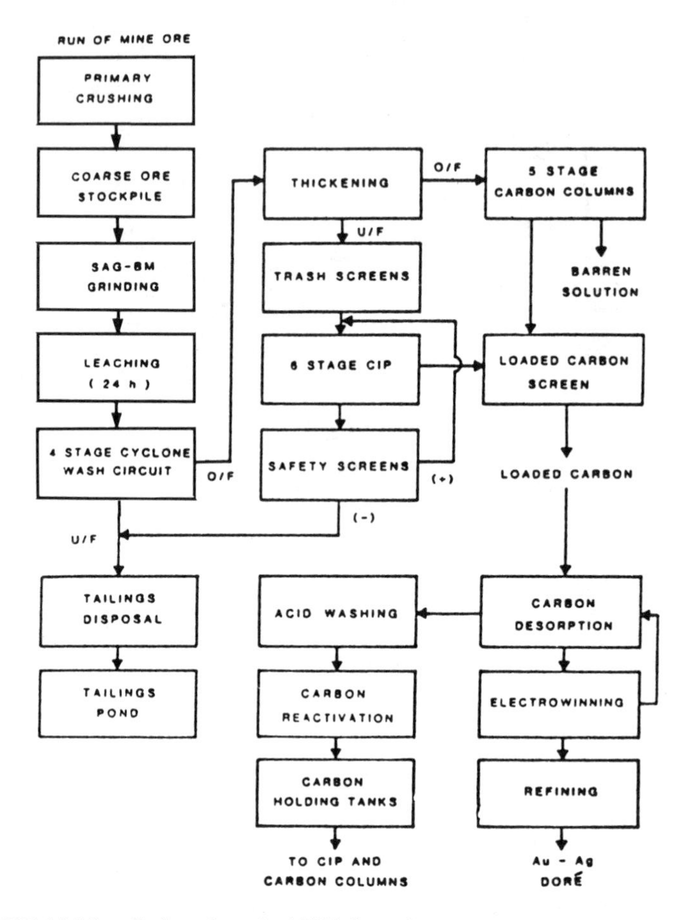

FIGURE 10-20. Carbon-in-pulp (CIP) flow sheet.
Reprinted by permission. S. D. Hills, The carbon-in-pulp process. U.S.B.M. IC 40–43, 1986.

developed an electrolytic cell for the continuous electrowinning of gold from the strip liquor. The first commercial CIP plant was operated from 1954 to 1960 at Golden Cycle Corporation's mill in Cripple Creek, Colorado. The Zadra process was applied with coarse carbon in screen baskets. Homestake Mining Company—in cooperation with U. S. Bureau of Mines metallurgists—started a large Zadra CIP plant (2,000 tons/day) in 1973. The South African Council for Mineral Technology (MINTEK) started work on CIP in 1976. CIP plants based on the Zadra process have proliferated all over the world since the late 1970s.

The *CIP flow sheet* (Figures 10-20 and 10-21) includes cyanide leaching of the fine ore, counter-current carbon-pulp contact, carbon-pulp separation, carbon stripping and regeneration, and gold electrowinning. The

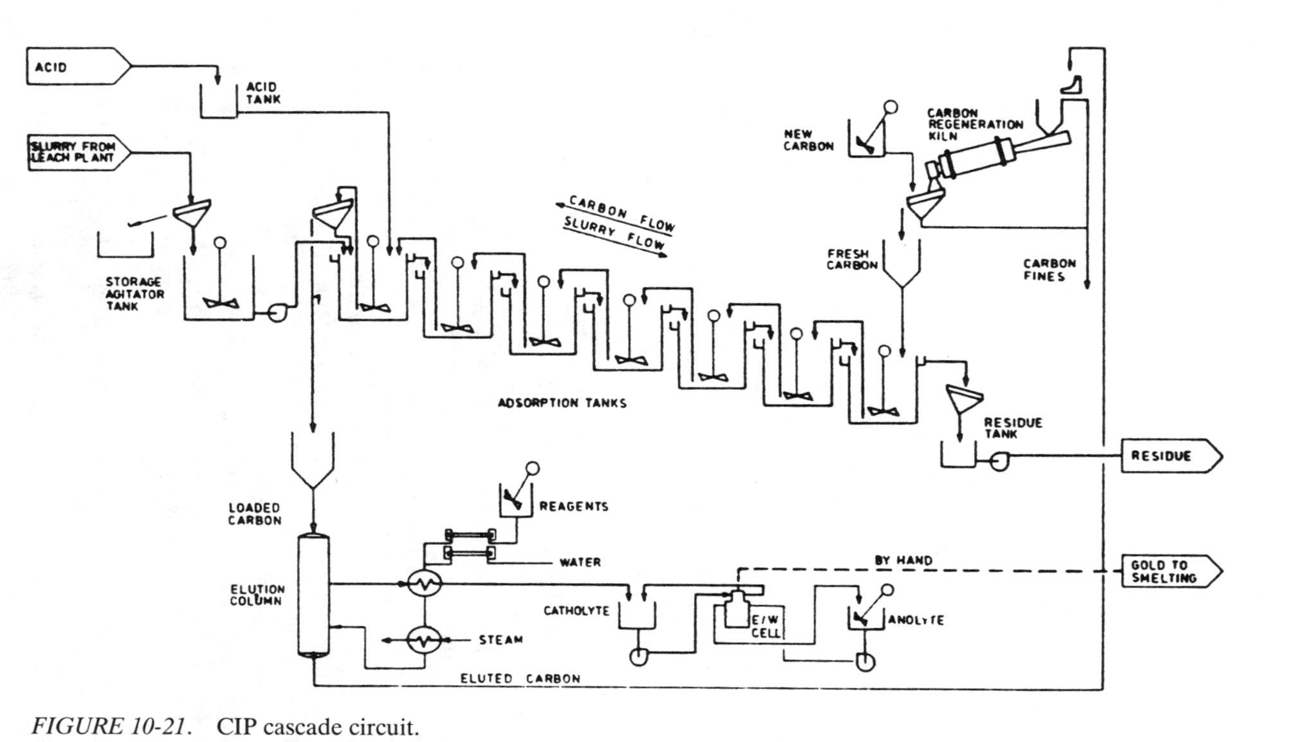

FIGURE 10-21. CIP cascade circuit.

Reprinted by permission. G. C. Young, W. D. Douglas, and M. J. Hampshire, in *Mining Engineering*. Society for Mining, Metallurgy, and Exploration (Littleton, CO), March 1984.

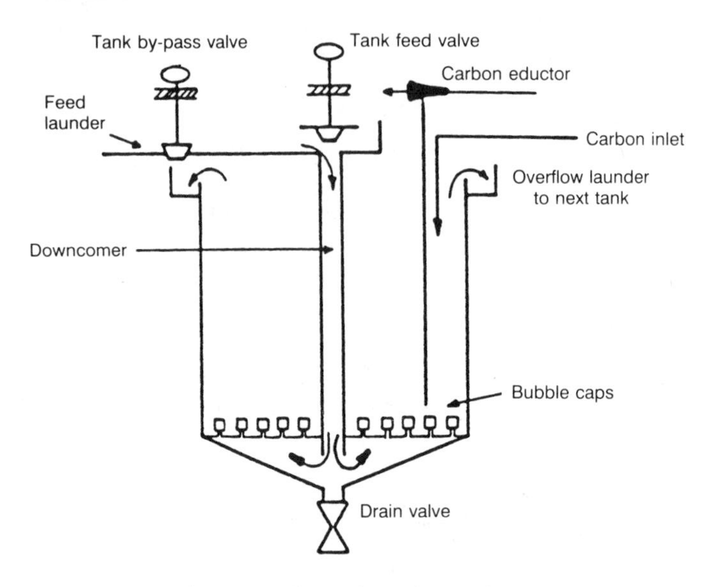

FIGURE 10-22. Cascade-type carbon adsorption tank.

Reprinted by permission. R. Pyper and S. G. Allard, Consideration in choice of carbon adsorption systems for heap leaching. Paper XXIII, *First International Symposium of Precious Metals Recovery* (Reno, NV), June 1984.

leaching is usually accomplished in four to six stages (using agitated tanks) to allow for the required cyanidation reaction. The ore, ground to -100 mesh or finer, is leached with cyanide solution as a thick (45–50% solids) slurry. Lime is added to maintain a protective alkalinity (pH of greater than 10); the pulp is vigorously aerated and agitated to promote the solid-liquid reaction. The required retention time depends on the maximum size of the gold grains, and may vary from 2 to 24 hours (see Chapters 4 and 8). Deep tanks with turbine agitators of low tip speed are preferred. Pachuca tanks have also been successful in leaching gold ores. The dissolution of gold and silver must be essentially completed before the leached pulp flows to the CIP adsorption circuit.

The gold-recovery circuit may include counter-current cyclone washing and/or thickening to supply a 40–50% slurry to the CIP circuit. The underflow pulp flows through trash screens (to remove wood chips and coarse grains) and is fed to the CIP circuit.

The CIP adsorption circuit is a cascade of six to eight large (e.g., 9 m dia. $\times$ 9.25 m) agitated tanks, each of which is equipped with a number of equalized pressure-air cleaned internal launder screens (Figure 10-22). The screens retain the granular carbon while allowing the fine ore slurry to advance to the next stage. Carbon is moved counter-currently to the flow of the ore pulp, from stage to stage, by vertical, recessed impeller pumps.

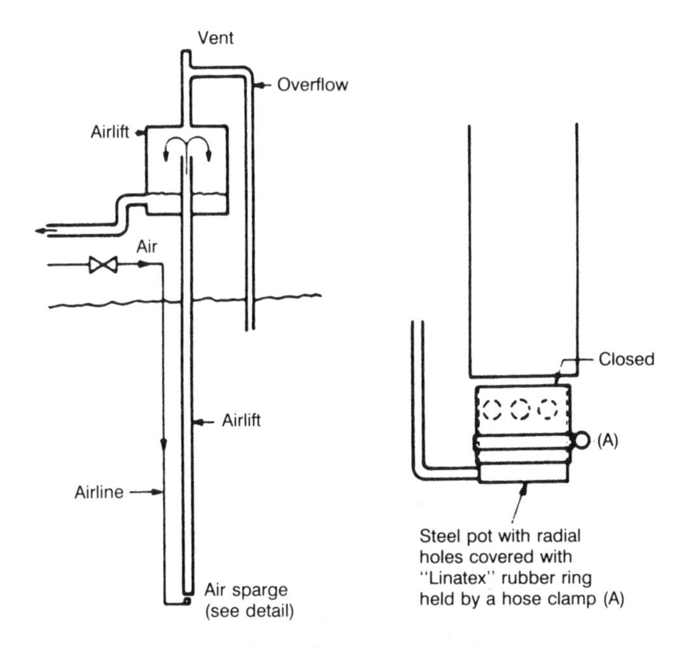

FIGURE 10-23. Air injector for carbon/slurry transfers.

Reprinted by permission, G. C. Young, W. D. Douglas, and M. J. Hampshire, in *Mining Engineering*. Society for Mining, Metallurgy, and Exploration (Littleton, CO), March 1984.

Carbon concentration in each tank is maintained, roughly constant, at 15 to 40 g/l, depending on the operation. Carbon can be loaded with 3,000–4,000 g Au/ton when removed from the screen of the first CIP tank (which is the last tank for the carbon flow).

Two types of interstage screening devices have been used for the separation of carbon from pulp. In the Homestake separation circuit, there is a 20-mesh vibrating screen above each tank. The pulp and the entrained carbon are raised by means of an outside airlift onto the vibrating screen above the tank. The pulp flows through the screen to the next CIP tank in line; the coarse granular carbon is retained on the screen and is discharged into the tank from which it came. Carbon is moved counter-currently between stages by vertical, recessed impeller pumps (Figure 10-23).

In the MINTEK concept of the CIP circuit, 20-mesh, air-cleaned screens are installed in the periphery launder of the adsorption tanks. The screens retain carbon within the tank while allowing the overflow slurry to advance from stage to stage.

A prescreening operation, such as that presented in Figure 10-24, can reduce or even eliminate screening problems between the CIP stages. Some advantages of the prescreening procedure include the following.

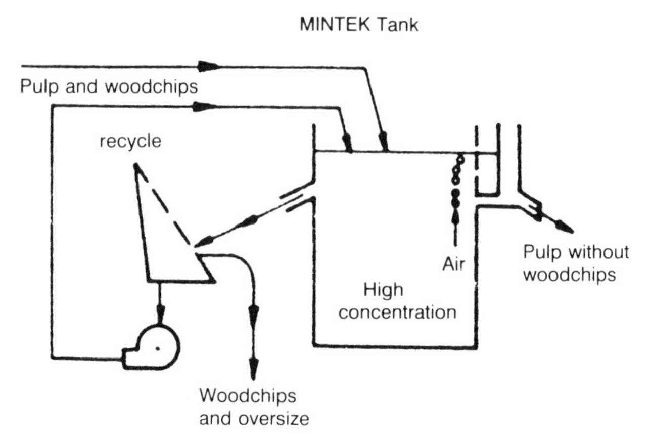

FIGURE 10-24. Flow diagram for prescreening woodchips.

- The tank can serve to absorb surges in the quantities of woodchips and in the volume of pulp.
- The volume of pulp passing through the vibrating screen is relatively small.
- Washing the high concentration of woodchips on the vibrating screen results in more efficient operation.
- The efficiency of screening woodchips by the almost-vertical screens is very much greater than the vibrating screen.
- The pulp passing through this 28-mesh vertical screen passes very readily through similar 20-mesh screens on the periphery of the absorption tanks.

The optimum parameters for CIP are

pH	10–11
Pulp density	40–45% solids
Free cyanide	0.05% into circuit; more than 0.015% out of circuit

The pulp-density limits are important for the intermixing of the carbon with the ore suspension (Figure 10-25). If solids make up less than 40% of the pulp the carbon tends to sink; if the pulp is more than 45% solids, the carbon floats.

With the proper conditions of CIP above, gold will load in preference to copper. If there is mercury in the ore, it will follow the gold and must be retorted from the final product. Lead, zinc, arsenic, antimony, cobalt, and nickel will also be adsorbed, but they usually do not create problems.

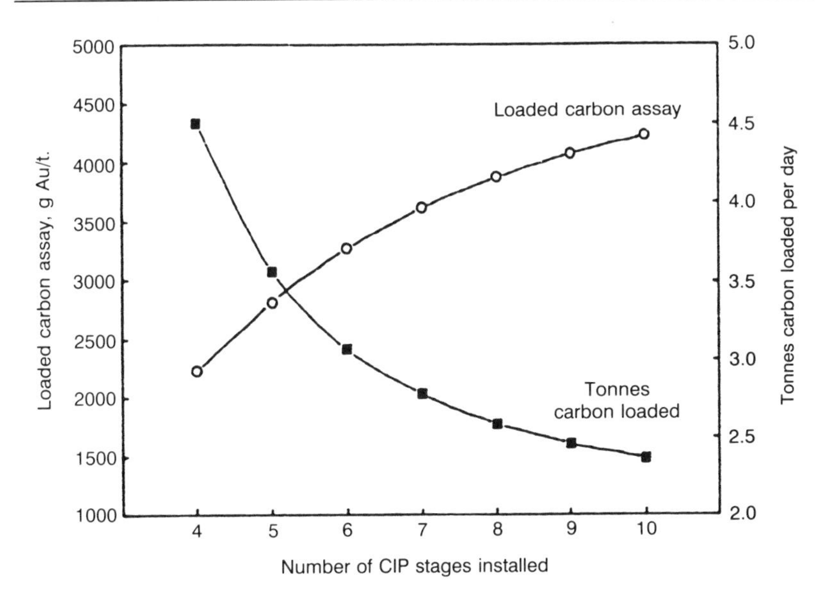

FIGURE 10-25. Effect of number of CIP stages on the assay and tonnage of loaded carbon.

Reprinted by permission. J. A. King, in *Proceedings of the International Symposium on Gold Metallurgy*. Canadian Institute of Mining and Metallurgy (Montreal), 1987: 66.

Pachucas can be used as CIP tanks and can handle somewhat coarser particles than agitated tanks. The impeller in agitated CIP tanks must be large and sweeping and have low tip speed. There is always some loss of gold with the carbon, but this can be minimized by screening the carbon after regeneration.

The optimum loading of the carbon has to be determined taking the economics of the operation into account. Frequent stripping and handling is costly, generates losses through fines, and may lower the capacity of the carbon (Figure 10-26). It is recommended that the gold loading on the carbon be kept moderately low, 150–300 oz./ton, to maintain a low gold inventory and control losses through uncollected carbon fines.

The carbon should be pre-abraded before use. Use of low-tip-speed agitators, plastic pipes with long-radius sweeping curves, and a minimum number of fittings is helpful in maintaining the integrity of the carbon grains. Some gold mills use dual impellers in the CIP tanks to lower the impeller tip speed. Coconut carbon should be used in CIP, since it resists abrasion. Depending on the relevant economics, a number of CIP plants use soda ash rather than lime to avoid lime scale buildup on both carbon and screens.

On-Line Analysis of Carbon-in-Pulp for Gold. An on-line analysis for gold can contribute to the smooth operation and control of a CIP plant.

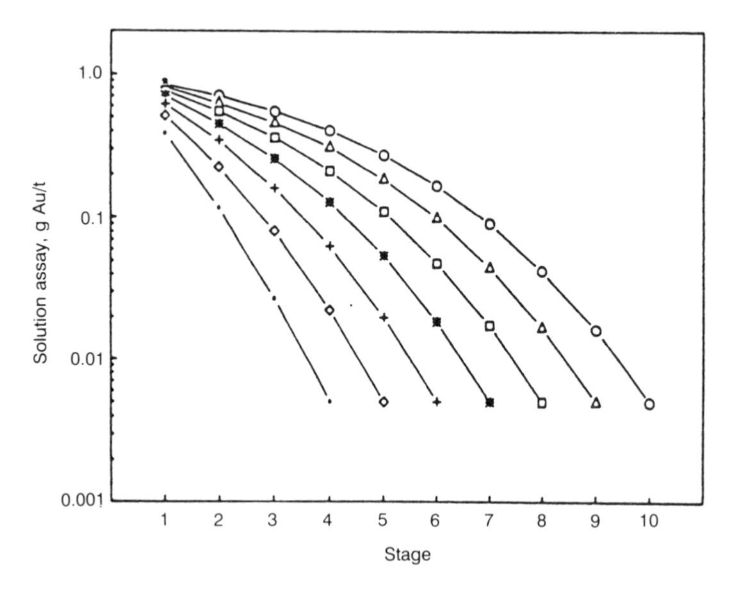

FIGURE 10-26. Effect of number of CIP stages on solution profiles for 0.005 g Au/t barren solution. (10,000 t/day feed solution with 1.0 g Au/t.)

Reprinted by permission. J. A. King, in *Proceedings of the International Symposium on Gold Metallurgy*. Canadian Institute of Mining and Metallurgy (Montreal), 1987: 66.

MINTEK has developed an on-line gold analyzer that uses atomic absorption spectrometry (AAS), allowing gold determination down to 0.001 g/ton in solution. Sample solutions are obtained automatically from the CIP tanks with properly designed sampling probes. In some South African gold mills, a profile of gold in solution is obtained at predetermined time intervals in order to optimize gold extraction.

Magnetic Carbon in CIP Systems. Magnetic carbon—made by cementing or incorporating magnetic particles onto the grains of activated carbon during its manufacture—was proposed as an alternative for the CIP systems. This "Magchar Process," patented in 1949 (Herkenhoff & Hedley, 1949), has been defunct since its inception mainly due to the high preparation cost of magnetic carbon and the high capital cost of magnetic separators (as compared to screens). The expected advantages of the Magchar Process are twofold.

1. The magnetic carbon can be fine and still permit treatment of coarse-ground ore pulps.
2. The adsorption of gold can be very fast due to fine magchar particles.

Carbon-in-Leach

In the carbon-in-leach (CIL) process, cyanidation of the ore and adsorption of soluble gold on carbon proceed simultaneously (Figure 10-27). Since adsorption is significantly faster than leaching gold from its ores, the number and size of the required CIL tanks is determined by the leaching properties of the ore. The rate of adsorption is generally proportional to the gold concentration in solution; hence it is advantageous to use the initial vessels of the circuit for plain preleaching. A convincing reason for selecting CIL technology is the presence of organic carbon in the ore feed, which robs soluble gold from the pregnant solution and thus increases the gold loss in the tails. The CIL process substantially prevents this gold robbing by adsorbing the solubilized gold, mostly on the added activated carbon.

The fine ($-660\ \mu$) ore pulp in the CIL tanks contains 6–10 g of coarse ($3300 \times 1200\ \mu$) carbon per liter. In CIL, pH is maintained at approximately 11.0, and cyanide concentration is maintained at 0.25 g/l (consistent with the cyanidation requirements). The fine slurry is transferred from tank to tank counter-currently to the activated carbon, as in the CIP process. Each tank is coupled with one vibrating screen. The ore slurry and the carbon are moving from tank to tank with airlifts. There is a loss of fine carbon in the tails due to abrasion. To minimize the fine-carbon loss, the carbon is screened after every regeneration cycle.

Carbon-in-Leach with Oxygen (CILO)

Elmore et al. (1984) introduced the use of oxygen blowing rather than air in the CIL process. The solubility of oxygen in water is 4.8 times greater when the water is in contact with oxygen instead of air. The solubility of gold, when there is adequate cyanide present, is controlled by the oxygen concentration in solution. Hence, gold extraction is significantly faster, and cyanide consumption is impressively lower, when oxygen is blown into the ore slurry under cyanidation.

Carbon-in-Pulp Versus Carbon-in-Leach

CIP and CIL are currently the principal routes for recovering gold and/or silver from pregnant solutions. As already mentioned, leaching generally requires a much longer pulp residence than adsorption. Therefore, it is possible to reduce the total required capacity (volume) of CIP equipment by using the leaching vessels for cyanidation and simultaneous adsorption (and ending up with a CIL circuit).

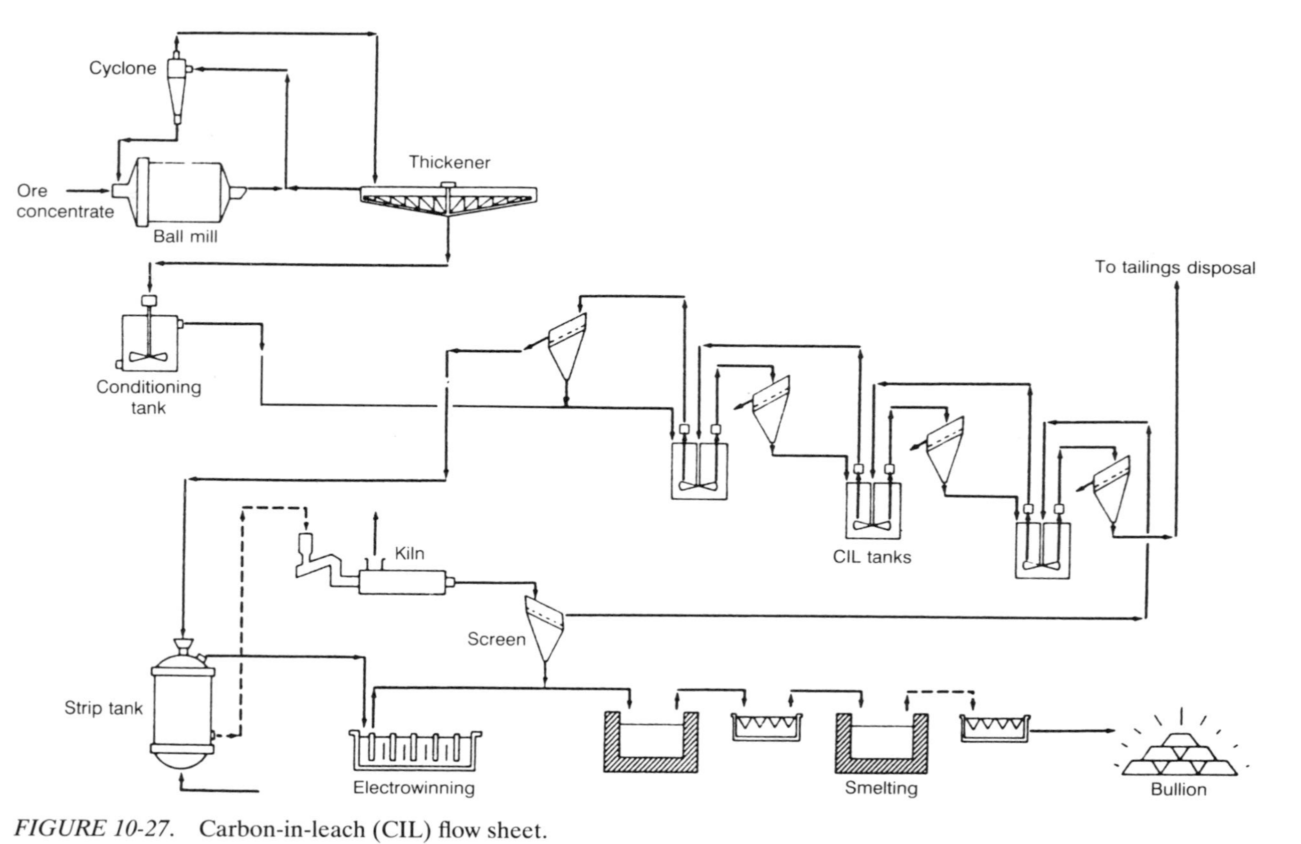

FIGURE 10-27. Carbon-in-leach (CIL) flow sheet.

Reprinted by permission. C. A. Fleming in *Extraction Metallurgy* 85: 757–87. I.M.M. (London), 1985.

TABLE 10-2. Comparison of CIL and CILO treatments with a relative time factor of 4.8.

	Leach time, hrs		*Au extraction, %*		*NaCN consumed, g/t*	
Sample	*CIL*	*CILO*	*CIL*	*CILO*	*CIL*	*CILO*
Ore A	24	5	90.2	89.5	65	15
B	24	5	95.4	95.7	60	10
C	24	5	90.1	89.6	100	25
D	24	5	79.4	76.8	145	80
E	24	5	83.1	78.4	240	160
F	24	5	79.0	84.2	260	190
G	24	5	59.1	61.9	90	20
H	24	5	98.3	97.2	205	10
Tailing A	96	20	68.1	68.3	260	130
Test avg.			82.5	82.5	160	70

Reprinted by permission. C. L. Elmore et al., Development of a carbon-in-leach with oxygen (CILO) process for gold ores. Paper distributed by Kamyr Inc., Ridge Center (Glens Falls, NY), 1984.

Very limited tests comparing CIP and CIL performances with identical ores has been carried out. Newrick et al. (1983) carried out a conceptual comparison of CIP *vs*. CIL by using computer programs based on a mathematical model for CIP developed by MINTEK's Laxen and Fleming 1982. This conceptual comparison is summarized in Table 10-3.

With increased tonnage and/or gold and silver values, the amount of carbon to be handled increases substantially, and large carbon-handling facilities are required. At some high grade of ore (1.5 oz. Au + Ag/ton), it becomes less economical to operate a CIP plant than a Merrill-Crowe zinc-precipitation process. However, every gold plant—even with Merrill-Crowe—should be using carbon columns to scavenge precious metals from bleed streams and tailings return water. It is worth noting that there are operations with a zinc-precipitation plant followed by a carbon-in-column installation, which minimizes the gold loss from the zinc-circuit barren solution.

Carbon-in-Column/Fluidized-Bed Contactors

The adsorption of gold from unclarified solution by activated carbon in a multistage carbon-packed column was investigated by Mehmet et al. (1986) (Figure 10-28). Satisfactory gold extractions—up to 99.6% recovery—can be achieved, with an average concentration of 0.0077 ppm in the barren solution from a feed averaging 1.89 ppm. Excessive loading of $CaCO_3$ on the carbon can be prevented by adjusting the pH of the pregnant feed to about 7.5.

TABLE 10-3. Comparison of Merrill-Crowe cementation *vs*. carbon adsorption.

	Merrill-Crowe
Advantages	• Low labor costs for operation, maintenance • Precious-metal concentration in the leach liquor has little effect on chemical requirements • Low capital expenditure for installation • Can handle large silver-to-gold ratios in pregnant liquor
Disadvantages	• Pregnant solution needs pretreatment prior to precipitation • Process is sensitive to interfering ions
	Carbon systems
Advantages	• No pretreatment of pregnant liquor required • Process handles slimy and carbonaceous ores • Very efficient recoveries, irrespective of incoming precious-metal concentration • Higher gold recovery (up to 99.9%) than Merrill-Crowe, since tail-solution values of gold and silver are lower than with zinc; carbon systems get virtually all the soluble values that would be lost with filter cakes or CCD final underflow
Disadvantages	• High silver grade in pregnant liquor results in high carbon movement • Carbon is susceptible to fouling by calcium and magnesium salts • Carbon regeneration and stripping are labor intensive • Adsorption processes are more expensive to put on line than zinc-cementation operations • High tie-up of gold in inventory • Loss of gold with fine carbon • CIP sends more cyanide to the tailings pond than CCD and there is no recycling of solution as with CCD.

The pregnant solution should flow upward, mildly fluidizing the carbon in the column. Excellent liquid-solid contact is achieved by the flow of solution through the fluidized carbon. A wire mesh screen of the proper size should be installed at the barren-solution outflow to trap carbon-entrained particles. A series of columns or tanks can be used. The solution enters the first column containing the most gold-loaded carbon, cascades to the following column, and leaves as barren solution from the "last" column containing fresh (or the least-loaded) carbon.

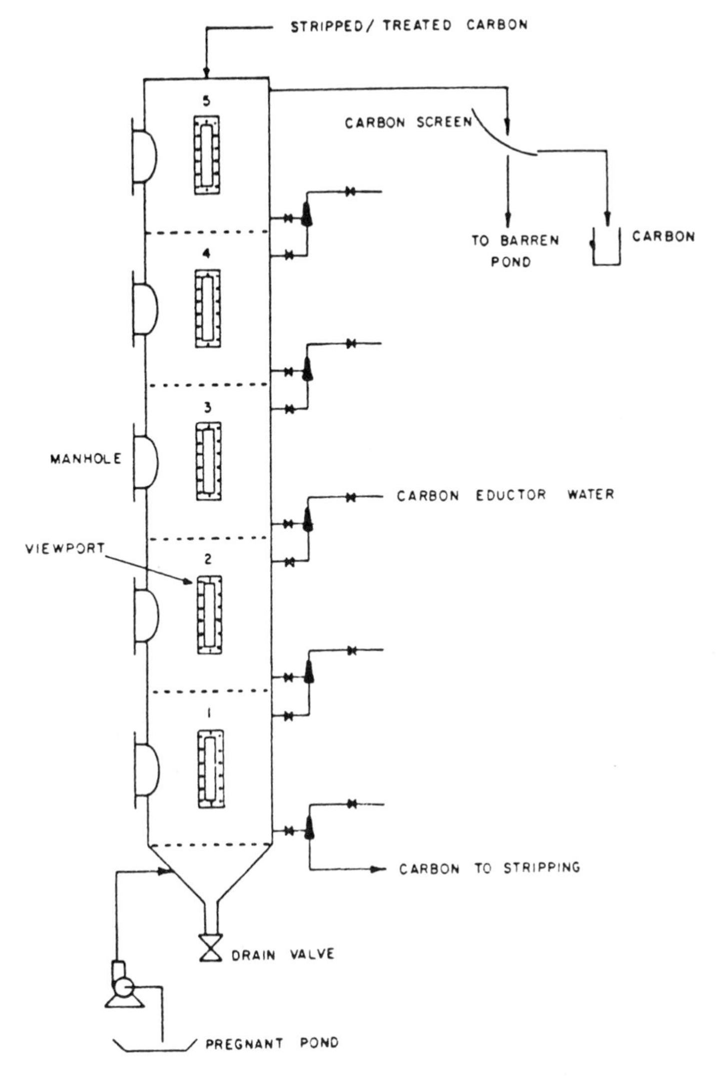

FIGURE 10-28. Typical five-stage adsorption column.

Reprinted by permission. R. Pyper and S. G. Allard, Consideration in choice of carbon adsorption systems for heap leaching. Paper XXIII, *First International Symposium on Precious Metals Recovery* (Reno, NV), June 1984.

Activated carbon in a series of fluidized-bed contactors is suitable for the treatment of unclarified pregnant solutions from heap-leaching operations. It constitutes, in practice, the method of gold recovery in such operations.

Ion-Exchange Resins

The use of activated carbon has tremendously improved the efficiency of gold recovery from pregnant solutions. However, several significant drawbacks are inherent in the use of activated carbon (see page 228).

An alternative technology studied since the late 1940s is the application of anion-exchange resins for the recovery of gold from cyanide solutions. Ion-exchange resins have been used successfully in the uranium industry. Attempts have been made to transfer this technology to the processing of gold ores since 1951. The number of publications on the potential use of ion-exchange resins in gold processing is impressively large but, in spite of this, there is no big-scale operation in the Western world.

Industrial-scale use of this technology appears limited to the U.S.S.R. The first reported use was at the very large Muruntau gold deposit in western Uzbekistan. This flow sheet incorporates semi-autogenous grinding, with gravity concentration of gold by tabling of the autogenous mill discharge. The classifier overflow is thickened, and the thickener underflow is leached with cyanide in Pachuka tanks. The partially leached pulp is then contacted with IX resin in a resin-in-leach operation. The resin is eluted and recycled to the leaching operation. Gold is precipitated from the eluate, cast into anodes, and purified electrolytically. Although no detailed information has been published on the Muruntau operation, it is surmised that IX-resin technology was selected for this operation on the basis of previous relevant experience in uranium technology.

The Western world has adopted carbon-in-pulp and carbon-in-leach technology due to difficulties in separating resins from pulps and eluting resins, and the generally poor resistance of the resins to abrasion. The most intensive effort on the development of IX resins for gold recovery has been going on at the South African Council for Mineral Technology (MINTEK).

There is need for a resin developed specifically for the recovery of gold from cyanide solutions. The ideal resin should be selective for the precious metals in alkaline solutions, have a high loading capacity, be easy to elute and regenerate, and be coarse and physically strong and durable.

IX resins are claimed to have the following advantages, as compared to activated carbon.

- Faster kinetics and higher equilibrium loadings
- Lower temperature and pressure of elution
- No requirement for thermal reactivation
- Less sensitive to fouling and poisoning

Some of the disadvantages cited for presently available IX resins, as compared to activated carbon, are as follows.

- Less selectivity
- Small particle size
- Poor physical strength
- Low bead density
- High cost

Some of the disadvantages of the IX resins can be overcome, but only at the expense of reducing the claimed advantages. IX resins can be eluted at low temperatures and pressures; however, the stripping is generally incomplete and, after a number of cycles, the resins must be incinerated to reclaim the contained gold values. With the high cost of resins, their incineration makes the IX process unattractive.

Anionic resins are used almost exclusively for the recovery of precious metals from solutions. Anionic resins, both weakly and strongly basic, are employed. The strongly basic resins are generally less selective and more difficult to elute, but they load more rapidly and to a higher level than weak-base resins.

Weak-base resins are, by definition, those having primary, secondary, or tertiary amine groups attached to the hydrocarbon matrix. The amine groups become deactivated and stop functioning as ion exchangers above a certain pH known as pKa. According to Riveros and Cooper (1987), "all commercial weak-base resins have a pKa lower than the normal pH of 10–11 of a cyanide (mill) leach liquor." Hence, the pH of mill solutions has to be adjusted to lower than pKa in order to maintain the integrity of the weak-base resin.

Green and Potgeiter (1984) synthesized polyvinyl resins with imidazoline and tetrahydropyridine as active groups and with pKas of 10.2 and 9.7, respectively, (i.e., about 2 pKa units higher than those of commercial weak-base resins). Gold can be eluted from weak-base resins more rapidly, more efficiently, and with cheaper chemicals than from strong-base resins. Therefore, the operating costs should be lower for a gold recovery process using a weak-base resin.

Strong-base resins adsorb metal-cyanide complex anions over a broad pH range, including the operating range (10–11) of gold-cyanide pregnant solutions. They have a higher loading capacity for the complex anions of precious metals than weak-base resins. Hence, a smaller inventory of strong-base resin is required for a given gold mill, and the gold elution and gold recovery equipment will be smaller. Strong-base resins are significantly cheaper than weak-base resins. The main problem with strong-base resins is the difficulty of their elution. In fact, the destruction of the resin has been proposed to recover the adsorbed gold.

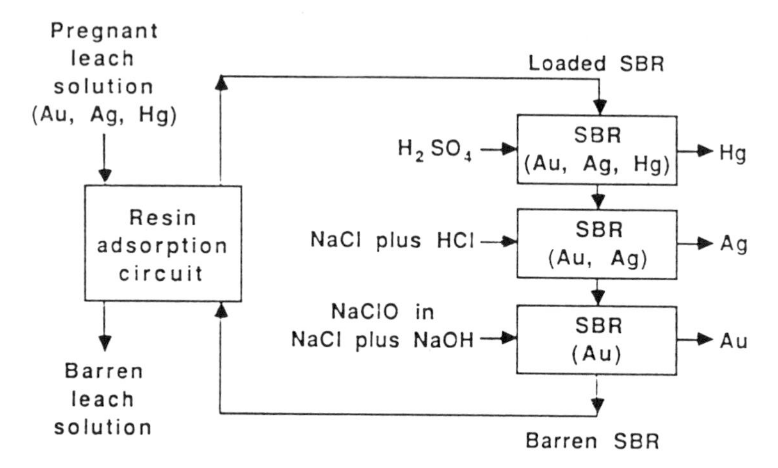

FIGURE 10-29. Ion-exchange flow sheet for recovery of mercury, silver, and gold by sequential elutions.

Reprinted by permission. P. A. Riveros and W. C. Cooper, *Gold Metallurgy*. Canadian Institute of Mining and Metallurgy (Montreal), 1987: 385–6.

Ionac ASB-1, a strong-base resin made by Sybron Chemicals, claims a capacity of 0.7 g of gold per gram of dry resin, and an operating capacity of 50–100 oz. Au/ft. Gold exchange up to 95–98% on the resin is claimed at flow rates of ten bed-volumes per hour through an ASB-1 resin bed. Zinc cyanide solution can regenerate the ASB-1 resin, yielding as eluate a concentrated gold cyanide solution (Dayton, 1987).

Palmer (1986) has demonstrated the feasibility of sequential elution of mercury, silver, and gold from strong-base ion-exchange resins with successive eluants. Sulfuric acid (2N H_2SO_4) is followed by hydrocloric acid (1N HCl with 200 g/l NaCl) and sodium chloride solution (NaCl 150 g/l, 0.5% NaClO and 5 g/l NaOH); see Figure 10-29.

Fleming (1985) proposed elution with thiocyanate anion and continuous removal of gold cyanide, by electrolysis, from the recirculating eluate (Figure 10-30). The effectiveness of this method is due to the affinity of anion-exchange resins for the highly polarized thiocyanate anion. However, this strong affinity creates the problem of how to displace the thiocyanate anion from the resin after gold elution.

Resin-in-Pulp Process

Commercial-size plants using ion-exchange resin-in-pulp (RIP) have been in operation for the recovery of uranium since the early 1950s. Such plants treated "thin" pulps; attempts to treat high-density pulp by "floating resin" failed.

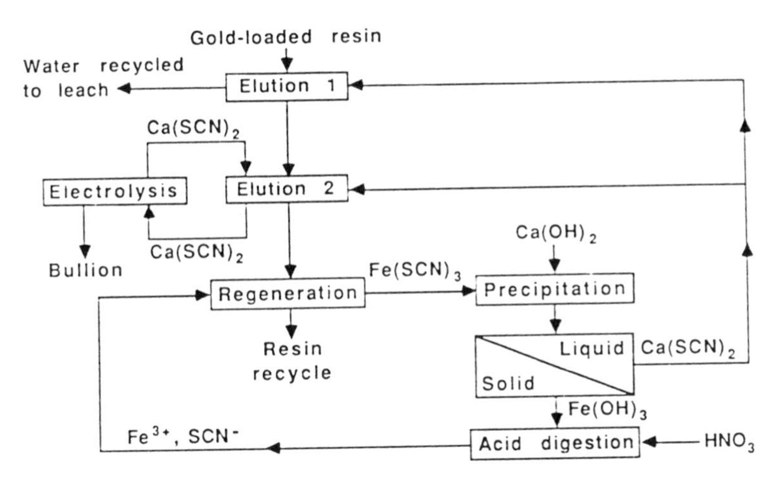

FIGURE 10-30. Aurocyanide elution from a strong-base resin, followed by regeneration of the resin with ferric ions.

Reprinted by permission. P. A. Riveros and W. C. Cooper, *Gold Metallurgy*. Canadian Institute of Mining and Metallurgy (Montreal), 1987: 385–6.

Attempts to apply resin-in-pulp in cyanided gold ore slurries indicated that the resin has a strong tendency to float in pulps with 40–45% solids (as employed in CIP). The expected cost advantages of RIP in "thick" gold ore slurries (as compared with CIP) would disappear completely if the resin-in-pulp process had to operate in "thin" (6–8% solids) slurries.

Ion-Exchange Fibers

In resin-in-pulp technology, the resin should be separated from the ore pulp by screening, as in CIP and CIL operations. Resins coarse enough for this purpose can be manufactured. However, loading rates decrease rapidly as the diameter of the resin particles is increased, due to slow diffusion of ions toward the center of the bead.

MINTEK suggested the use of a fibrous ion-exchange material that would provide a large surface area for the attachment of functional groups capable of extracting metals from solutions. The kinetics should be fast, as the diameter of the fibers would be about 10 microns. However, ion-exchange fibers could break during RIP operations, and make the separation of the IX fibers from wood fibers resulting from mine-timbering operations difficult. Fibrous IX materials could be produced in the form of filaments, cloth, or nonwoven material.

Elution of Ion-Exchange Resins

Elution methods for gold-loaded IX resins include the use of sodium hydroxide, ammonium thiocyanate, sodium perchlorate, zinc cyanide, thiourea and acids, dimethylformamide, and electro-elution. Weak-base resins can be eluted by treatment with sodium hydroxide solution. This points up a weakness of these resins in cyanide-gold metallurgy, as cyanide systems normally operate at a relatively high pH and elution can occur. Regeneration is not required after elution with sodium hydroxide.

Ammonium thiocyanate is reported to elute strong-base resins effectively, and it allows for relatively easy regeneration of the resin before recycling. The addition of sodium hydroxide to thiocyanate allows zinc, iron, and copper to be selectively eluted from IX resins. Zinc cyanide is an effective eluant for strong-base resins, but regeneration of those resins is difficult after this treatment. Dimethylformamide has been used in the plating industry. Electro-elution was studied in both South Africa and the U.S.S.R. during the 1950s, but there is no recent mention of its use.

Claims as to the ease and/or completeness of IX-resins elution should be examined carefully. Most elution tests reported do not involve multicycle (loading-elution-regeneration) experimentation, and the results can be misleading. Ashing avoids problems with elution and regeneration, but probably can be justified economically only with high-grade, clean solutions that allow high loadings of gold on the resin.

The incineration of IX resins is not easy, since it produces environmentally hazardous fumes, which are mainly heterocyclic amines and have a very unpleasant odor. In small-scale experiments, these malodorous amines can be removed by bubbling the off-gases through dilute aqueous acid solutions.

Any resin employed in the recovery of precious metals will ultimately become poisoned and exhausted. Incineration will probably be required, at some time, to recover the precious-metals content of the exhausted resins.

In view of the problems associated with the elution and regeneration of conventional IX resins, some attention has been given to the production of low-cost resins that could be used once and incinerated to recover the gold. The Australian CSIRO studied the combustion of polydiallylamine resins, which are reportedly cheap, easy to make, and selective for gold from cyanide solutions.

Ion-exchange resins have failed, so far, to achieve widespread acceptance in the gold extraction industry. Even ion-exchange resins in columns encounter problems when treating gold pregnant solutions from heap leaching containing more than traces of copper and other nonferrous metals. The lack of a selective-for-gold ion-exchange resin with appropriate density for RIP, the absence of an efficient resin elution process, and the small size of resin beads prevent the extensive use of ion-exchange resins in the gold extraction industry.

Solvent Extraction

Mooiman and Miller (1983) and Mooiman et al. (1984) presented interesting laboratory results on the solvent extraction of gold and silver from cyanide solutions. Uranium and gold are extracted by solvent extraction from the pregnant leach solution of a tailings treatment plant in South Africa (von Michaelis, 1987). Uranium is first stripped from the pregnant organic phase; gold is electrowon from the aqueous solution after solvent regeneration.

Wan and Miller (1986) have shown that gold can be recovered from waste alkaline cyanide-electroplating solutions by solvation extraction with alkyl phosphorus esters, such as tri-n-butyl phosphate (TBP) and di-n-butyl n-butyl phosphate (DBBP). Stripping of these loaded organic extractants with acids or bases is not feasible. However, gold can be recovered by direct electrolysis of the organic phases.

An interesting development from the U.S.S.R. (Ivanoyski et al., 1975) is the use of liquid ion exchanger (solvent extractant), contained in porous plastic slabs and immersed in the ore pulp, in order to extract gold cyanide as in a resin-in-leach system. Obviously, plugging of the pores of the plastic could be a very serious problem.

Metal Chelating Agents

A patented metal chelating agent, called "Vitrokele," bounded to a porous substrate such as polystyrene, was reported to have been tested for gold adsorption (Holbein et al., 1988). Vitrokele is claimed to be more selective for gold and also physically stronger than activated carbon. According to the inventors, Vitrokele can be eluted at significantly lower temperatures than carbon, and it does not need thermal regeneration.

Vitrokeles are complex compounds made to be selective for different metals. They are used in the atomic-energy and water-treatment fields. The Vitrokele complexes contain anionic ligands, not ion-exchange sites.

References

Note: Sources with an asterisk (*) are recommended for further reading.

Adamson, R. J. 1972. *Gold Metallurgy in South Africa.* pp. 120–121. Johannesburg: Chamber of Mines of S. Afr.

Arnold, J. R., and W. J. Pennstrom. 1987. The gold electrowinning/replate circuits at Gold Fields' Ortiz and Mesquite operations. *Min. & Metal. Proc.*, May: 65–67.

Barin, I., H. Barth, and A. Yaman. 1980. Electrochemical investigation of the kinetics of gold cementation from cyanide solutions. *Erzmetall.* 33(7/8): 399–403.

*Briggs, A. P. W. 1983. Problems encountered during the commissioning of the carbon-in-pulp plant at Beisa Mine. *J. S. Afr. I.M.M.*, October: 246–53.

Chapman, T. G. 1939. Cyanidation of gold bearing ores. U.S. Patent 2,147,009.

Cole, P. M., D. S. von Broemsen, and P. A. Laxen. 1986. A novel process for the regeneration of carbon. In *Gold 100. Proceedings of the International Conference on Gold* Vol. 2. Johannesburg: S. Afr. I.M.M., pp. 133–56.

Crowe, T. B. 1918. Effect of oxygen upon the Precipitation of Metals from Cyanide Solutions. *Bull. Am. Inst. Mining Eng.* 1279-82.

Davidson, R. J. 1974. The mechanism of gold adsorption on activated charcoal. *J. S. Afr. I.M.M.*, November: 67–76.

Davidson, R. J., and V. Veronese. 1979. Further studies on the elution of gold from activated carbon using water as the eluant. *J. S. Afr. I.M.M.*, October: 437–45.

Davidson, R. J., W. D. Douglas, and J. Tumilty. 1982. Aspects of laboratory and pilot plant evaluation of CIP with relations to gold recovery. Paper presented at the XIV Internat. Mineral Processing Congress, 17–23 October, at Toronto, Canada.

Davis, W.N. 1880. *U.S. Patent* 227,963.

Dayton, S. H. 1987. Gold processing update. *Eng. Min. J.*, June: 25–29.

*Eisele, J. A., M. D. Wroblewski, M. D. Elges, and G. E. McCleland. 1986. Staged heap leaching and direct electrowinning. *U.S.B.M.* IC 9059.

Elmore, C. L., R. J. Brison, and C. W. Kenney. 1984. Development of a carbon-in-leach with oxygen (CILO) process for gold ores. Paper distributed by Kamyr Inc., Ridge Center, Glens Falls, NY.

Filmer, A. O. 1982. The electrowinning of gold from carbon-in-pulp eluates. In *Carbon-In-Pulp Seminar*. The Australian I.M.M., Perth and Kalgoorlie Branches, and Murdoch University, 49–66.

Finkelstein, N. P. 1972. Recovery of gold from solutions. In *Gold Metallurgy in South Africa*, edited by R. J. Adamson. Johannesburg: Chamber of Mines of S.A., 327–41.

Fisher, J., and D. E. Weimer. 1964. *Precious Metals Plating*. Teddington: R. Draper Ltd.

Fleming, C. A., 1982. Recent developments in carbon-in-pulp technology in South Africa. In *Hydrometallurgy: Research, Development, and Plant Practice*, edited by K. Osseo-Assare and J. D. Miller. New York: The Metallurgical Society of A.I.M.E., 839–57.

Fleming, C. A. 1985. Novel process for recovery of gold cyanide from strong-base resins. In *Extraction Metallurgy 85*. London: I.M.M., 757–87.

Fuersteneau, M. C., C. O. Nebo, J. R. Kelso, and R. M. Zaragosa. 1987. Rate of adsorption of gold cyanide on activated charcoal. *Min. & Metal. Proc.*, November: 177–81.

Green, B. R., and A. H. Potgeiter. 1984. Unconventional weak-base anion exchange resins, useful for the extraction of metals, especially gold. In *Ion-Exchange Technology*, edited by D. Naden and M. Streat. Chichester, U.K.: Horwood, 626–36.

Habashi, F. 1987. One hundred years of cyanidation. *C.I.M. Bull.*, 80(905):108–114.

Heinen, H. J., D. J. Peterson, and R. E. Lindstrom. 1976. Gold desorption from activated carbon with alkaline alcohol solutions. In *World Mining and Metals Technology*, Vol. 1. New York: A.I.M.E., 551–64.

Herkenhoff, E., and N. Hedley. 1949. Magnetic Activated Carbon. U.S. Patent 2,479,930.

Herkenhoff, E. 1982. Magchar: An alternative for gold plants using carbon-in-pulp systems. *Eng. & Min. J.* August: 84–87.

Hills, S. D. 1986. The carbon-in-pulp process. *U.S.B.M.* IC 40–43.

Holbein, B. E., D. K. Kidby, and A. L. Huber. 1988. Integrated gold and cyanide recovery with Vitrokele and Cyanosave. Randol Gold Forum 88. 22–24 January, at Scottsdale, AZ.

Ivanoyski, M. D., M. A. Meritvhev, and V. D. Potekhin. 1975. Recovery of gold from ore pulps by means of pore carrier fluid (solvent) extractant. *Soviet J. Non-ferrous Metals* 16(1): 88–91.

Johnson, W. D. 1894. Method of Abstracting Gold and Silver from their Solutions in Potassium Cyanides. U.S. Patent 522,260.

King, J. A. 1987. Gold isotherms. In *Gold Metallurgy*, edited by R. S. Salter, D. M. Wyslouzi, and G. W. McDonald. New York: Pergamon Press, 59–75.

Laxen, P. A., and C. A. Fleming. 1982. A review of pilot-plant testwork conducted on the carbon-in-pulp process for the recovery of gold. Paper presented at the CMMI 12th Congress. 3–7 May, at Johannesburg.

Lazowski, M. 1848. On some properties of carbon. *Chem. Gaz.* (6): 43.

Lowenheim, F. A. 1977. *Electroplating.* New York: McGraw-Hill, 270.

Marques and Nell Inc. 1984. Furnace economically regenerates spent carbon. *Chem. Eng.*, February 6: 35–36 (U.S. Patent 4,374,092).

*McDougall, G. J., R. D. Hancock, M. J. Nicol, O. L. Wellington, and R. G. Copperthwaite. 1980. The mechanism of adsorption of gold cyanide on activated carbon. *J. S. Afr. I.M.M.*, September: 293–305.

Mehmet, A., W. A. M. te Riele, and B. W. Boydell. 1986. *J. Met.*, June: 23–28.

Milligan, D. A., O. A. Muhtadi, and R. B. Thorndycraft. 1988. Metal production. In *Introduction to Evaluation, Design and Operation of Precious Metal Heap Leaching Projects*, edited by D. J. A. van Zyl et al. Littleton, CO: S.M.E.-A.I.M.E., 137–51.

Mooiman, M. B., and J. D. Miller. 1983. The solvent extraction of gold from aurocyanide solutions. *Proceedings ISEC*. New York: AIChE., 530–31.

Mooiman, M. B., J. D. Miller, J. D. Hiskey, and A. R. Hendriks. 1984. Comparison of process alternatives for gold recovery from cyanide leach solutions. In *Gold and Silver Heap and Dump Leach Practice*. New York: A.I.M.E.-S.M.E., 93–108.

Muhtadi, O. A. 1988. Metal extraction (recovery systems). In *Introduction to Evaluation, Design and Operation of Precious Metal Heap Leaching Projects*, edited by D. J. A. van Zyl et al. Littleton, CO: Soc. Mining Eng.-A.I.M.E., 124–36.

Newrick, G. M., G. Woodhouse, and D. M. G. Dods. 1983. Carbon-in-pulp versus carbon-in-leach. *World Min*., June: 48–51.

Nicol, D. I. 1979. The adsorption of dissolved gold on activated charcoal in a NIMCIX contactor. *J. S. Afr. I.M.M.*, December: 497–500.

Nicol, M. J., E. Schalch, P. Balestra, and H. Hegedus. 1979. A modern study of the kinetics and mechanism of the cementation of gold. *J. S. Afr. I.M.M.*, February: 191–98.

Palmer, G. R. 1986. Ion-exchange research in precious metals recovery. *U.S.B.M.* IC 9059: 2–9.

Paul, R. L., A. O. Filmer, and M. J. Nicol. 1982. The recovery of gold from concentrated aurocyanide solutions. In *Hydrometallurgy: Research, Development, and Plant Practice*, edited by K. Osseo-Assare and J. D. Miller. New York: The Metal. Society of A.I.M.E., 689–704.

Pyper, R., and S. G. Allard. 1984. Consideration in the choice of carbon adsorption systems for heap leaching. In *First International Symposium on Precious Metals Recovery*. Paper XXIII. 10–14 June. Reno, NV.

Raub, C. J., H. R. Khan and M. Baumgartner. 1986. High temperature gold deposition from acid cyanide baths. *Gold Bull.* 19(3): 70–74.

Riveros, P. A., and W. C. Cooper. 1987. Ion exchange recovery of gold and silver from cyanide solutions. In *Gold Metallurgy*, edited by R. S. Salter, D. M. Wyslouzil, and G. N. McDonald. New York: Pergamon Press, 379–93.

Ross, J. R., H. B. Salisbury, and G. M. Potter. 1973. Pressure stripping gold from activated carbon. Paper presented at the Annual A.I.M.E. meeting, at Chicago, IL.

Stone, A. M. 1985. Elution process and apparatus for extraction of particulate material from a vessel. U.S. Patent 4,555,385.

von Michaelis, H. 1987. Recovering gold and silver from pregnant leach solutions. *Eng. Min. J.*, June: 50–55.

Wan, R. Y., and J. D. Miller. 1986. Solvation extraction and electrodeposition of gold from cyanide solutions. *J. Met.*, December: 35–40.

Wilkinson, P. 1986. Understanding gold plating. *Gold Bull.* 19(3): 75–81.

Young, G. C., W. D. Douglas, and M. J. Hampshire. 1984. Carbon-in-pulp process for recovering gold from acid plant calcines at President Brand. *Min. Eng.*, March: 257–64.

*Zadra, J. B. 1950. A process for the recovery of gold from activated carbon by leaching and electrolysis. *U.S.B.M.* RI 4672.

Zadra, J. B., A. L. Engel, and H. J. Heinen. 1952. Process for recovering gold and silver from activated carbon by leaching and electrolysis. *U.S.B.M.* RI 4843.

CHAPTER 11

Melting and Refining of Gold

Refining of gold comprises the following sequence of operations: melting, refining, de-golding, and electrorefining. *Melting* gold precipitates and/or gold cathodes is required for homogenization, reliable sampling, and pyrorefining. The crucible—which contains the charge—must be able to resist the chemical action of the fluxes used (mixtures of borax, niter, and silica at varying proportions). A borax-silica-bone ash mixture is often used to cover the melt.

There are four types of furnaces used by gold producers to melt precipitates and cathodes to dore.[1]

- Graphite crucibles heated externally with oil or propane burners
- Reverberatory furnaces fired with gas or propane
- Blast furnaces
- Induction furnaces

All four types of furnaces are fired and operated on an intermittent basis, and are very seldom kept warm all the time. The graphite crucible and the induction furnace are, as a rule, tilting furnaces. The reverberatory furnace is used by large-scale producers, usually in silver refining or in treating copper refining slimes. Smelting of secondary materials can be conducted in a blast furnace specifically designed for this purpose (Embleton, 1981).

The induction furnace is preferable, since it has high speed of melting. It is also the cleanest furnace to operate and the one with the smallest volume of flue gas and the minimum dust losses (only 2% of the reverbatory furnace gas volume). The efficiency, simplicity, and high speed of

[1]Gold-silver alloy, typically containing less than 5% base-metal impurities.

operation achieved by induction equipment more than justify its high capital cost. Induction furnaces tend to give well-mixed melts due to the stirring action of the electromagnetic field. Crucible or reverberatory furnace melts should be stirred well before sampling.

Sampling of molten dore can be conducted in two ways. The most reliable and accepted sampling is with a glass vacuum tube. Special pyrex-glass tubes, sealed under vacuum, can be purchased for sampling. The glass at one end of the sampling tube is thin, and it cracks when immersed in the molten metal; the tube is immediately filled with a dore sample, due to its vacuum. Alternatively, a small ladle can be used to take a deep sample.

Gold refining can be achieved by high-temperature chlorination of the molten metal (Miller process) followed by electrorefining (Wohwill process). *The Miller process* consists of chlorine injection into the molten bullion by means of an immersed clay tube. This process has three stages:

1. Slow reaction of chlorine gas with the base-metal impurities forms volatile chlorides.
2. Rapid production of nonvolatile chlorides, the "Miller salt," and skimming of those salts from the surface of the melt—the molten Miller salt is transported to a holding furnace.
3. Potential formation of volatile gold chlorides may result in furnace losses.

A borax-silica-bone ash mixture is often used to cover the melt and control the loss of volatile gold chlorides. Chlorination of the bullion has to stop close to 99 fineness to restrict the loss of volatile gold chlorides. The Miller process gases are vented through a milk-of-lime scrubbing system, followed by a filter-bag.

In *de-golding*, the nonvolatile silver and base-metal chlorides (Miller salt) contaminated with the borax/silica/chloride slag have to be treated in order to recover any contained gold. Gold in the Miller salt may be as particles mechanically carried over from the molten metal phase, or as gold chloride formed at the end of the chlorination process.

The Miller salt in the holding furnace is treated with sodium carbonate. Three reactions take place to precipitate molten silver.

$$2AgCl + Na_2CO_3 \rightarrow Ag_2CO_3 + 2NaCl$$
$$Ag_2CO_3 \rightarrow Ag_2O + CO_2$$
$$2Ag_2O \rightarrow 4Ag + O_2$$

As the fine silver droplets settle, percolating through the molten chlorides, they scavenge any existing fine free gold. This silver-gold molten alloy is recycled to the Miller process. Base-metal chlorides react with soda ash, forming oxides that float to the surface, and with the borax/silica flux.

The Wohlwill Electrorefining Process

The Miller process can produce 99.9% fine gold, if volatile loss can be collected, but it still contains platinum group metals (PGM) and traces of impurities, mainly silver and copper. The bullion from the Miller process has to be cast into anodes and be submitted to electrorefining in order to produce high-purity gold. Cathode starter sheets are made from titanium metal or from thin gold strips. The electrolyte is an aqueous chloride solution prepared by dissolving gold with chlorine gas in the presence of hydrochloric acid.

$$2Au + 3Cl_2 + 2HCl \rightarrow 2HAuCl_4$$

Gold dissolves from the anodes and is selectively plated at the cathodes, while silver forms insoluble chloride slime. Copper, platinum, and palladium form soluble chlorides and are removed by means of a bleed stream from the electrolyte.

The main ionization reaction is

$$HAuCl_4 \rightarrow H^+ + AuCl_4^-$$

(Ionization of the form $AuCl_3 \rightleftharpoons Au^{3+} + 3Cl^-$ is known to be low.) The following reactions take place at the electrodes.
At the anode

$$Au \rightarrow Au^{3+} + 3e$$
$$Au^{3+} + 4Cl^- \rightarrow AuCl_4^-$$

At the cathode

$$AuCl_4^- \rightarrow Au^{3+} + 4Cl^-$$
$$AuCl_4^- + 3e \rightarrow Au$$

The Wohlwill process has, of course, a very high inventory cost. In order to accelerate this electrolytic process and minimize its inventory cost, high current densities are employed (sometimes exceeding 100 A/ft^2); the electrolyte is heated and its gold content is maintained at a high level; and air is sparged under the cathodes to enhance the diffusion of the gold ions. The Wohlwill cathodes have at least 999^+ fineness.

Gold Refining by Dissolution/Precipitation

Gold bullion can be dissolved in *aqua regia*, and after filtration of the auriferous chloride solution it can be reprecipitated. This dissolution is used in most small refineries, and it is used occasionally in electrolytic refineries to make fresh electrolyte for Wohlwill cells.

Gold bullion in granular form is charged in a reactor lined with acid resistant brick. *Aqua regia*—one part nitric acid with four parts hydrochloric acid—is then metered into the reactor.

$$Au + HNO_3 + 4HCl \rightarrow H[AuCl_4] + 2H_2O + NO$$

Gold and some platinum group metals (Pt, Os, and Pd) dissolve, whereas any silver contained in the bullion is precipitated as silver chloride. Nitrogen oxide that evolves during the dissolution has to be meticulously scrubbed, and this adds a major cost to the operation.

Silver chloride is filtered from the pregnant solution, and gold is precipitated by reducing reagents such as sulfur dioxide gas or sodium metasulfite. The nature of the reductant, and especially the rate of its addition, affect the size and form of the precipitated gold. Very fine gold may create losses during filtration; coarse gold may entrap solution. The precipitated gold, after careful washing, is melted to a bullion with .9995 to .9999 fineness.

PGM separation and recovery is critical to most refiners as an important additional revenue. The group includes:

- Platinum, Pt (II, IV oxidation states)
- Palladium, Pd (II, IV)
- Iridium, Ir (III, IV, VI)
- Rhodium, Rh (II, III, IV, VI)
- Osmium, Os (IV, VIII)
- Ruthenium, Ru (III, IV, VIII)

Platinum and palladium, the most important of the group, are soluble in *aqua regia*, and they follow gold through the refinery. Gold is preferentially reduced, either in the Wohlwill cell or during the SO_2 precipitation, whereas platinum and palladium remain as chlorides in solution. Platinum can be precipitated as ammonium chloroplatinate by the addition of ammonium chloride to the solution. This precipitate is filtered, dried, and ignited to form platinum sponge. Palladium is then precipitated by the addition of dimethylglyoxime and the precipitate, recovered by filtration, is ignited and forms palladium sponge. Alternatively, palladium can be oxidized to its highest state by nitric acid or sodium chlorate, and can be precipitated as ammonium chloro salt. Multiple washes, dissolutions, and precipitations may be necessary to produce high-purity palladium.

Reference

Embleton, F. T. 1981. A new gold refining facility. *Gold Bull.* 14(2): 65–68.

CHAPTER 12

Gold Mill Tailings

Tailings are, by definition, fine-particle residues of milling operations that are devoid of metal values. Particle-size distribution is one of the most important ways of characterizing tailings. The mining industry distinguishes "sands" and "slimes" as components of tailings (usually +200 mesh, 74 μ, for sands). Gold losses to tailings is the most important parameter in deciding whether a deposit or milling process is economically viable or not.

Disposal of Tailings

Tailings may be discarded on land, into a water course, or in a sizable body of water (a lake or sea). In the case of underground mining, at least part of the tailings may have to be pumped back into the mine to backfill excavated space, but usually a significant proportion of the tailings remains to be discarded. In gold milling operations, tailings should not be stored underground, if possible; it is prudent to dispose of gold milling tailings on the surface and have them easily available when higher gold prices or/and more efficient extraction processes exist. Storing tailings in active open-pit mines is obviously impossible, and abandoned open-pit quarries are, as a rule, far from active gold mills. Disposal of gold mill tailings in rivers and lakes is frowned upon for environmental reasons. Factors influencing the design of tailings impoundment are listed in Table 12-1.

Site selection for tailings disposal has to be based on economic and environmental considerations. The tailings impoundment site has to be close to the mill, for economic reasons and to conform with the following three requirements:

TABLE 12-1. Factors influencing impoundment design.

Site characteristics	Topography, drainage area Geology/seismicity Precipitation, evaporation
Tailings characteristics	Gradation (sand-slimes) Clay content Chemical
Effluent characteristics	pH Metallic cations (Pb, Cd, etc.) Anions (SO_4, Cl, etc.) Other (CN, Ra, etc.) Oxidation/leaching potential
Mine/mill characteristics	Tailings output Effluent output Recirculation capacity Sand backfilling

Reprinted by permission. S. G. Vick, in *Min. Eng.* S.M.E. (Littleton, CO), June 1981: 653–7.

1. Be mineralogically barren
2. Have strong structural geology to bear the weight of the impoundment
3. Have a geomorphology that allows surface waters to bypass the dam or drain through it

Tailings dam designs are usually categorized as upstream, downstream, or centerline. These dam types, along with their respective advantages and disadvantages, are described in Figure 12-1. Dam wall designs may be valley dams constructed across dry river valleys; lagoons, normally constructed on flat or sloping grounds, with walls on all sides of the impoundment (usually built from coarse tailings); or valley-side dams, with a curved or multisided wall toward the valley side (avoiding the valley river flow).

Tailings dams can be built with mine overburden and waste rock covered with tailings, or the whole dam may consist of tailings (sands with superimposed fines, separated for construction purposes) (Figure 12-2). A summary of constraint and disposal options compatibility for different tailings disposal options is presented in Table 12-2.

Water/Cyanide Recovery from Tailings Slurries

Cyanide/water recovery from tail slurries may be required in areas where water is scarce or where water is very abundant due to high rainfall. Water should be recycled to the mill, and any excess water should be disposed of after rigorous treatment, for removal of cyanide and metal contaminants. Tailings are partially de-watered before being pumped to the tailings dam

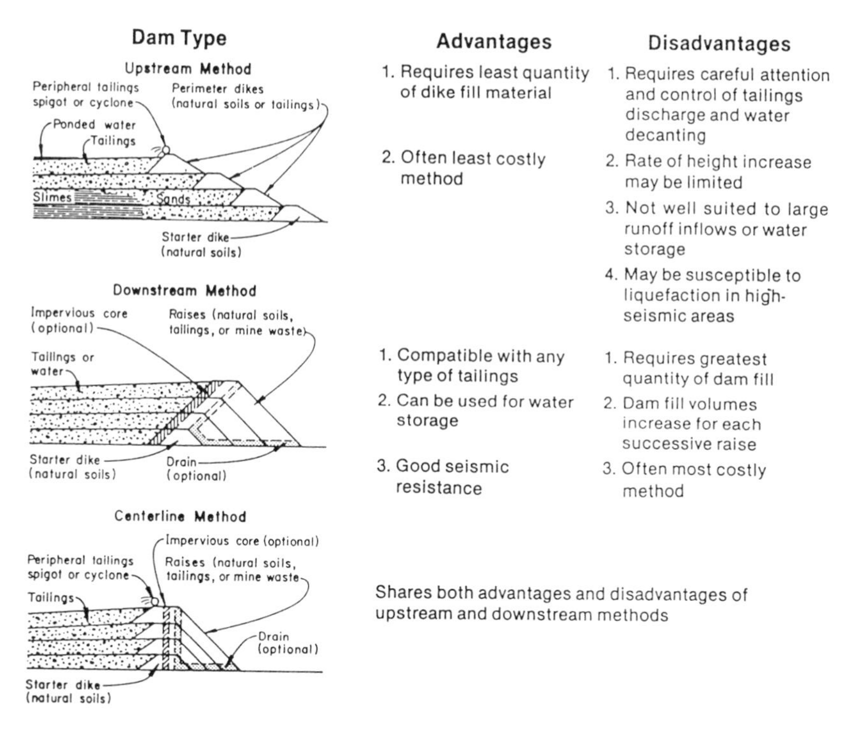

FIGURE 12-1. Tailings dam designs.

Reprinted by permission. S. G. Vick, in *Min. Eng.* S.M.E. (Littleton, CO), June 1981: 653–7.

either by counter-current decantation (CCD) in a series of thickeners (where the tailings are separated from the gold pregnant solution and washed), or by filtration (Young et al., 1986).

Properly designed tailings dams have seepage-collection trenches (see Figure 12-2) from which the collected water is recycled to the tailings or to the mill (depending on the plant water balance), or is disposed of elsewhere after treatment (Figure 12-3). Cyanide should be considered a "transient pollutant," since its toxic properties decrease rapidly over time, without the need of destroying it with chemical agents. Natural degradation of cyanide is caused by volatilization, photodecomposition, biodecomposition, and conversion to thiocyanite (Schmidt et al., 1981).

Excess solution from the tailings dam should be recycled to the mill, to the extent that the dissolved (base-metal) impurities allow such recycling. If excess water has to be rejected, environmental regulations apply to the bleed stream. The bleed stream may have to be discharged through activated carbon or ion-exchange columns, or be submitted to a process that destroys the residual cyanide.

TABLE 12-2. Summary of constraint and disposal options compatibility.

Design constraints	*Disposal option*						
	Upstream dam	Downstream dam	Centerline dam	Stope backfilling	"Below grade"	Offshore disposal	Thickened discharge
Site characteristics							
Narrow valley, high buildup rate	x	•	?		x		?
High runoff, large drainage area	x	•	•		x		x
High seismicity	x	•	?		•		x
Coastal site, deep nearshore water						•	
Tailings characteristics							
Principally sands	•	•	•	•	•	•	•
Principally slimes	x	•	?	?	•	x	x
Effluent characteristics							
Low pH, metallic cations	?	•	•	?	•	x	?
Neutral pH, anions	•	•	•	•	•	•	•
Radiological[1]	x	?	x	?	•	x	x
Other requiring strict seepage limitations or liners	x	•	•	?	•	x	?
Mine/mill characteristics							
Underground mine				•			
High effluent discharge, low recirculation	x	•	?	?	x	?	?

• *usually compatible*
x *usually incompatible*
? *compatibility depends on specific circumstances*
[1]*based on US regulatory requirements*

Reprinted by permission. S. G. Vick, in *Min. Eng.* S.M.E. (Littleton, CO), June 1981: 653–7.

Destruction of Cyanide in Gold-Mill Effluents

In spite of the natural degradation of cyanide, sometimes it has to be destroyed in order to preserve wildlife and comply with environmental regulations. The following processes lead to quick and efficient destruction of cyanides.

- Alkaline chlorination
- INCO's sulfur dioxide-air process
- DEGUSSA's hydrogen peroxide process
- Biological treatment

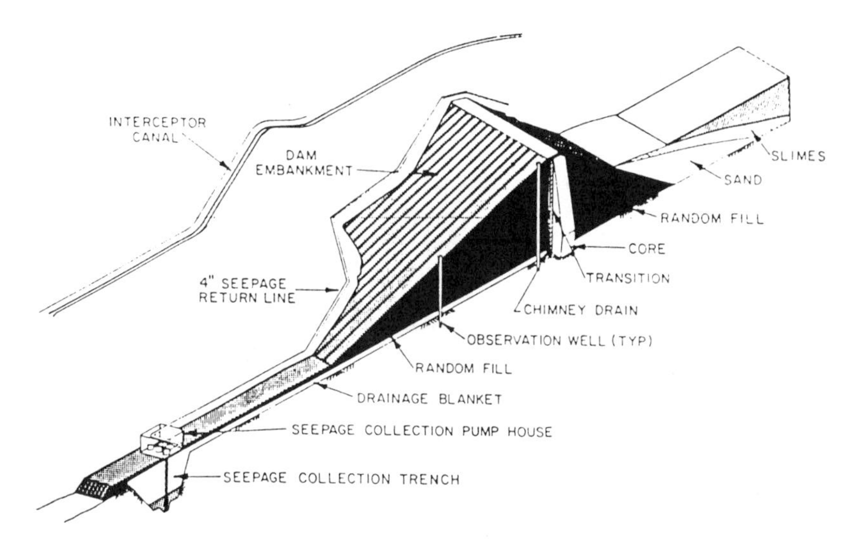

FIGURE 12-2. Section of downstream tailings impoundment.

Reprinted by permission. M. C. Fuersteneau and B. R. Palmer, *Gold Silver, Uranium, and Coal: Geology, Mining, Extraction and Environment*. A.I.M.E. (Littleton, CO), 1987: 483.

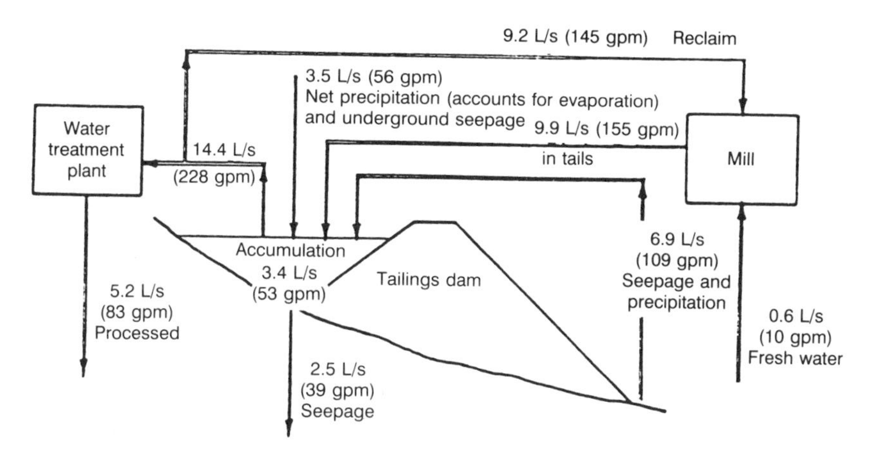

FIGURE 12-3. Average water balance in Noranda's Grey Eagle Mines mill.

Reprinted by permission. G. L. Simmons et al., Norand's carbon-in-pulp gold/silver operation at Happy Camp, CA. *Minerals and Metallurgical Processing*. A.I.M.E. (Littleton, CO), May 1985: 73–79.

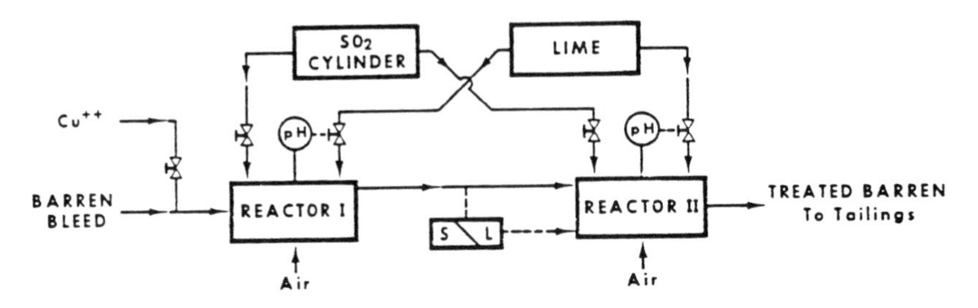

FIGURE 12-4. INCO's SO_2/air process of cyanide destruction.

Reprinted by permission. E. A. Devuyst, B. R. Conrad, and W. Hudson, Industrial application of the INCO SO_2-air cyanide removal process. Paper XXVI, *First International Symposium on Precious Metals Recovery*, (Reno, NV) June 1984.

The alkaline chlorination process oxidizes free cyanide-metal complexes in tailings pond water to carbon dioxide (CO_2) and nitrogen (N_2), and causes the precipitation of heavy metals as hydroxides.

$$CN^- + Cl_2 \rightarrow CNCl + Cl^-$$
$$CNCl + 2OH^- \rightarrow CNO^- \; Cl^- + H_2O$$
$$2CNO^- + 3Cl_2 + 6OH^- \rightarrow 2HCO^- + N_2 + 6Cl^- + 2H_2O$$

The destruction of free cyanide (NaCN), thiocyanate (SCN^-), and copper cyanides is achieved by oxidation, as depicted in the following reactions.

$$2NaCN + 6NaOH + 4Cl_2 + 0.5O_2 \rightarrow 8NaCl + 2CO_2 + 3H_2O + N_2$$
$$2Na_2Cu(CN)_3 + 28NaOH + 16Cl_2 \rightarrow 2Cu(OH)_2 + 6CO_2 + 32NaCl + 12H_2O + 3N_2$$
$$NaSCN + 4NaOCl + 2NaOH \rightarrow NaCNO \rightarrow Na_2SO_4 + 4NaCl + H_2O$$

Water is pumped to the water treatment plant from the tailings pond into the first agitated reactor, where sodium hydroxide is added to maintain pH in the range of 7.8 to 8.5. The water is then circulated through a Venturi-type chlorine injector and back to the first reactor until it is saturated with chlorine. Additional sodium hydroxide is dropped into the second agitated reactor to raise the pH of the water to the 10.8 to 11.5 range and thus prevent the evolution of toxic gas and cause the precipitation of heavy metals (as hydroxides).

INCO's SO_2-air process for cyanide removal (Figure 12-4) is based on the oxidation of cyanide (free or complexed with transition metals other than iron and cobalt).

$$CN^- + SO_2 + O_2 + H_2O \rightarrow CNO^- + H_2SO_4$$

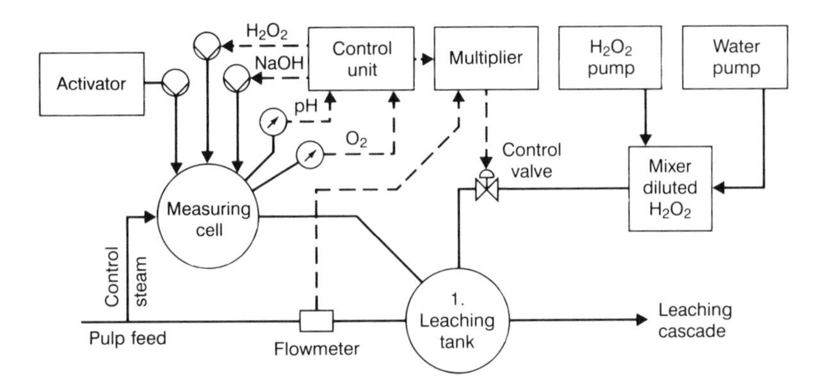

FIGURE 12-5. Flowchart of cyanide destruction with hydrogen peroxide.
Reprinted by permission. Degussa Corporation (Allendale, NJ) 1987.

Copper ions catalyze this reaction and, when the copper concentration in the tailings water is insufficient, some copper sulfate solution has to be added. The optimum operating pH is in the range of 8 to 10, whereas temperature has no significant effect in the range of 5° to 60° C. The suggested concentration of SO_2 in air is at or below 2%. Dilute roaster gases or dilute sulfite solution can be used. Metal ions in the tailings water precipitate either as ferrocyanide compounds (e.g., $Zn_2Fe(CN)_6$) or as hydroxides. Soluble arsenic and antimony contaminations can also be precipitated in the presence of iron.

In the presence of sufficient copper catalyst, the rate-limiting step in the oxidation of cyanide is the rate of oxygen transfer to the solution. Hence, well-agitated reactors, enhancing the dispersion of air, are preferred.

Detoxification of cyanide effluents with hydrogen peroxide started being applied in the mid-1980s (Figure12-5). Hydrogen peroxide is a "clean" chemical, and it oxidizes cyanide without the formation of toxic intermediates.

- Oxidation of CN^-

$$CN^- + H_2O_2 \rightarrow CNO^- + H_2O$$

- Hydrolysis of CNO^-

$$CNO^- + 2H^+ + H_2O \rightarrow CO_2 + NH_4^+, \text{ or}$$
$$CNO^- + OH^- + H_2O \rightarrow CO_3^{2-} + NH_3,$$
depending on the pH.

If excess hydrogen peroxide is left in the treated water, it rapidly decomposes.

$$2H_2O_2 \rightarrow 2H_2O + O_2$$

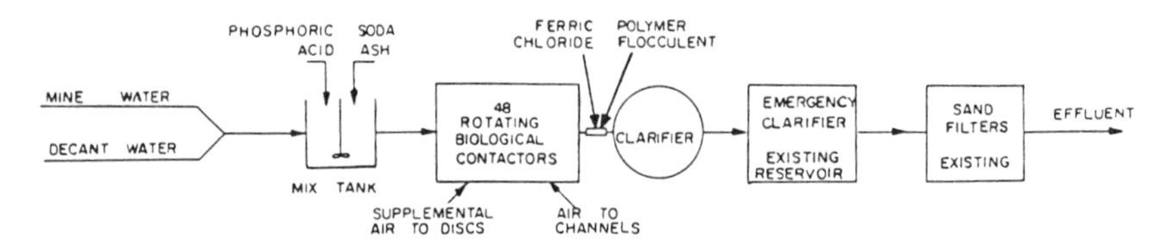

FIGURE 12-6. Biodegradation of cyanide in waste water.

Reprinted by permission. T. I. Mudder and J. L. Whitlock, Biological treatment of cyanidation waste water. *Minerals and Metallurgical Processing.* A.I.M.E. (Littleton, CO), August 1984: 161–72.

Hydrogen peroxide oxidizes both free and weakly bound complex cyanides (normally measured as "weak acid dissociable" compounds), such as CN^-, HCN, $Cd(CN)_4^{2-}$, $Zn(CN)_4^{2-}$, $Cu(CN)_2^-$, $Cu(CN)_3^{2-}$, and $Cu(CN)_4^{3-}$. For example

$$2Cu(CN)_3^{2-} + 7H_2O_2 + 2OH^- \rightarrow \underline{2Cu(OH)_2} + 6CNO^- + 6H_2O$$

On the contrary, aurocyanide and ferro- and ferricyanides are not oxidized (not destroyed) by hydrogen peroxide. The ferrocyanide ion can be removed from solution, if required, by precipitation with copper or iron ions.

$$Fe(CN)_6^{4-} + 2Cu^{2+} \rightarrow \underline{Cu_2Fe(CN)_6}$$

Hydrogen peroxide is an ideal reagent for the destruction of cyanides, due to its excellent environmental properties; the unused reagent (H_2O_2) decomposes to oxygen and water.

Biological treatment of cyanidation waste waters has been used. An attached-growth, aerobic biological treatment of cyanidation waste waters, developed at Homestake Mining Company, oxidizes free and complexed cyanides, thiocyanates, and the ammonia that is produced during the primary oxidation (Figure 12-6) (Mudder & Whitlock, 1984). The bacterial seed was found indigenous to Homestake waste waters. The only chemical requirements are an inorganic carbon source to assist nitrification (e.g., soda ash) and phosphorus (H_3PO_4) as a trace nutrient. The two-stage process employs *Pseudomonas paucimobilis* bacteria, in the first stage, to oxidize metal-complexed cyanides and thiocyanates to ammonia, carbonates, and sulfates while absorbing heavy metals on the biofilm. In the second stage, ammonia is converted to nitrites and finally to nitrates with the assistance of *Nitrobacter* and *Nitrosomonas* bacteria.

The process employs "Rotating Biological Contactors" (RBC), each with an active surface area of 100,000 to 150,000 sq. ft., allowing either plug flow or mixing by rotation. After treatment in the RBCs, the biomass (in the form of sludge) is dropped in a clarifier. Ferric chloride and/or a flocculant are added to enhance precipitation of residual metals and clarification.

The products of the biological degradation include nonpolluting anions, sulfates, nitrates, and carbonates. Accumulation of toxic metals within the biomass may cause slow microbial growth and failure of the system. If pretreatment for metal ions removal were required, its cost would make the bioprocessing of cyanides impractical.

Arsenic Removal from Gold-Mine Wastes

Low concentrations of arsenic may be found in the water of arseniferous tailings; considerable to high concentrations of arsenic are usually found in the quench-waters of roasting operations (see Chapter 5). Potential methods of arsenic removal from gold mill effluents include the following:

- Chemical precipitation
- Adsorption processes (activated carbon and ion-exchange resins)
- Reverse osmosis

The large volumes of mine and mill effluents that have to be treated make the costs of reverse osmosis and ion-exchange resins prohibitive. Activated carbon is not very effective in adsorbing arsenic. If the contaminated stream contains low concentrations of gold, the activated carbon will adsorb the gold along with the arsenic, and the adsorbed gold could pay for the arsenic treatment process.

Chemical precipitation of arsenic from roasting quench solution (132 ppm As) and from water (0.5 ppm As) of a lake where gold mill tailings were disposed were investigated by Rosehart and Lee (1972). Their precipitation results, obtained with different reagents, indicate that pH adjustment at 12, with lime, achieves the highest removal of arsenic (95%) and heavy metals, at the lowest cost per pound of arsenic removed.

Recovery of Gold from Accumulated Old Tailings

High gold prices and improved extraction technology have kindled interest in reworking old tailings. Retreatment of old tailings presents a number of incentives.

- The tailings can be easily drilled and sampled.
- Reserves and expected gold extractions can be accurately assessed.
- The tailings can be recovered, for reprocessing, at very low cost.
- An operating (or abandoned) mill is nearby.
- Crushing and grinding costs, if any, are minimal.

Tailings can be recovered for reprocessing from a number of faces with high-pressure (300 lb/in^2) water monitors. Three treatment alternatives usually have to be tested and evaluated.

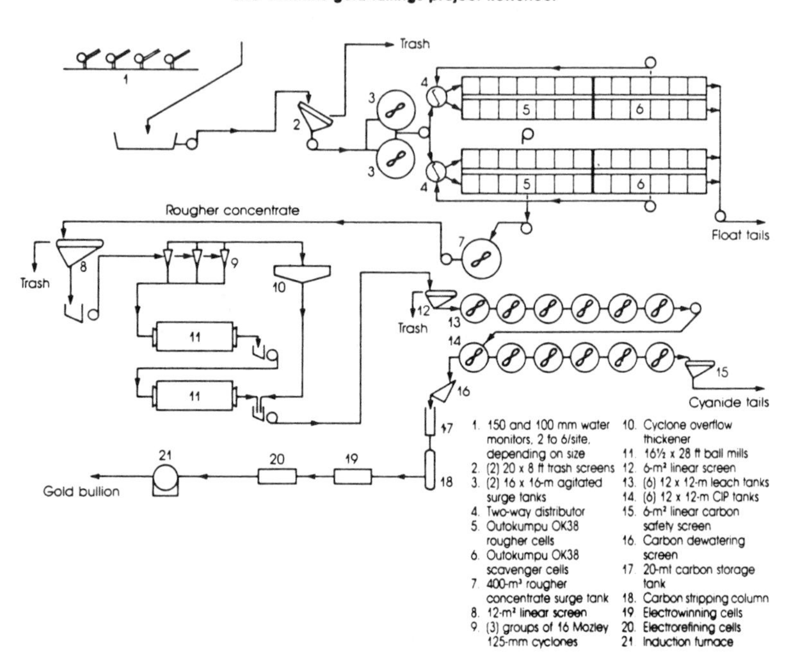

FIGURE 12-7. Recovery of gold from old tailings by flotation and cyanidation for the concentrate.

Reprinted by permission. K. R. Suttil, ERG Timmins gold tailings project flow sheet. *Eng. Min. J.* (Chicago), September 1988: 59.

1. Direct cyanidation of tailings
2. Upgrading the tailings by flotation
3. Flotation followed by cyanidation of the concentrate

Thorough testing and optimization of the proposed treatment are strongly recommended. Gold that was not recovered during the initial treatment tends to be either extremely fine or refractory. Flotation followed by cyanidation (Figures 12-7 and 12-8) is the preferred flow sheet in most cases, since it restricts cyanidation to the upgraded auriferous concentrate.

High ferrous iron content in the old tailings results in very high oxygen and cyanide consumptions. In many cases, desliming of the tailings may have a beneficial effect, since ferrous iron tends to concentrate in the slimes. Finally, carbon-in-pulp is the preferred process for recovering gold from the pregnant tailings slurry, thus avoiding tedious de-watering and filtration.

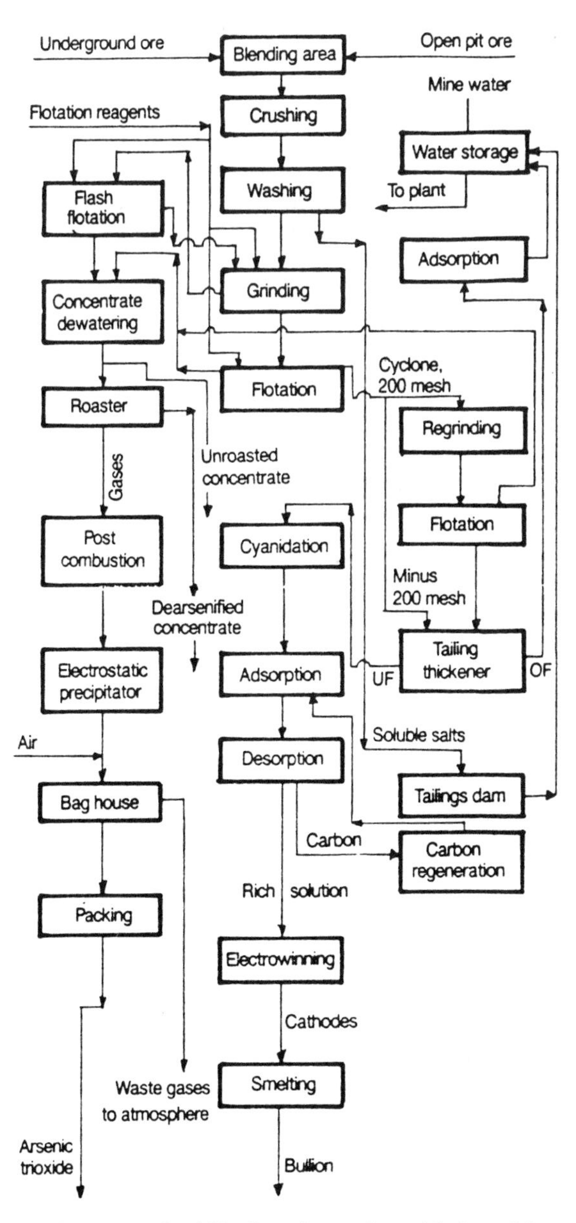

FIGURE 12-8. Recovery of gold by flotation and cyanidation of the tailings of an arsenic ore concentrator.

Reprinted by permission. R. J. M. Wyllie, *Eng. Min. J.* (Chicago), March 1988:41.

References

Note: Sources with an asterisk are not cited but are recommended for further reading.

Degussa Corporation. 1987. *Hydrogen Peroxide*. Allendale, NJ: company brochure.

Devuyst, E. A., B. R. Conrad, and W. Hudson. 1984. Industrial application of the INCO SO_2-air cyanide removal process. In *First International Symposium on Precious Metals Recovery*. Paper XXVI. 10–14 June, at Reno, NV.

*Griffiths, A., H. Knorre, S. Gos, and R. Higgins. 1987. The detoxification of gold-mill tailings with hydrogen peroxide. *J. S. Afr. I.M.M.*, September: 279–83.

Mudder, T. I., and J. L. Whitlock. 1984. Biological treatment of cyanidation waste waters. *Min. & Metal. Proc.*, May: 73–79.

Rosehart, R., and J. Lee. 1972. Effective methods of arsenic removal from gold mine wastes. *Can. Min. J.*, June: 53–57.

Schmidt, J. W., L. Simovic, and E. Shannon. 1981. *A Progress Report*. Wastewater Technology Center—Environment Canada.

Simmons, G. L., D. L. Blakeman, J. W. Trimble, and S. W. Banning. 1985. Noranda's carbon-in-pulp gold/silver operation at Happy Camp, Ca. *Min. & Metal. Proc.*, August: 161–65.

Vick, S. G. 1981. Siting and design of tailings impoundments. *Min. Eng.*, June: 653–57.

Wyllie, R. J. M. 1988. El Indio. *Eng. Min. J.*, March: 34–71.

Appendix A

Conversion Factors

Mass

1 kilogram (kg)	= 2.2046 pounds
1 tonne (t)	= 1000 kilograms
	= 0.9842 tons (long)
	= 1.1023 tons (short)
1 ounce (oz.)	= 28.35 grams (g)
1 troy ounce (tr. oz.)	= 31.103 grams

Length

1 meter (m)	= 39.3701 inches
	= 3.2808 feet
	= 1.0936 yards
1 inch (in.)	= 2.54 centimeters (cm)
1 foot (ft.)	= 30.48 centimeters
1 mile	= 1.6094 kilometers (km)

Area

1 square meter (m^2)	= 10.7639 square feet (sq. ft.)
1 acre	= 0.405 hectares
	= 4050 square meters

Volume

1 cubic meter (m^3)	= 35.3147 cubic feet
1 liter (l)	= 0.2642 gallons (US)
	= 0.2200 gallons (Imp.)
1 cubic foot (ft^3)	= 0.02832 cubic meters
1 cubic yard (yd^3)	= 0.7646 cubic meters
	= 764.6 liters
1 fluid ounce (oz.)	= 0.02957 liters

Force

1 newton (N)	= 1 $kg.m/s^2$
	= 0.2248 pounds force
	= 100,000 dynes

Pressure	
1 pascal (Pa)	= 1 N/m^2
	= 1.4504×10^{-4} pounds force/ square inch
	= 7.5006×10^{-3} mm Hg (0° C)
	= 0.9869×10^{-5} atmospheres
Density	
1 kilogram/cubic meter (kg/m^3)	= 0.001 gram/cubic centimeter
	= 0.06243 pound/cubic foot
Energy (Work and Heat)	
1 joule (J)	= 1 N.m
	= 0.7376 foot pound force
	= 9.478×10^{-4} British thermal units (Btu)
	= 2.388×10^{-4} calories
Power	
1 watt (W)	= 1 J/s
	= 1.341×10^{-3} horsepower
Viscosity	
1 pascal second	= 1 $N.s/m^2$
	= 1 kg/m.s
	= 1000 centipoise
Magnetic Flux	
1 weber (Wb)	= 1 V.s
	= 10^8 maxwell
Magnetic Flux Density	
1 tesla (T)	= 1 Wb/m^2
	= 10^4 gauss
1 ampere/meter (amp/m)	= 0.01256 oersted
Gas Constant	
8.314 (N/m^2) (m^3)/mol K	= 1.987 cal/mol K
	= 1.987 Btu/lb mole °R
	= 0.7302 atm ft^3/lb mole °F
Gravitational Acceleration	
9.8066 m/s^2	= 32.174 ft/s^2

Appendix B

Gold Production Statistics

TABLE B-1. Estimates of total world gold production, by time period.

Time period	*Total world gold production, 10^6 tr oz*	*Areas of major production ranked by size*
3900 B.C.–A.D. 1492	400–500	Africa, Europe, Asia
1493–1600	23	South America, Africa, Europe
1601–1700	29	South America, Africa, Europe
1701–1800	61	South America, Europe, Africa, Mexico
1801–1900	374	United States, Australia, Soviet Union, South Africa, Asia
1901–30	585	South Africa, United States, Soviet Union Australia, Canada, Asia
1931–83	2,286	South Africa, Soviet Union, Canada, United States, Australia
Total (rounded)	3,800–4,000	

Reprinted with permission. P. R. Thomas and E. H. Boyle, Jr. 1986. Gold availability—world: A minerals availability appraisal. *U.S.B.M. IC 9070.*

TABLE B-2. Gold production, 1971, 1981, and 1983 for the 10 largest producing countries, thousand troy ounces.

Country	*1971*	*1981*	*1983*	*Change 1971–81*	*Change 1981–83*
South Africa	31,389	21,121	21,847	−10,268	+762
Soviet Union	6,700[a]	8,425	8,600	+ 1,725	+175
Canada	2,243	1,673	2,274	−570	+601
United States	1,495	1,379	1,957	−116	+578
China	50[a]	1,700	1,900	+1,650	+200
Brazil	157	1,200	1,600	+1,043	+400
Australia	672	591	1,035	−81	+444
Philippines	640	758	817	+118	+59
Papua New Guinea	24	553	582	+529	+29
Chile	64	401	571	+337	+170
Total	43,434	37,794	41,183	−5,640	+3,389
8 largest market economy countries[b]	36,684	27,669	30,683	−9,015	+3,014
Total world	46,494	41,250	44,533	−5,244	+3,283

[a]Estimated. [b]Less the Soviet Union and China.

Reprinted with permission. P. R. Thomas and E. H. Boyle, Jr. 1986. Gold availability—world: A minerals availability appraisal. *U.S.B.M. IC 9070.*

Appendix C

Comparative Long-Term Total Production Costs in Selected Countries

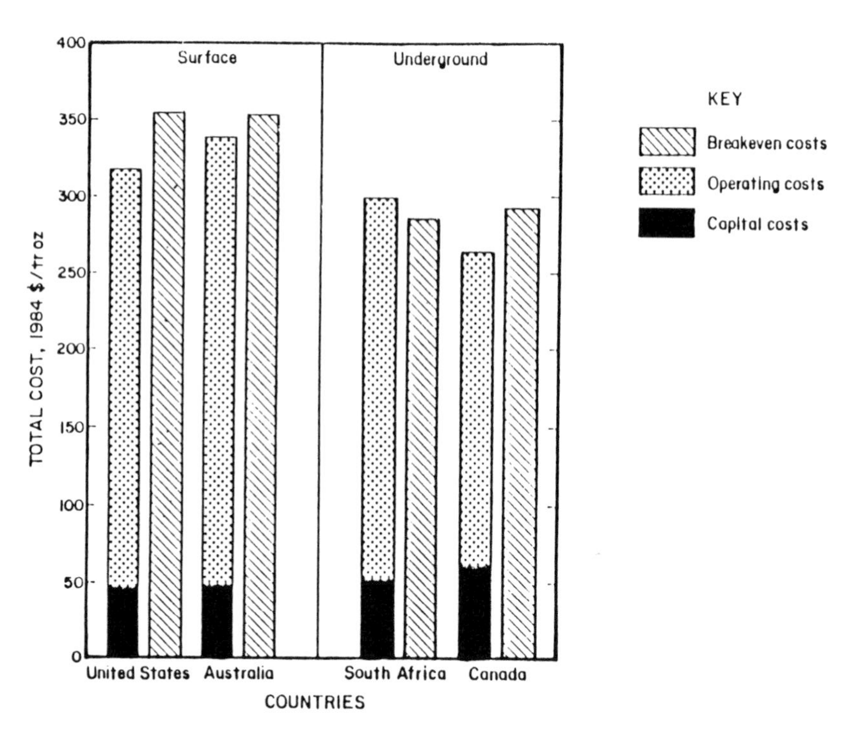

Reprinted with permission. P. R. Thomas and E. H. Boyle, Jr. 1986. Gold availability—world: A minerals availability appraisal. *U.S.B.M. IC 9070.*

Appendix D

Primary Gold Deposits and Mines in the U.S.

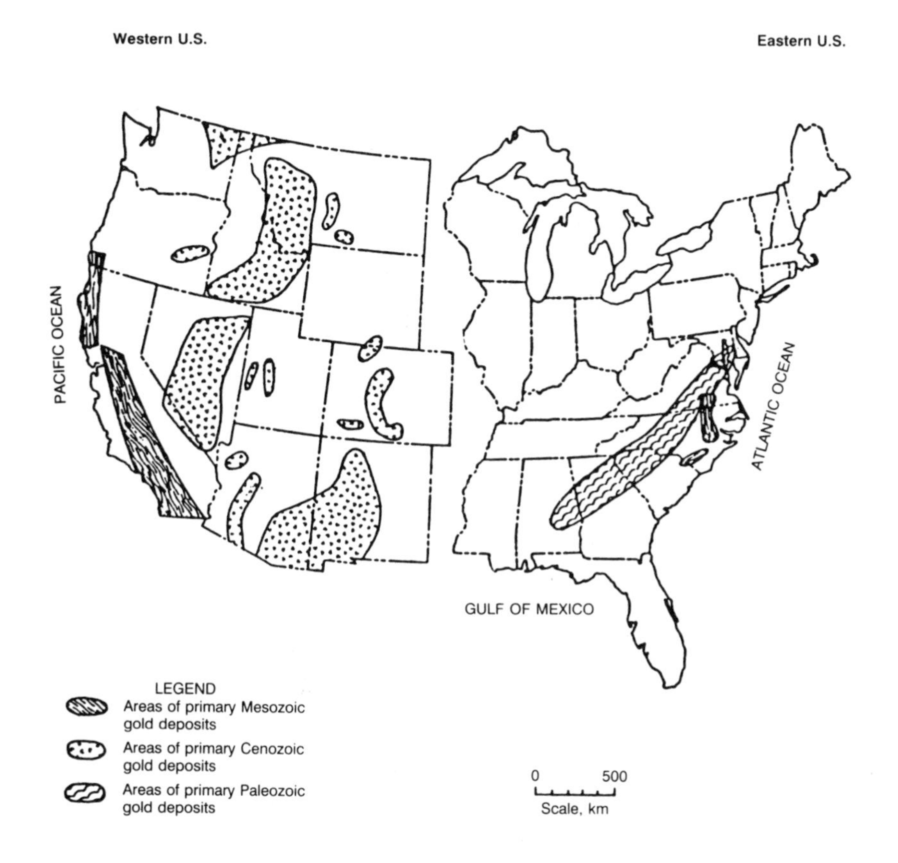

FIGURE D-1. Primary gold deposits in the U.S.

Reprinted with permission. P. R. Thomas and E. H. Boyle, Jr. 1986. Gold availability—world: A minerals availability appraisal. *U.S.B.M. IC 9070.*

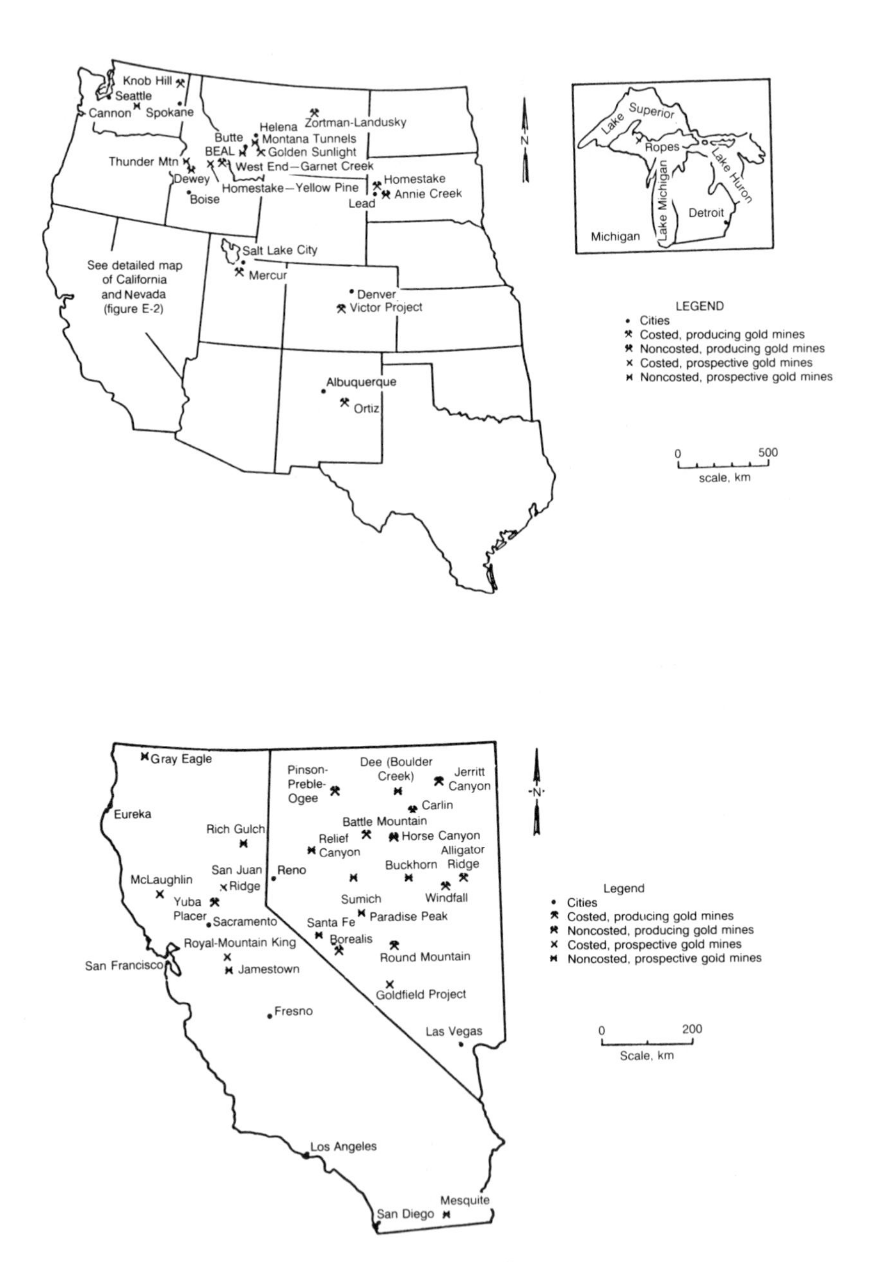

FIGURE D-2.. Primary gold mines in the U.S.

Reprinted with permission. P. R. Thomas and E. H. Boyle, Jr. 1986. Gold availability—world: A minerals availability appraisal. *U.S.B.M. IC 9070*.

Appendix E

Gold Mill Sampling and Metallurgical Balance (Five CIP tanks; five carbon adsorption columns)

Daily

The following daily samples should be taken throughout the plant.

I. The mill feed (minus ⅝ inch) as it discharges from the belt feeding the grinding circuit:
   A. Grab sample of 0.5 to 1.0 kg of material every two hours to calculate moistures.
   B. Twenty-four hour composite using an automatic straight-path sampler collecting approximately 3 kg of material every 20 minutes.

II. The cyclone overflow:
   A. Grab sample of approximately 100 ml of pulp is taken. The solids are settled and the clear solution assayed directly by AA for gold. This is repeated every two hours to indicate gold extraction.
   B. Twenty-four hour composite consisting of 1 l grab samples taken every two hours.

III. The pond reclaim solution:
   A. Twenty-four hour composite using a Clarkson feeder; this composite is 3 to 4 gal solution.

IV. The carbon adsorption columns:
   A. Wire sampler collecting a 24-hour solution composite of the thickener overflow feeding Column Number 1.
   B. Wire sampler collecting a 24-hour solution composite of the discharge from Column Number 5.
   C. Solution dip samples from Tanks 1 through 5 at 7 A.M.; Sampled using a 1 liter beaker connected to a wooden stick.
   D. The carbon in Tank Number 1 is sampled at 7 A.M. using a 1 l plastic beaker attached to a wooden stick.
   E. Carbon level measured using a rope calibrated in feet with a weight attached to the end.

V. The leach feed (thickener underflow):
   A. Denver straight-path sampler collecting approximately 1 l pulp every 20 minutes; composite samples are per shift.

VI. Carbon-in-pulp tanks:
   A. One liter grab sample of pulp every two hours of the CIP feed (leach tank discharge); this is a 24 hour composite.
   B. One liter dip sample from Tanks 2, 3, and 4 at 7 A.M.; the samples are filtered and the clear solutions assayed.
   C. The carbon-to-pulp ratio in Tanks 1 through 5 is calculated at 7 A.M. by dipping out 10 liter pulp (2 l/dip) and screening at 20 mesh. The +20 mesh carbon is washed thoroughly, and the carbon volume measured using a plastic cone calibrated in milliliters. The carbon from Tanks 1 and 2 is dried and fire assayed for Au in triplicate.
   D. The CIP tails from Tank Number 5 are sampled using a 3-stage Vezin sampler. Composite samples are per shift.

VII. The loaded carbon:
   A. Grab samples of 500–600 g are taken before and after acid washing, and after stripping.

VIII. Carbon strip solutions:
   A. Samples are taken before and after the electrolytic cell. These are taken using a Cole Parmer Masterflex multichannel pump, collecting approximately 2 gallons per 500 tons of solution.

Sample Preparation Procedures

Solution samples are pressure filtered, with the filtrate being assayed by AA and fire (10 assay ton) for gold. The filter paper is discarded.

Before pulp samples are taken, an initial 50 ml store-purchased clorox (NaOCl) is added to each bucket. This addition destroys any CN present and stops the leaching process.

Pulp samples are given to the chemical laboratory and prepared as diagrammed in Figure E-1. A graph for determining the minimum sample mass is shown in Figure E-2.

Solution composites are:

1. Pond reclaim
2. Leach feed
3. Thickener overflow
4. Column barren
5. CIP feed
6. CIP tails

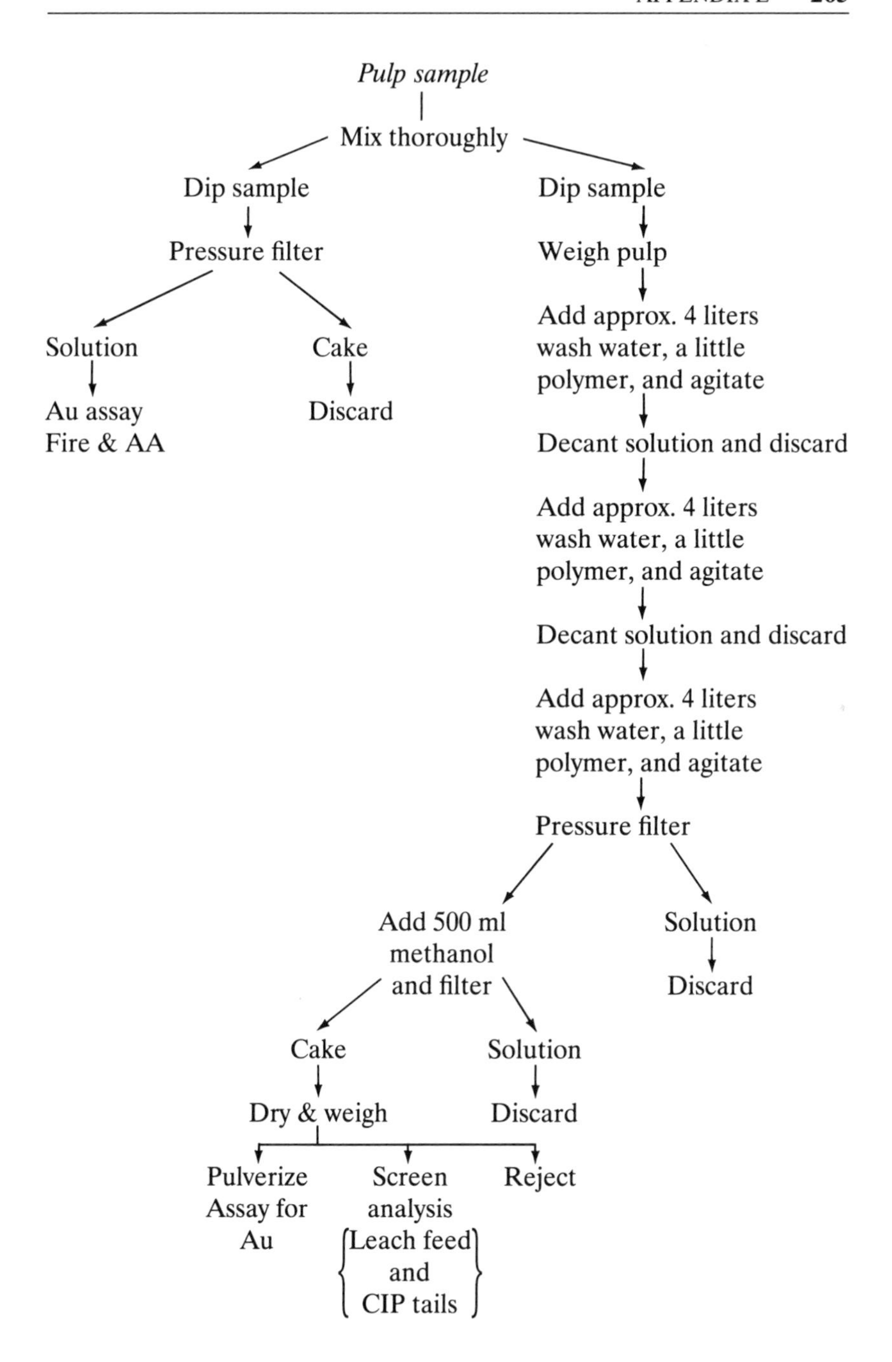

FIGURE E-1. Preparation of pulp samples.

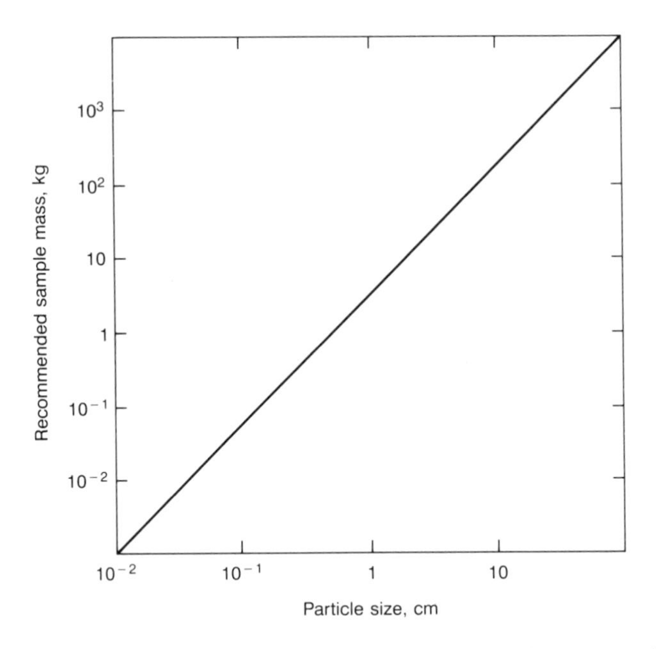

FIGURE E-2. Recommended minimum sample mass based on top particle size.

Solid composites are:

1. Mill feed
2. Leach feed
3. CIP feed
4. CIP tails

All carbon samples are washed over a 20 mesh screen. Approximately 50 g carbon is sampled across the screen and placed in a conical beaker. Added to the beaker is 200 ml of 50% HCl, and the mixture is heated until all action ceases. The acid is decanted and the carbon is washed thoroughly, dried, and fire assayed for gold. A screen analysis of the carbon is not performed.

Metallurgical Balance and Metal Inventory

The daily gold extraction balance is calculated on the basis of "Au In − Au Out = Au Extracted"

	Head assay from previous day (oz/ton Au) × Corrected dry tonnage
Plus	Reclaim solution assay (oz/ton Au) × Reclaim solution tonnage (mag flowmeter)
Minus	Tail solution assay (oz/ton Au) × Tail calculated solution tonnage

Minus	Tail solids assay (oz/ton Au) × Tail calculated solid tonnage
Equals	Ounces of gold extracted

Gold extraction is calculated by dividing the ounces of Au extracted (equation shown above) by the total ounces of Au going into the plant [(Head assay × Dry ore tonnage) + (Reclaim solution assay × Reclaim solution tonnage)].

This is the only balance made on a daily basis. Weekly or semimonthly balances are not performed.

The monthly balance is an accounting of gold produced that month, and includes a gold inventory throughout the plant. This inventory includes the gold content in the grinding circuit, thickening, leach and CIP tanks, adsorption columns, any loaded carbon, and the refinery.

Appendix F

Data Commonly Collected During Baseline Studies for an Environmental Assessment of a Heap Leaching Operation

Physical factors

1. Location
2. Geomorphic/physiographic
   a. Geological hazards
   b. Unique land forms
3. Climate
4. Soils
   a. Productivity
   b. Capability
   c. Hazard
      (1) Erosion characteristics
      (2) Mass failure
5. Minerals and energy resources
   a. Locatable minerals
   b. Leasable minerals
   c. Energy sources
6. Visual resources
7. Cultural resources
   a. Archaeological
   b. Historical
   c. Architectural
8. Wilderness resources
9. Wild and scenic rivers
10. Water resources
    a. Water quality
    b. Stream-flow regimes
    c. Flood plains
    d. Wetlands
    e. Groundwater recharge areas
11. Air quality
12. Noise
13. Fire
    a. Potential wildfire hazard
    b. Role of fire in the ecosystem
14. Land use including prime farm, timber, and rangeland
15. Infrastructure improvements
    a. Roads
    b. Trails
    c. Utility corridors and distribution
    d. Water collection, storage, and distribution
    e. Communications system
    f. Solid waste collection
    g. Sanitary waste collection

Reprinted by permission. J. W. Thatcher, D. W. Stuthsacker, and J. Keil, Regulatory aspects and permitting requirements for precious metal heap leach operations. In *Introduction to Evaluation, Design and Operation of Precious Metal Heap Leaching Projects.* Society for Mining, Metallurgy, and Exploration (Littleton, CO), 1988: 40–58.

Biological factors

1. Vegetation
   a. Forest, including diversity of tree species
   b. Rangeland, including conditions and trends
   c. Other major vegetation types
   d. Threatened or endangered flora
   e. Research natural area (RNA) potentials
   f. Unique ecosystems (other than RNAs)
   g. Diversity of plant communities
   h. Noxious weeds
2. Wildlife
   a. Habitat
   b. Populations
   c. Threatened or endangered fauna
   d. Diversity of animal communities
   e. Animal damage control
3. Fish/aquatic biology
   a. Habitat
   b. Populations
   c. Threatened or endangered fish, including state-listed species
4. Recreation resources (usually a combination of physical and biological factors)
5. Insects and diseases
6. Exotic organisms; for example, Russian thistle, Siberian ibex

Social Factors

1. Population dynamics
   a. Size (growth, stability, decline)
   b. Composition (age, sex, minority)
   c. Distribution and density
   d. Mobility
   e. Military
   f. Religious
   g. Recreation/leisure
2. Special concerns
   a. Minority (civil rights)
   b. Historic/archaeological/cultural
3. Ways of life, defined by
   a. Subcultural variation
   b. Leisure and cultural opportunities
   c. Personal security
   d. Stability and change
   e. Basic values
   f. Symbolic meaning
   g. Cohesion and conflict
   h. Community identity
   i. Health and safety
4. Land tenure and land use
5. Legal considerations

Appendix G

Flowchart of Gold Recovery from Copper Refinery Slimes

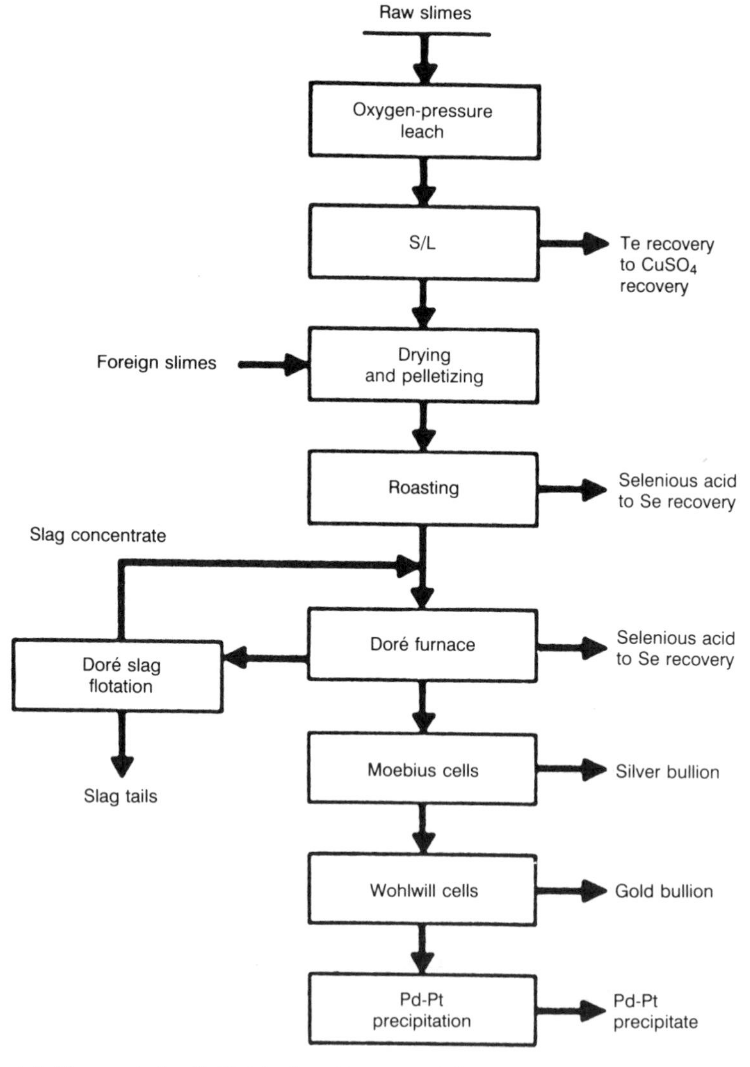

Reprinted by permission. B. H. Morrison, Recovery of silver and gold from refinery slimes at Canadian copper refiners. In *Extraction Metallurgy '85*. Institution of Mining and Metallurgy (London), 1985: 254 (Figure 1).

Appendix H

Solubility of Minerals and Metals in Cyanide Solutions

	Mineral		% Dissolved in 24 hours	Reference
Gold minerals	Calaverite	$AuTe_2$	Readily soluble	Johnston (1933)
Silver minerals	Argentite	Ag_2S	Readily soluble	Leaver, Woolf, and Karchmer (1931)
	Cerargyrite	$AgCl$		
	Proustite	Ag_3AsS_3		
	Pyrargyrite	Ag_3SbS_3	Sparingly soluble	
Copper minerals	Azurite	$2\,CuCO_3 \cdot Cu(OH)_2$	94.5	Leaver and Woolf (1931)
	Malachite	$CuCO_3 \cdot Cu(OH)_2$	90.2	
	Chalcocite	Cu_2S	90.2	
	Cuprite	Cu_2O	85.5	
	Bornite	$FeS \cdot 2\,Cu_2S \cdot CuS$	70.0	
	Enargite	$3\,CuS \cdot As_2S_5$	65.8	
	Tetrahedrite	$4\,Cu_2S \cdot Sb_2S_3$	21.9	
	Chrysocolla	$CuSiO_3$	11.8	
	Chalcopyrite	$CuFeS_2$	5.6	
Zinc minerals	Smithsonite	$ZnCO_3$	40.2	Leaver and Woolf (1931)
	Zincite	ZnO	35.2	
	Hydrozincite	$3\,ZnCO_3 \cdot 2\,H_2O$	35.1	
	Franklinite	$(Fe,Mn,Zn)O \cdot (Fe,Mn)_2O_3$	20.2	
	Sphalerite	ZnS	18.4	
	Gelamine	$H_2Zn_2SiO_4$	13.4	
	Willemite	Zn_2SiO_4	13.1	
Iron minerals	Pyrrhotite	FeS	Readily soluble	Hedley and Tabachnick (1958)
	Pyrite	FeS_2		
	Hematite	Fe_2O_3	Sparingly soluble	
	Magnetite	Fe_3O_4		
	Siderite	$FeCO_3$	Practically insoluble	
Arsenic minerals	Orpiment	As_2S_3	73.0	
	Realgar	As_2S_2	9.4	
	Arsenopyrite	$FeAsS$	0.9	

(continued)

Mineral			% Dissolved in 24 hours	Reference
Antimony minerals	Stibnite	Sb_2S_3	21.1	
Lead minerals	Galena	PbS	Soluble at high alkalinity	Lemmon (1940)

Reprinted by permission. F. Habashi, Kinetics and mechanism of gold and silver dissolution in cyanide solution. Montana Bureau of Mines and Geology, *Bull. 59.*, 1967.

Index

ABC grinding circuit, 60–62
Acid-forming constituents, 116
Acid pressure oxidation, 103
Activated carbon, 84, 133, 185, 230–231
  adsorption, 133, 193–206
  chemical treatment, 199–200
  columns, 133, 227–228, 229
  desorption, 133
  elution, 196, 198, 204
    continuous, 204
    industrial uses, 216–230
  regeneration, 204–206
  stripping, 194, 196, 198
  systems, 72–74, 195
  thermal reactivation, 204
Aeration, 107, 108, 157, 159
  tanks, 109
Agglomeration, 116, 117, 126, 127, 131, 134
  binders, 125, 127, 128
  effects, 119, 123–125
Agglomerators, pan type, 128
Agitation, intensity of, 158
Alchemists, 141, 142
Alcohol stripping process, 201
Alkalies, 156
Alkali sulfide, 155
Alluvial gold, 25
Aluminum
  foil, as cathode, 204
  as gold precipitant, 192–193
Amalgam, 67
Amalgamation, 44, 65, 67, 142, 165
Amenable gold ores, 55, 83
American Cyanamid Company, 98
Ammonia, 139
Anglesite, 156
Antifreeze, in leaching solution, 133
Antimony, 1, 4, 6, 87, 103, 156, 161, 164
  pentasulfide, 85
  sulfate, 156
Aqua regia, 1, 12, 21, 138, 182, 244
Arsenate, 88, 92, 93
Arsenic, 1, 4, 6, 80, 84, 87–88, 92, 93, 96, 99, 103, 105, 155–156
  oxidation to arsenate, 99, 103
  removal, 253
Arsenides, 7, 92
Arsenites, 84, 92, 98
Arseno process, 103
Arsenopyrite, 7, 84–85, 87, 88, 90–91, 98
  removal of, 98, 103
Arsine, 84
Auric chloride, 17, 90
Auric ions, 148
Auricyanide
  complex, 14
  ions, 194
Aurous
  chloride ion, 14, 90
  complex ions, 14
  cyanide, 145, 149
  thiosulfate ion, 14
  thiourea ion, 14
Australia, 30, 51
Autoclave
  cyanidation, 164
  materials of construction, 100–101, 103

Autoclave *continued*
  oxidation, 99–102
Autogenous grinding, 59–61

Bacteria, nutrients, 106, 107
Bacterial leaching, 105–107, 108
Bacterial oxidation, 105–107
Barren solution, 133, 210
Base metals, 137, 154
  cyanides of, 161
  cyanogen compounds of, 161
Bentonite, 125
Berlin blue, 141–142
Bicarbonates, 151
Bio-heap leaching, 107
Bioleaching, 105
Biological oxidation, 87, 105–107
Bismuth, 1, 80, 162, 164
Bisulfite
  addition, 109
  precipitation with, 136
"Black gold," 7
Blast furnaces, 241
Blue acid, 142
Bottle tests, 118
Boundary layer, 163
Brass, 132, 137
Brazil, 6
Briquetting, of concentrates, 93
Bromine, 146, 182
Bucket-line dredgers, 32–34

Calcines
  fusion, 97
  lockup in, 97–98
  mill, 92
  surface area, 91
  washing, 98
Calcite, 98
Calcium, 197
  aurocyanide, 197
  chromate, 153
  hydroxide, 68, 152, 149
  hypochlorite, 109
  ion effect, 152
  peroxide, 152
  sulfate, 98
  sulfide, 98
Calmet Process, 164
Carat gold, 137
Carbon
  in columns, 196
  in-leach (CIL), 196, 197
  in-pulp (CIP), 196, 197, 216–224
Carbonaceous materials, 81, 82, 83, 84, 87, 101, 107, 108, 109, 116, 193
Carbonaceous ores, 83, 107
Carbonates, calcium, 98
Carbonatites, 4
Carbon columns, 73, 194
Carbon dioxide, 146, 163
Carbon-in-leach (CIL), 66, 73, 225, 226, 230
  with oxygen (CILO), 225
  comparison CILO vs. CIL, 227
Carbon-in-pulp (CIP), 73
Carbonites, 4
Carbon loading, 107
Carborundum, 139
Carlin Gold Mining Company, 115
Carlin type
  gold deposits, 7
  gold ores, 107, 116
Cashman process, 103
Cationic reagent, 157
CCD. *See* Counter-current-decantation
Cement, 128
Cementation of gold, 133, 141, 186–192
  effects of solution composition, 187–189
Chalcocite, 84
Chalcopyrite, 84, 101
Charcoal, 185, 196
Chemical oxidation, 107–110
Chlorides, effect on zinc consumption, 109
Chloridizing roasting, 110
Chlorination, 67, 87, 107, 109
  rate, 109
  roasting, 89–90
  solutions, 185

tanks, 109
Chlorine, 90, 109, 141, 182, 243
addition rate, 109
discovery, 141
in situ, 107
oxidation, 87
Chlorauric acid, 16
Clarification
precoat pressure, 190
of solution, 189, 190, 191
Classifiers, 62
Clay, well-packed pads, 122, 123
Clays, 82, 117, 126
adsorbing gold, 101-102
Coinage metals, numerical properties, 2
Colluvial subenvironment, 30
Column leach tests, 118-119
Comminution circuits, 55-63
Copper, 141, 243
alloys with gold, 137
cyanide, 185
electrolytic refinery slimes, 80, 138
heap leaching, 115
numerical properties, 2
recovery from solution, 94
Corrosion theory of cyanidation, 144
Counter-current-decantation (CCD), 70-72, 158, 228
thickeners, 155
Covellite, 84
Crocoite ($PbCrO_4$), 157
Crowe vacuum tower, 190, 192
Crushing, 57-59, 115, 116, 117, 126
degree of, 123
Cuprocyanide ions, 151
Cuprous cyanide, 151
Current density, 243
Cyanate, 142, 144
Cyanicides, 81, 85, 92, 98, 116, 155
Cyanidation, 44, 59, 67-68, 92, 98, 116, 155, 185
chemical enhancement of, 162-163
effect of
Eh, 148-150
flotation reagents, 157
foreign ions, 154-157
oxygen, 146-148
particle size, 162
temperature, 160-162
of gold ores, 145-175
high-pressure, 102
high-pressure, low-alkalinity, 102-103
inhibitors, 155
intensive, 164, 165, 166
kinetics, 158-165
mechanism, 145-157
mill, 62
with oxygen, 160
particle size, 198
pH, 145, 150, 198
physical enhancement, 163-165
in pipe reactor, 103, 169
plant size, 63
pressure, 103, 164
process, 67-68
rate, 158-160
limiting, 159
reactor, 163
residence time, 158
solution, 92, 115, 144
acidification, 162
barren, 129
stripping by aeration, 162
theories, 143-145
with ultrasonic treatment of the pulp, 163-165
Cyanide, 93, 142, 155
analyzers, 157
biodegradation, 164
complex ions, 162
concentration, 162, 197
consumption, 98, 99, 119, 157, 165
decomposition, 150, 158
destruction, 163-164, 248-253
hydrolysis, 68
ions, 144
leaching, 67, 142
loss, 155, 158
regeneration, 165, 168
solubility of minerals and metals in, 272-273

Dump leaching, 120, 124, 126-129
ore preparation, 126-129

Edward's roaster, 94
Egyptians, ancient, 25, 41
Eh, 142, 145–146
Electrochemical
  order of metals, 186
  reaction, 145
Electromotive force (EMF), 161, 162
Electronics industry, 138
  gold scrap, 138
Electroplating, 138
Electrorefining, 56
Electrowinning, 56, 69, 73–75, 185, 204, 209–214, 218, 219, 226
  flow of solution, 214
  operating parameters, 215–216
  sandwich, type of cell, 212
  on steel wool cathodes, 213
Electrum, 79, 80, 101
Elements associated with gold, 2, 3
Elkington's process, 142
Elsner's equation, 142, 144
Eluvial environment, 30
EMF. *See* Electromotive force
Emitters, of solution, 132
Environmental regulations for leaching pads, 122, 123
Ethers, 157
Expanding pad method, 123
Exploration,
  for gold, 8–9
  techniques, 28, 29

Fanning concentrators, 49
Feasibility study, 120
Ferric
  arsenate, 99
  chloride, 90
  hydroxide, 102, 153
  ions, 102
  sulfate, 98, 99, 152, 155, 193
Ferrocyanide, 156, 183, 187
Ferrous
  ions, 102
  sulfate, 152, 155
Filter bags, 134
Filters
  candle, 192
  Stellar, 191, 192
Fines, 116, 117, 126
Flash chlorination, 109–110
Flotation, 86
  collectors, 154
  of silica, 110
Flow sheet
  carbon-in-leach (CIL), 226
  carbon-in-pulp (CIP), 218, 219
  continuous leaching and electrolysis of gold-laden activated carbon, 211
  grinding and gravity concentration, 65
  gold elution and electrolysis, 212
  gold milling, 66, 68–74
  ion exchange elutions, 232, 233
  Merrill-Crowe plant, 190
Fluosolids roaster, 94, 96–97
  temperature control, 96
Fluvial subenvironment, 30
Freezing of solutions, 133
Furnaces. *See* Melting furnaces
Fusion, of calcines, 97

Galena, 84, 156
Gantry cranes, 126
Gas superficial velocity, 159
Geochemical diagram, 5
Geochemistry of gold, 1–4
Geomembrane, 125
Gilding baths, 137
Glacial subenvironment, 30
Gold
  abundance of
    in lithosphere, 1
    in sea water, 8
  adsorption, 197–198, 199
  alloys, 11, 13, 79, 81, 137, 138
  alluvial, 7, 25, 31
  amalgam, 67
  amenable, ores, 55
  anodes, 243
  antimonides, 80
  assaying, 21–22
    atomic absorption spectography, 22

detection, 21
gravimetric methods, 21–22
neutron activation, 22
sample preparation, 21
spectrophotometric methods, 22
volumetric methods, 22
X-ray fluorescence, 22
associated with sulfides, 3–4
-bearing minerals, 3–4, 80
biological systems, effect on, 7–8
bromides, 16, 17
bullion, 243
in carbonaceous materials, 4
cathodes, 241
chlorides, 16–17, 89
coarse free, 158
complex ions, 1, 18–20
in copper refinery slimes, 80
cyanidation, 141–170
cyanide, 18
deposits, 3, 4–7
dissolution of, 143, 145, 162, 243
rate of, 162
electroplating, 1, 206–209
electrowinning, 66, 72–75, 133, 196, 206–216
current density, 210
current efficiency, 210
elements associated with, 2, 3
exploration, 8–9
fluorides, 16
geochemistry, 1–4
gravity concentration, 41–53
halides, 16
host materials, 84
"invisible," 80
melting and refining, 241–244
mill
sampling, 264–268
tailings, 244–255
Miller process, 242, 243
milling flow sheets, 68–78
minerals, 3–4, 80
native ores, 3, 79, 80, 116, 142
nuggets, 31
parting, 138
plating baths, 137
precipitates, 109, 241
pressure leaching, 164
production
costs, 261
statistics, 259–260
recovery, 38, 133
from copper refining slimes, 271
from solution, 70–76
from tailings, 253–255
refining, 76, 242, 243–244
small scale, 139
scrap, 137–139
secondary, 137
in sea water, 8
selenides, 4, 80
silica-locked, 110
in sulfides, 80
tellurides, 4, 79, 80
in veins, 80
volatilization loss, 93
"Golden fleece," 41
Gravity concentration, 32, 41–43, 62, 63–65, 66, 162, 230
Green gold, 138
Grinding, 58, 59–63
ABC, 60–61
autogenous, 57, 59
circuit, 62
fine, 110
media, 60, 62
semi-autogenous, 230
Guiana, 30

Halides of gold, 16
Heap, 115, 116
construction, 129–133
flooding, 131
operation, 129–133
ponding, 131
spraying, 131
sprinkling, 131
Heap leaching, 115, 120
advantages, 134
environmental assessment, 269–270
methods, 120–123
pilot test, 119–120
water circuits, 121

Hematite, 91, 99
High-pressure stripping, 201
Homestake Mining Company, 211
Humic acid, 82
Hydraulic mining, 37–38
Hydrochloric acid, 157, 182, 232, 243
Hydrocyclones, 44, 51
Hydrofluoric acid, 110
Hydrogen, 161
  cyanide, 142
  peroxide, 132, 143, 144, 152, 163
    sulfide, 193
    theory, 143
Hydrogen Theory, Janin's, 144
Hydronium ions, 99
Hydrophobic, 157
Hypochlorite *in situ*, 107

Improved Mass Transfer (IMT) cell, 213
Induction furnaces, 241
In-situ leach, 181
Ion-exchange
  fibers, 233
  resins, 185, 230–235
    elution, 231, 232, 233, 234
    flow sheet, 232
    incineration, 231
    strong base, 231
    weak base, 231
Iron, 99, 141
  oxides, 84
  vitriol ($FeSO_4$), 141

Jarosites, 99
Jewelry, 138
Jigs, 43, 44, 45–47

Ketones, 157
Kinetics, of gold cyanidation, 158–185

Laterites, 30–31
Lateritization, 30
Leaching,
  circuit, 70
  dump, 117
  heap, 117, 120–133
    ore preparation, 126–129
    pad preparation, 124–126
  pulverized gold ores, 67–68
  with pure oxygen, 160
  valley, 122
  vat, 133–135
Leaching solution
  pressure emitters, 132
  sprinklers, 137
Lead, 99, 103, 157
  effect on gold dissolution, 162, 163
  nitrate, 68
  solder, 138
Lime, 84, 85, 128
  addition, 70, 126, 132, 157
  consumption, 98, 119
Linear screen, 63
Liner systems, 125
Liners, mill, 60–62
Litharge (PbO), 155, 156
Lithium, 197

MacArthur-Forrest process, 143
Magchar Process, 224
Magnesium, 197
Marine placer
  deposits, 32
  environment, 26, 27
Mass transfer, 158
Melting furnaces, 241
Mercury, 99, 162
  extraction of, 75
  selective electrowinning, 75–76
Merrill-Crowe process, 72, 189–192
Micron desorption system, 196–197
Mill liners, 60–62
Milling, 55–76
  flow sheets, 68–74
MINTEK, 205, 212–213, 224, 230
  steel wool cell, 212–213, 233
Multi-hearth roaster, 94, 95

Neptunian theory, 25
Nickel, 139
  arsenides, 84
  sulfides, 84
Nitric acid, 138

Ore
  amenable, 55
  mineralogy, 118, 126
  run-of-mine, 126
  testing, 116-120
Orpiment (As_2S_3), 155
Oxidation treatment
  acid pressure, 103
  conditions, 99
Oxygen, 142, 144, 155, 161, 165
  absorption in solution, 145
  consumers, 84, 99, 154, 155
  mass transfer rate, 159
  theory of cyanidation, 143
Oxygen Theory, Elsner's and Bodlander's, 144

Pachuca tanks, 70, 164, 230
Pactolus River, 7
Pad, impervious, 115, 117, 120
  expanding, method, 122
  liner systems, 120, 122, 125
Palladium, 80, 139, 243
  pallado-arsenide (Pd_2As), 80
Permeability, 116, 120
pH, 100, 109, 152, 231
  cyanidation optimum, 106, 119
Pilot plant testing, 40-41, 119-120
Pinched sluices, 49
Pipe reactor, 103, 160, 164
Placers
  environments, 25-31
  exploration techniques, 28-29
  gold in, 25, 30, 38-53
  gravity concentration, 41-53
  mining methods, 28-29
Platinum, 139, 243, 244
  group of metals (PGM), 243, 244
Plumbite, 156
Polythionate, 155
Ponding, 120
Ponds, in gold mills, 117, 122, 13
Potassium, 197
  cyanide, 142
  ferrocyanide, 142
  hydroxide, 142
  ions, 99
  permanganate, 146
Potential-pH diagram, Zn-H_2O-CN, 188
Pourbaix diagram, Au-H_2O, 15
Preaeration, of slurry, 70
Pregnant solution, 121, 133, 194, 230, 231
Preheating
  ore slurry, 108
Pressure
  cyanidation, 164
  leaching, 102
  oxidation, 101-102
Process feasibility study, 120
Pyrargyrite (Ag_3SbS_3), 101
Pyrite, 3, 6, 85, 86, 88, 89, 99, 101, 102, 105, 151, 154
  roasting, 90-98
Pyrolysis, 102
Pyrrhotite, 84, 88, 89, 155

Quartz, 3, 6

Realgar (As_2S_2), 155
Redox potential diagrams, 149, 150, 151
Reducing agents, 193
Refractory gold ores, 79-110
  classification, 81
  flotation of, 80, 86
  mineralogy, 79-85
  pyritic, 88
  treatment of, 87-110
Reichert Cones, 44, 49-51
Resin-in-leach, 230
Resin-in-pulp, 232-233
Reverberatory furnaces, 241
Reynolds number, 42, 43
Riffled sluices, 48
Riffles, 44
Rintoul furnace, 207
Roasting, 87-98
  with additives, 93
  chemistry of, 88-90
  chloridizing, 89-90, 93
  conditions affecting, 90-97
  control of the gaseous phase, 92
  flash, 88

Roasting *continued*
  furnaces, 94-97
  optimum temperature, 96
  problems, 97-98
  reactions, 88
  with sodium carbonate, 93
  temperature effect, 91
Rotary kiln, 94
Rubber mill liners, 62

SAG grinding, 61
Salsigne process, 156
Sampling, 39-40, 117-118
  dore, 242, 264-265
Scale inhibitors, 132
Screening, 57, 58, 59
  linear screen (Delcor), 63
Scrubbers, wet, 38
Scrubbing, 44
Secondary gold, recovery of, 137-139
Selenium, 164
Semi-autogenous grinding (SAG), 59
Settling velocity, of particles, 42
Shaking tables, 43, 44, 51-53
Shape factor, Corey, 42
Silicates, 84, 110, 157
Silver, 1, 2, 4, 5, 6, 7, 99, 142
  dissolution of, 143
  electroplating, 142
  in gold veins, 6
  leaching of, 158
  numerical properties of, 2
  sterling, 137
Sizing, 44
Skarn-type deposits, 4
Sluices, 43, 44, 49
Soda ash, 86
Sodium, 197
  bisulfite, 139
    ions, 99
  hydroxide, 132
  hypochlorite, 107, 109
  peroxide, 146
  sulfide, 86
Solid-liquid separation, 70-74
Solubility of minerals and metals in
    cyanide solutions, 272-273
Solvent extraction, 185, 235
Sphalerite, 101
Spiral concentrators, 44, 47-48
Sprinkling, 120
Stability diagram Fe-O-S, 89
Stannous chloride, 139
Steel, 137
  wool, 210
Stellar filter, 191, 192
Sterling silver, 137
Stibnite, 85, 98, 155, 156
Suction-cutter dredging, 34-36
Sulfate, 155, 189
Sulfide ion, 155, 156
Sulfides, 80, 84, 98, 106, 164
Sulfites, 98, 155
Sulfocyanides, 183
Sulfolobus, 106
Sulfur, 98, 99, 106, 155
Sulfuric acid, 98, 232
Svyagintsevite (Pd_3Pb), 80

Tailings dam, 98, 158, 246-247
Tellurides, 84
Tellurium, 87, 164
Testing
  bacterial oxidation, 106-107
  laboratory, 40
  pilot plant, 40
Thermocouples, 96
Thermophilic bacteria, 106
Thio-antimonate, 85
Thio-antimonite, 85, 156
Thio-arsenite, 156
Thiobacillus
  ferrooxidans, 106
  thiooxidans, 106
Thiocyanite ions, 154, 155, 232
Thiol-type collectors, 157
Thiosulfate, 154, 155, 232
Thiourea leaching of gold, 171-180
Tin, 137
Titanium, 100, 101, 103, 243
Transitional placer
  deposits, 32
  environment, 26
Transmutation, apparent, 141

Tumbling grinding mills, 60–62

Ultrasonic treatment, of pulp, 163–164
Uranium minerals, 84
U.S.S.R., 230

Valley-leach method, 122
Van der Waals forces, 194
Vat leaching, 134–135
Vitox oxygenation, 147, 160

Washing, of calcines, 98
Waterflush Technology, 58
White gold
  with nickel, 139
  with palladium, 139
Wigglers, 131
Wobblers, 131, 132
Wohlwill electrorefining process, 243

Xanthates, 157

Zadra process, 201
  electrolytic cell, 210
    cathode compartment, 210–212
Zinc, 133
  cementation process, 186–192, 194
  consumption, 109
  cyanide, 232
  domains of solubility in cyanide, 189
  dust, 189
  precipitates, 190
  shavings, 186